AF614749

Methods in Molecular Biology™

Series Editor
John M. Walker
School of Life Sciences
University of Hertfordshire
Hatfield, Hertfordshire, AL10 9AB, UK

For further volumes:
http://www.springer.com/series/7651

Metastasis Research Protocols

Second Edition

Edited by

Miriam Dwek

Molecular and Applied Biosciences, School of Life Sciences, University of Westminster, London, UK

Susan A. Brooks

Department of Biological & Medical Sciences, Oxford Brookes University, Oxford, Headington, UK

Udo Schumacher

Institute Anatomie und Experimentelle Morphologie, Universitätsklinikum, Hamburg-Eppendorf, Germany

Humana Press

Editors
Miriam Dwek
Molecular and Applied Biosciences
School of Life Sciences
University of Westminster
London, UK

Susan A. Brooks
Department of Biological & Medical Sciences
Oxford Brookes University
Oxford, Headington, UK

Udo Schumacher
Institute Anatomie und
Experimentelle Morphologie
Universitätsklinikum
Hamburg-Eppendorf
Germany

ISSN 1064-3745 ISSN 1940-6029 (electronic)
ISBN 978-1-61779-853-5 ISBN 978-1-61779-854-2 (eBook)
DOI 10.1007/978-1-61779-854-2
Springer New York Dordrecht Heidelberg London

Library of Congress Control Number: 2012936663

Printed on acid-free paper

Humana Press is part of Springer Science+Business Media (www.springer.com)

Preface

Diverse molecular, cellular, and environmental events must all come together to allow the successful formation of secondary cancers, metastases. The vast amount of knowledge that has been amassed in relation to the biology underlying cancer formation is being applied to enable a better understanding of the metastatic process. It is well accepted that elucidation of the key events in this process will lead to the next generation of laboratory tests for early-diagnosis of metastases and for the treatment of occult as well as more clinically advanced disease.

This second edition of *Metastasis Research Protocols* brings together updated versions of the seminal techniques that were presented in the first edition and also includes new techniques that have recently been shown to be important in illuminating the processes underlying this important area of biology.

The first volume: *Analysis of Cells and Tissues* takes the reader through key cellular and molecular techniques relevant to the exploration of cancer cells and tissues, the focus is on the tools that have been shown to be helpful in unravelling the molecular processes important in cancer metastasis. The second volume: *Models of Metastasis* is concerned with the interaction between cancer cells and host/environment. This volume focuses on the diverse range of in vitro and in vivo models of metastasis.

In keeping with the first edition, volume one includes techniques such as immunochemistry (Chaps. 1–4), PCR (Chap. 5), and SDS-PAGE (Chap. 6), the mainstay of many laboratories. As before, we make aim for each of the techniques to stand-alone and for these to be valuable to the newcomer as well as the experienced researcher. Volume one has been extended to include newer techniques, for example, affinity measurement of biomolecular interactions (Chap. 11), methylation analysis of microRNA (Chap. 15), and RNAi technology (Chap. 16). Similarly, volume two retains chapters that describe in detail the methods often used to assess cancer cell adhesion to extracellular matrix components and endothelial cells (Chaps. 1–5), syngeneic and xenograft models of metastasis (Chaps. 9, 10, 16, and 17). In addition, updated methods are presented, for example, for the production of in vivo double-knock out models (Chap. 14) and the application of fluorescent imaging techniques for monitoring the development of metastases in vivo (Chaps. 11 and 12).

We have sought to keep the important key methods in both volumes and to introduce new methods which are making an impact in the area of metastasis research. One of the much-loved aspects of the *Methods in Molecular Biology* series is that it aims to impart knowledge of complex methodology to the end-user in an accessible manner. The "Notes" section found at the end of each chapter serves to demystify the techniques in a handy "hints and tips" format, this enables researchers who may be hesitant to adopt a new procedure to try it out, thereby adding to their repertoire of laboratory techniques. We have tried to maintain this key element in the chapters presented in these two volumes and we hope that you find this to be a continued useful aspect of this series.

Finally, we would like to thank all our contributors who have worked tirelessly to master their techniques, for sharing these with us and you, the reader. We hope you find that in these two volumes, methods that will assist in helping you to make new observations that may in turn lead to the development of new treatments aimed at combating cancer metastasis.

Acknowledgements

Miriam Dwek acknowledges the invaluable contribution of HRMD, thank you for all your input.

Contents

Contributors

BABAK AFROUGH • *Department of Molecular and Applied Biosciences, School of Life Sciences, University of Westminster, London, UK*

CORD ALBAT • *Institute of Pathology, Medizinische Hochschule Hannover, Hannover, Germany*

VITALY BALAN • *Tumor Progression and Metastasis, Karmanos Cancer Institute and Wayne State University, Detroit, MI, USA*

MATTHIAS BALLMAIER • *Department of Pediatric Hematology, Medizinische Hochschule Hannover, Hannover, Germany*

CHRISTINE BLANCHER • *Wellcome Trust Centre for Human Genetics, Headington, Oxford, UK*

STEPHAN BRAUN • *Universitätsklinik für Gynäkologie und Geburtshilfe Medizinische Universität Innsbruck, Innsbruck, Austria*

SUSAN A. BROOKS • *Department of Biological & Medical Sciences, Oxford Brookes University, Oxford, Headington, UK*

IAN D. BULEY • *Department of Cellular Pathology, Peninsula Medical School, Torbay Hospital, Torquay, UK*

MATTHIAS CHRISTGEN • *Institute of Pathology, Hannover Medical School, Hannover, Germany*

MICHAEL R. DASHWOOD • *Department of Clinical Biochemistry, Royal Free and UCL Medical School, London, UK*

DEREK DAVIES • *FACS Laboratory, London Research Institute, Cancer Research UK, London, UK*

MIRIAM DWEK • *Department of Molecular and Applied Biosciences, School of Life Sciences, University of Westminster, London, UK*

RAFAEL FRIDMAN • *Department of Pathology, Wayne State University, School of Medicine, Detroit, MI, USA*

SIMON FRY • *Department of Molecular and Applied Biosciences, School of Life Sciences, University of Westminster, London, UK*

DEBBIE M.S. HALL • *Department of Biological & Medical Sciences, Oxford Brookes University, Oxford, UK*

BJÖRN INGEMARSSON • *Attana AB, Stockholm, Sweden*

LYNDAL KEARNEY • *Haemato-Oncology Research Unit, Division of Molecular Pathology, The Institute of Cancer Research, Surrey, UK*

HANS KREIPE • *Institute of Pathology, Medizinische Hochschule Hannover, Hannover, Germany*

ANTHONY LEATHEM • *Department of Molecular and Applied Biosciences, University of Westminster, London, UK*

ULRICH LEHMANN • *Institute of Pathology, Medizinische Hochschule Hannover, Hannover, Germany*

MARILENA LOIZIDOU • *Division of Surgery & Interventional Science, Royal Free and UCL Medical School, London, UK*
ROB MC CORMICK • *Cancer Research UK, Weatherall Institute of Molecular Medicine, John Radcliffe Hospital, Oxford, UK*
PRATIMA NANGIA-MAKKER • *Tumor Progression and Metastasis, Karmanos Cancer Institute and Wayne State University, Detroit, MI, USA*
KLAUS PANTEL • *Institut für Tumorbiologie, Universitätsklinikum Hamburg-Eppendorf, Hamburg, Germany*
DILUKA PEIRIS • *Department of Molecular and Applied Biosciences, School of Life Sciences, University of Westminster, London, UK*
AVRAHAM RAZ • *Tumor Progression and Metastasis, Karmanos Cancer Institute and Wayne State University, Detroit, MI, USA*
DEREK E. ROSKELL • *Department of Cellular Pathology, University of Oxford, Oxford Radcliffe Hospitals NHS Trust, Level 1, John Radcliffe Hospital, Oxford, UK*
JULIEN SAINT-GUIRONS • *R&D department, DHR Finland Oy, Innotrac Diagnostics, Turku, Finland*
CAROLINE SCHREIBER • *Max-Delbrück-Center for Molecular Medicine, Free University Berlin, Berlin, Germany*
JANET SHIPLEY • *Sarcoma Molecular Pathology Team, Divisions of Molecular Pathology and Cancer Therapeutics, The Institute of Cancer Research, Surrey, UK*
ANJUM SOHAIL • *Department of Pathology, School of Medicine, Proteases and Cancer Program, Karmanos Cancer Institute, Wayne State University, Detroit, MI, USA*
MARTA TOTH • *Department of Pathology, School of Medicine, Proteases and Cancer Program, Karmanos Cancer Institute, Wayne State University, Detroit, MI, USA*
KIRSTEN VORMBROCK • *Max-Delbrück-Center for Molecular Medicine and Institute for Biochemistry, Free University Berlin, Berlin, Germany*
HAZEL M. WELCH • *Division of Surgery & Interventional Science, Royal Free and UCL Medical School, London, UK*
DANIEL WICKLEIN • *Centre for Experimental Medicine, Institute for Anatomy and Experimental Morphology, Hamburg, Germany*
ULRIKE ZIEBOLD • *Marie Curie Excellence Group, Max-Delbrück-Center for Molecular Medicine, Berlin, Germany*

Chapter 1

Basic Immunocytochemistry for Light Microscopy

Susan A. Brooks

Abstract

Immunocytochemistry, the identification of cell- or tissue-bound antigens in situ, by means of a specific antibody–antigen reaction, tagged microscopically by a visible label, has a remarkably wide range of applications. The basic techniques are straightforward and can be adapted to explore the localisation of virtually any molecule of interest to the researcher in samples of normal and/or malignant cells. Heterogeneity can be mapped and loss or gain of immunoreactivity with tumour progression can be visualised. In this chapter, methodologies are given for appropriate preparation of cells and tissues, including cells cultured on coverslips (which can be used for live cell imaging), cell smears, frozen (cryostat) and fixed, paraffin wax-embedded tissue sections. Heat- and enzyme-based antigen retrieval methods are covered. Basic detection methods, which can be readily adapted, are given for direct (labelled primary antibody), simple indirect (labelled secondary antibody), avidin–biotin (biotinylated primary antibody), avidin–biotin complex (ABC), peroxidase–anti-peroxidase or alkaline phosphatase–anti-alkaline phosphatase (PAP or APAAP), and polymer-based methods. The use of enzyme labels including horseradish peroxidase and alkaline phosphatase, and fluorescent labels, are considered.

Key words: Immunochemistry, Antibody, Avidin–biotin complex, Direct, Indirect

1. Introduction

1.1. Immunocytochemistry

Immunocytochemistry may be defined as the identification of a cell- or tissue-bound antigen in situ, achieved by means of a specific antibody–antigen reaction, viewed microscopically by virtue of a visible label. Successful immunocytochemistry therefore requires (a) preservation of the antigen in a form recognisable by the antibody, (b) a suitable antibody, and (c) an appropriate label. The basic immunocytochemistry technique was first described by Coons et al. (1–3) who employed antibody directly labelled with a fluorescent tag to identify antigen in tissue sections. Since that time the technique has been refined and expanded enormously. Some significant developments include the use of horseradish peroxidase

Miriam Dwek et al. (eds.), *Metastasis Research Protocols*, Methods in Molecular Biology, vol. 878,
DOI 10.1007/978-1-61779-854-2_1, © Springer Science+Business Media, LLC 2012

(HRP) (4) and alkaline phosphatase (AP) (5) as label molecules, the development of many, increasingly sensitive, multi-layer detection methods, and exploitation of the strong binding between avidin and biotin in detection techniques (6, 7). In recent years, there has been a renaissance in the use of immunocytochemical approaches in conjunction with imaging of live cells using confocal microscopy and related techniques.

1.2. Range of Applications

Immunocytochemistry is appropriate for a wide range of applications. Any cell- or tissue-bound molecule to which an antibody can be raised can, theoretically, be detected in situ. It is a technique of particular interest in metastasis research as it facilitates the detection of virtually any molecule of interest to the researcher in samples of tumour, normal tissues, or cells. Of particular interest in this field is that of the heterogeneity of immunoreactivity by cells within a morphologically homogeneous tumour mass, or between normal and cancer cells. The gain or loss of certain antigens by tumour cells at different stages in the natural history of the disease and in relation to metastatic potential is also of great relevance. For example, loss of cell adhesion molecules by tumour cells may be instrumental in their breaking away from a primary tumour mass. Immunocytochemistry is the only technique that allows assessment of the expression of such molecules in situ.

1.3. Types of Cell and Tissue Preparations

The first requirement for successful immunocytochemistry is preservation of the antigen; the primary consideration must be what type of cell or tissue preparation to employ. Immunocytochemistry can be performed on a range of different cell and tissue preparations including living cell suspensions, cell smears, frozen (cryostat) sections and fixed, paraffin wax-embedded sections. Some of the advantages and disadvantages of these types of preparation are summarised in Table 1.

Of particular interest may be the application of immunocytochemistry to routinely formalin-fixed, paraffin wax-embedded archived tissue specimens, facilitating mapping of expression of molecules of interest retrospectively with the benefit of long-term patient follow-up. Many antigens are well preserved in such tissues, and retrospective analysis can be successfully carried out after years (sometimes even decades) of tissue storage, making this a powerful and informative approach. The only limitations are, firstly, that lipids are solubilised out and lost during processing to paraffin wax, and their antigenic structures are not present. Secondly, an antibody is required that successfully recognises antigen preserved in this manner. Some antigens are damaged, sequestered or altered by fixation and processing to paraffin wax, and many antibodies will therefore no longer recognise them and will only give successful results on fresh, frozen (cryostat) tissue sections or fresh cell preparations. Enzyme- and heat-mediated antigen retrieval techniques

Table 1
Advantages and disadvantages of different tissue preparations

Preparation	Suitable for	Advantages	Disadvantages
Cell suspensions	Living cells, e.g. blood cells, cultured cells, cells released from solid tissue masses Direct method using fluorescent-labelled antibodies most suitable Good for confocal imaging studies	Unaltered antigen immunolocalisation in the living cell seen. Excellent for cell surface antigens	Not suitable for demonstration of cytoplasmic antigens. Cells seen in isolation; no indication of tissue distribution of antigen
Cell smears	Any living cells in suspension, e.g. blood, cultured cells Any labelling method is suitable.	Quick and easy Good for cytoplasmic antigens	Morphology sometimes indistinct
Frozen sections	Any fresh solid animal or human tissue Any labelling method is suitable Good for confocal imaging studies	Relatively quick Fairly good morphology Spatial relationships of cells within tissues seen Good for cytoplasmic and cell surface antigens	Technically more demanding than suspensions or smears.
Paraffin wax sections	Any solid animal or human tissue Any labelling method suitable	Tissue preserved indefinitely. Excellent morphology Relationships between cells in tissues seen Cell surface and cytoplasmic antigens seen	More time consuming than other methods Glycolipids lost Fixation and processing may damage some antigens

are described later in this chapter and may be successful in partially, or fully, reversing the alterations caused by fixation and processing and thereby enable successful detection of otherwise undetectable antigens. The development of heat-mediated antigen retrieval (e.g. see refs. 8–11) has vastly expanded the repertoire of antibodies that can be used successfully on fixed and processed tissues.

Immunolocalisation of molecules in preparations of cultured cell lines or of cell suspensions from body fluids, such as ascites, blood, or pleural effusions may also be of interest in metastasis research, and immunocytochemistry on such preparations in the form of cytospins, cell smears, or cells cultured on coverslips, is generally very successful. These types of preparations labelled using immunocytochemistry involving a fluorescent label may yield stunningly beautiful and informative results when examined using confocal microscopy.

For the simpler, quicker immunocytochemical methods, cell and tissue preparations will adhere well to clean, dry glass microscope slides. For longer, multi-step techniques, the use of an adhesive is recommended. Silane is an effective adhesive which is widely available, cheap, and simple to use. Its use is essential if heat-mediated antigen retrieval methods are to be used subsequently.

1.4. Choice of Antibody

A requirement for successful immunocytochemistry is the availability of a suitable antibody directed against the antigen of interest. Detailed description of how to raise and purify antibodies lies beyond the scope of this volume. The main choice for the researcher lies between monoclonal and polyclonal antibodies. Both types of antibodies, directed against thousands of antigens of potential interest are available commercially, and can be produced "in house" if the appropriate facilities and expertise are available. Companies will produce "tailor-made" antibodies directed against peptide sequences requested by the customer at a relatively modest cost. It is essential to choose an antibody appropriate for immunocytochemistry specifically, as antibodies developed for other applications—for example, ELISA—may not work. It is also important to realise that different commercially available antibodies directed against the same molecule, or epitope on a particular molecule, may not be equally effective in immunocytochemistry and some "shopping around" may be helpful. Many companies will provide small samples of antibodies free of charge for researchers to evaluate.

The choice of monoclonal or polyclonal antibody depends largely on what antibodies are available that are directed against the antigen of interest. Each type of antibody has their own advantages and disadvantages and it cannot be assumed that either polyclonal or monoclonal antibodies are invariably superior. Briefly, polyclonal antibodies contain a cocktail of immunoglobulins directed against different epitopes of an antigen of interest, and often other, irrelevant, antigens too. They can sometimes cross react with molecules other than the one of interest and give spurious results or "dirty" background labelling. The cocktail of immunoglobulins present may, however, react with multiple epitopes on the antigen molecule of interest resulting in stronger and more effective labelling than is achieved with a comparable monoclonal antibody. Polyclonal antibodies are usually raised in rabbit (or sometimes large animals such as goat or sheep; chicken antibodies are also available) and tend to be cheaper to buy than monoclonal antibodies. Monoclonal antibodies are usually raised in mice or rats, and, as the name suggests, represent immunoglobulins produced by a single immortalised clone of cells, and therefore directed against a single epitope. These tend to be more expensive than polyclonal antisera; however, can lend themselves to extremely high working dilutions. The great advantage of monoclonal antibodies is their specificity which often gives rise to very clean labelling results. Many monoclonal antibodies are raised to synthetic peptide sequences, giving the advantage that the precise epitope they recognise is known. Cross-reactivity can sometimes occur even with monoclonal antibodies if the epitope they are directed against is shared by other, irrelevant molecules.

There are no hard and fast rules as to the choice of an antibody—for some applications, a particular monoclonal antibody may be ideal, whilst for others a polyclonal antiserum may produce better results. It is important to note the class of antibody being used. Most monoclonal antibodies used in immunocytochemistry are of the IgG class, however some may be IgM. This is an important consideration in many immunocytochemical methods, as detection of antibody binding to antigen may be achieved by subsequent reaction with a secondary antibody directed against the first. For example, to detect the binding of a monoclonal mouse IgG, a labelled secondary antibody raised in, for example rabbit, against the mouse IgG may be applied. A secondary antibody directed against mouse IgM would not be appropriate in this example.

For any application, the appropriate dilution of antibody must be determined. This can usually only be ascertained by performing a range of dilutions and checking which give optimum results in terms of strong specific labelling coupled with clean background. When using polyclonal antibodies, doubling dilutions are convenient, ranging from 1:50 to 1:3,200 as a rough guide. Monoclonal antibodies may be tested in the range of, possibly, 1–50 μg/ml. In the more complex multi-step techniques different dilutions of primary antibody and secondary labelling reagents may need to be titrated against each other in a "chequerboard of dilutions" to determine optimum working dilutions. An example of a typical "chequerboard of dilutions" is given in Table 2. One would expect the more concentrated solutions of primary and/or secondary antibody to give strong labelling but unacceptably high background; too high a dilution of either primary or secondary antibody will

Table 2
Sample chequerboard of dilutions used to determine the optimum dilution of primary and secondary antibody for use in, for example, an indirect detection method. The optimum combination of both will yield strong, specific labelling with clean background and no nonspecific labelling

Dilution of polyclonal secondary antibody e.g. swine antisera raised against rabbit immunoglobulins	Dilution of primary antibody e.g. polyclonal rabbit antisera raised against the molecule of interest						
1/50	1/50	1/100	1/200	1/400	1/800	1/1,600	1/3,200
1/100							
1/200							
1/400							
1/800							

yield low intensity, but probably very clean labelling. The optimum dilution of both in combination should yield deep, intense specific labelling with clean background and an absence of nonspecific labelling. It is worth taking time and care over titration experiments in order to achieve optimal experimental results.

1.5. Choice of Labels for Immunocytochemistry

The third requirement for successful immunocytochemistry is the presence of a visible label. The choice of label usually lies between a fluorescent label or the coloured product of an enzyme reaction, although other labels such colloidal gold, silver, or ferritin can also be employed (usually in immunocytochemistry for electron microscopy, which goes beyond the scope of this chapter). Traditionally, the most commonly used fluorescent label is fluorescein isothiocyanate (FITC) which fluoresces a bright yellow-green (12). Alternatives include tetrarhodamine isothiocyanate (TRITC) and Texas red, which fluoresce red (13). Many primary and secondary antibodies are available commercially, labelled with these compounds. There is also an ever increasing range of other fluorescent labels which open up the possibility of multiple labelling experiments and which are increasingly used in confocal imaging.

The most commonly used enzyme labels are HRP and AP. Many primary and secondary antibodies and other immunocytochemical reagents are available labelled with these compounds. Other less commonly used enzyme labels include glucose oxidase and beta galactosidase. The principle of using any enzyme label is the reaction with substrate on addition of a soluble chromagen to yield a precipitated or insoluble coloured product visible by light microscopy. HRP reaction with hydrogen peroxide and the chromagen diaminobenzidine (DAB) (14) yields a granular, brown, alcohol-insoluble product or with 3-amino-9-ethylcarbazole (AEC) (15) yields a granular, red, alcohol-soluble product. Other chromagens are also available, but are less commonly employed. For AP, reaction with naphthol phosphate as a substrate and a diazonium salt can yield a variety of coloured—most typically red or blue—alcohol-soluble azo dyes as products (16). Enzyme label detection kits, usually in the form of dropper bottles of concentrated reagents ready to be diluted in water or buffer, are commercially available as a convenient alternative to preparation of solutions "in house".

When enzyme labels are employed, the issue of endogenous, cell- or tissue-bound enzyme becomes an issue, and steps often need to be incorporated into the detection method to block endogenous enzyme prior to development of the final coloured label product. It is worth noting that endogenous AP is usually destroyed in processing to paraffin wax.

1.6. The Range of Detection Methods Available

A number of fairly standard immunocytochemical techniques exist, which vary in terms of complexity and sensitivity. They range from the simple "direct" technique where antigen is detected by the binding of a directly labelled antibody, through to much more complex, but highly sensitive multi-layer techniques. Examples of a range of detection methods are outlined in this chapter and their relative advantages and disadvantages are summarised in Table 3. They are also represented diagramatically in Figs. 1–5. For any particular application, the choice of technique depends, amongst other things, on the type of cell or tissue preparation to be labelled, the abundance and robustness of the antigen to be detected, and the affinity and efficacy of the antisera. Simpler "direct" techniques are often employed for labelling of living cells—for example, cultured cells or cells from body fluids—as they are least likely to damage delicate cells or antigens. These are commonly used in conjunction with fluorescent labels although enzyme labels can also be used. The more complex and sensitive multi-layer techniques are usually used in conjunction with more robust cell and tissue preparations and are particularly appropriate where antibody titres are low, or where antigen expression is scanty as the "layering" of reagent results in amplification of the final signal. Amplification is achieved because at every step, multiple reagent molecules have the opportunity to bind to the previous "layer", resulting in a much amplified "cloud" of label molecules marking the initial binding of antibody molecule to antigen. This important point is not shown in the figures illustrating the methods as, for the sake of clarity, the "layers" are represented in a simplified, linear manner. Many reagent manufacturers also market labelling reagents based on synthetic

Table 3
Advantages and disadvantages of different detection methods

	Advantages	Disadvantages
"Direct" method	Simplest method available. Quick. Limited number of reagents required. Works particularly well using fluorescent-labelled antibody and cell suspensions	Lacks sensitivity, therefore may not be appropriate for scantily expressed antigens. May suffer from high background
"Simple indirect" method	Increased sensitivity over direct method (~20× more sensitive). Relatively quick and straightforward	Requires more reagents than the direct method, therefore potentially more expensive. Extra step therefore takes longer
"Simple avidin–biotin" method	Relatively quick and simple, but highly sensitive and yields clean labelling	Endogenous biotin may confuse interpretation in some cases Glycosylated avidin may be recognised by tissue-bound lectins or bind charged sites nonspecifically
"ABC" method	Highly sensitive (at least 100× more sensitive than the direct method). Clean results	Large ABC complex sometimes causes steric hindrance. Tissue lectins may bind glycosylated avidin; avidin may attach nonspecifically by charge. Endogenous biotin may confuse interpretation. Cost implications of reagents
"PAP" or "APAAP" methods	Highly sensitive (~50× more sensitive than direct method)	Very time consuming. Cost implications of extra reagents. Largely superseded by ABC method
Labelled polymer methods	Highly sensitive. Reduced incubation times for faster results	Cost implications of specialised reagents

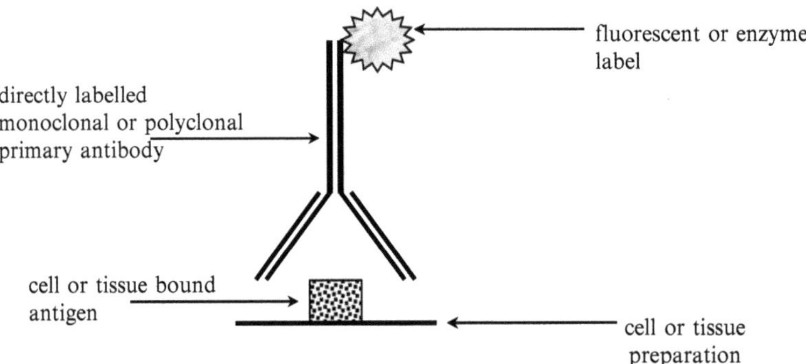

Fig. 1. "Direct" method—cell- or tissue-bound antigen is detected by binding of directly labelled primary antibody.

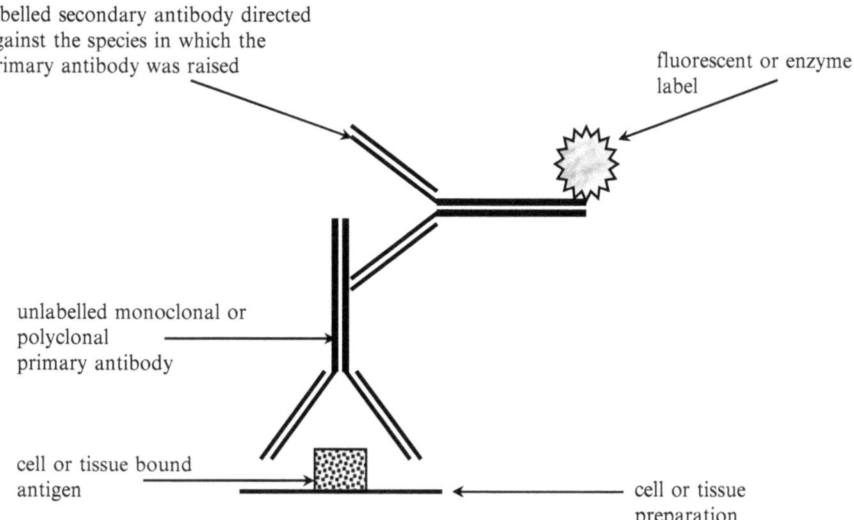

Fig. 2. "Simple indirect" method—cell- or tissue-bound antigen is detected by binding of unlabelled primary antibody, then labelled secondary antibody directed against the species in which the primary antibody was raised.

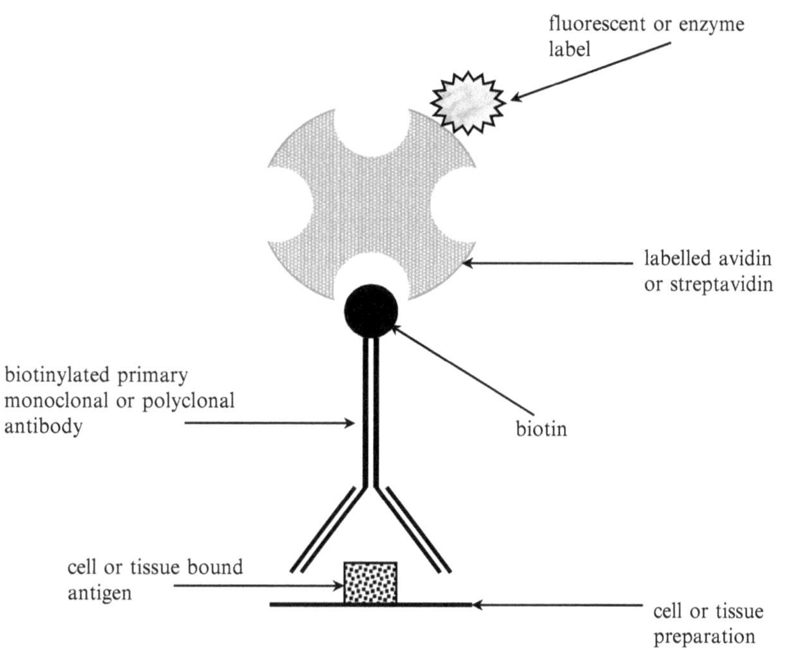

Fig. 3. "Simple avidin–biotin" method—cell- or tissue-bound antigen is detected by binding of a biotinylated primary antibody, then labelled avidin or streptavidin.

polymers that present a great concentration of labelling signal, and therefore improved sensitivity, allowing for the development of highly sensitive methods and the possibility of reduced incubation times, as illustrated in Fig. 6.

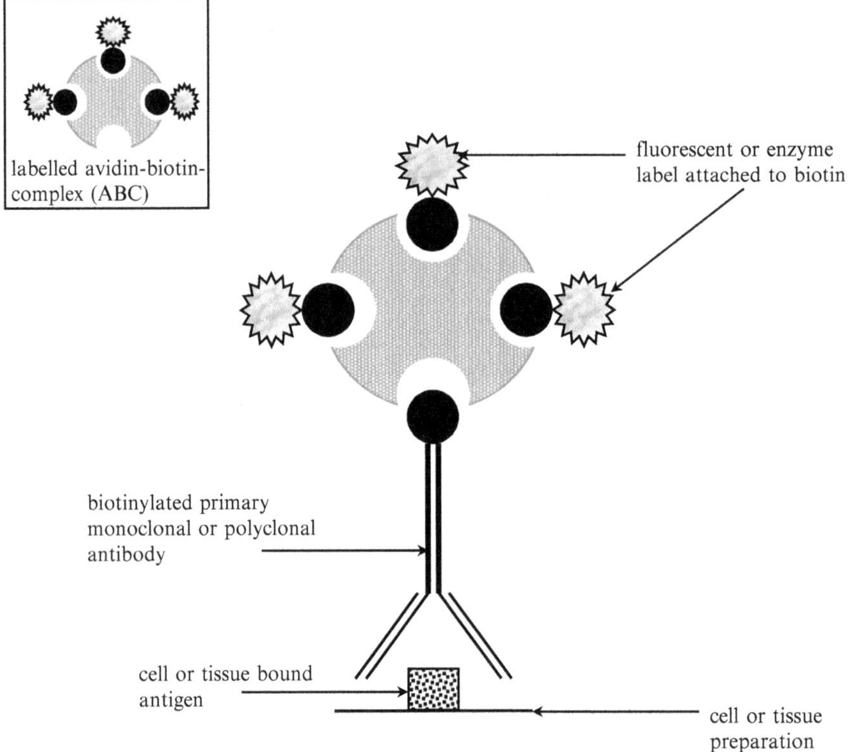

Fig. 4. "Avidin–biotin complex" (ABC) method—cell- or tissue-bound antigen is detected by binding of a biotinylated primary antibody, then labelled ABC.

In addition to the methods listed in this chapter, similar approaches are described in detail in the chapter by Brooks and Hall in this volume on the related technique of lectin histochemistry. Lectin histochemistry facilitates the detection of carbohydrate structures present in, for example, glycoproteins, glycolipids or glycosaminoglycans, in situ, by means of their recognition by a lectin.

1.7. Controls

The incorporation of appropriate positive and negative controls is of paramount importance. An appropriate positive control is a cell or tissue sample that is known to be rich in the antigen of interest and which should yield a labelling distribution consistent with its biology—for example, if the antigen of interest is known to be expressed by the endothelial cells lining human blood vessels, then a sample of highly vascular human tissue would be appropriate and labelling would be expected to be localised in the walls of blood vessels and nowhere else. Negative controls are often performed by omitting incubation with the primary antisera—perhaps replacing it with a similar dilution of an irrelevant antisera, non-immune sera or plain buffer. Under these circumstances, no labelling at all would

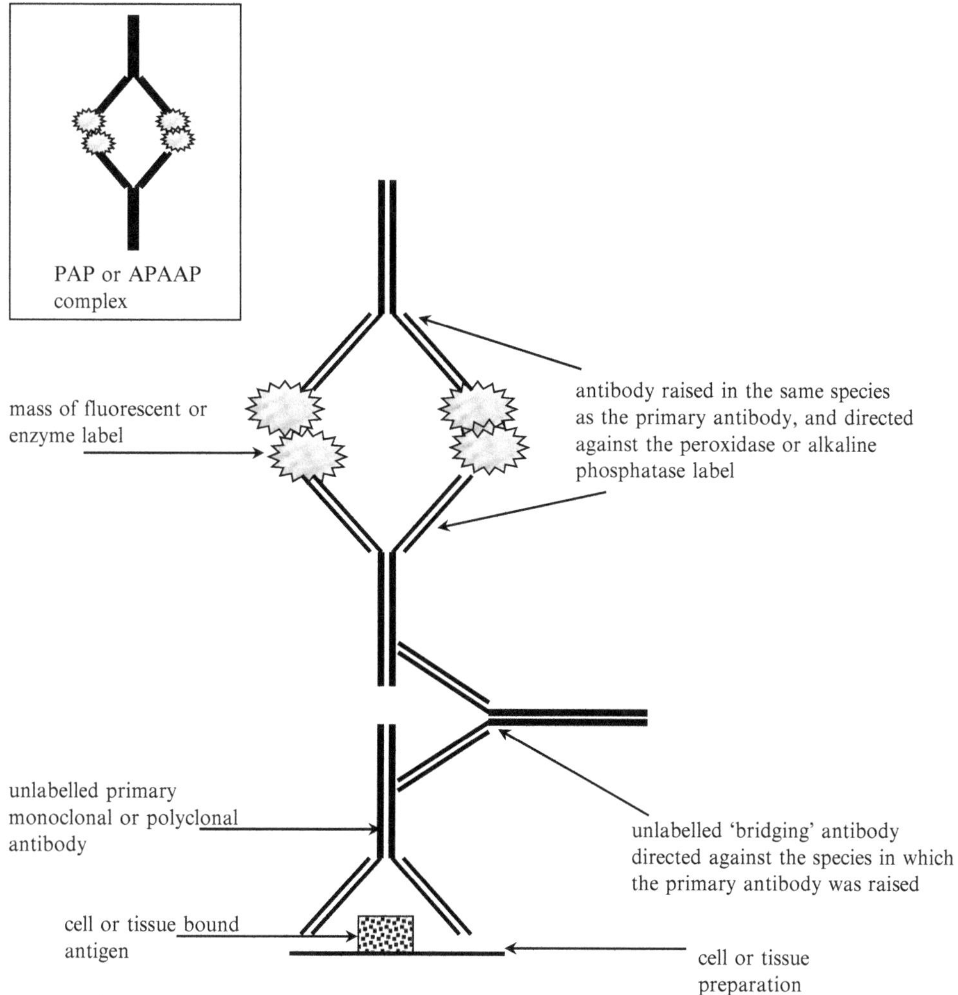

Fig. 5. "Peroxidase–anti-peroxidase" (PAP) or "alkaline phosphatase–anti-alkaline phosphatase" (APAAP) method—cell- or tissue-bound antigen is detected by, firstly, unlabelled primary monoclonal or polyclonal antibody, then a "bridging" antibody directed against the species in which the primary antibody was raised, and finally a labelled "PAP" or "APAAP" complex raised in the same species as the primary antibody. The labelled "PAP" or "APAAP" consists of a mass of enzyme label complexed with antibodies directed against it.

be expected. Any labelling that occurs would be as a result of other reagents reacting with the tissue nonspecifically. The specificity of labelling may be confirmed by incubating the cell or tissue sample with the primary antisera in the presence of a free antigen—for example, if labelling a sample using a primary antibody against a known peptide sequence, the antisera can be incubated in the presence of a free peptide. Under these circumstances, as the antisera will recognise antigen in solution as well as antigen in cell or tissue preparation, a diminution or even abolition of labelling would be expected.

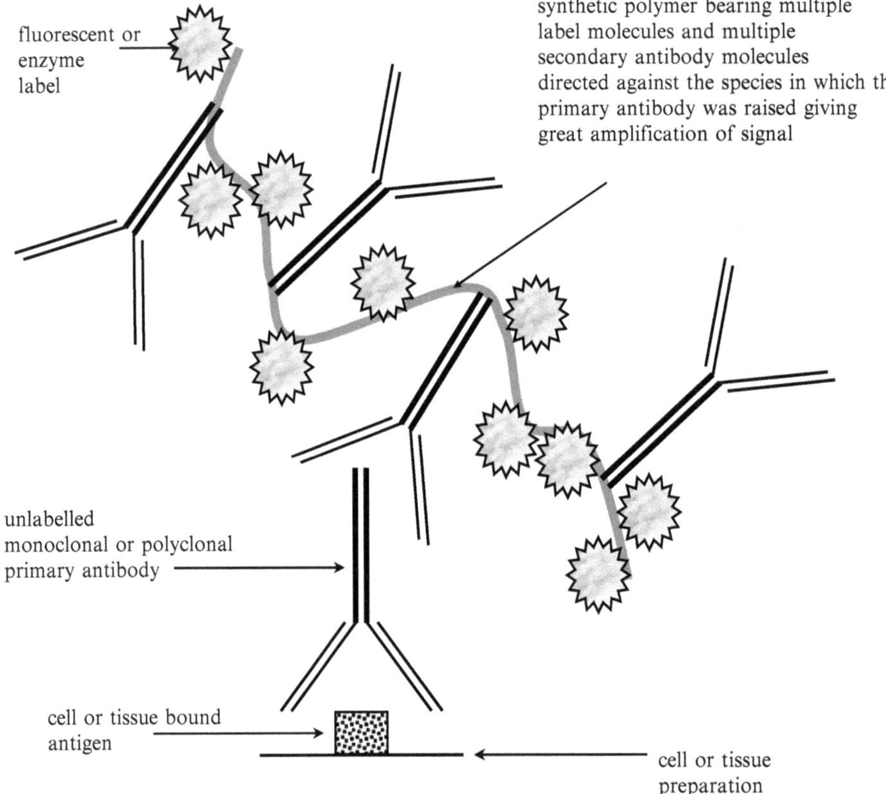

Fig. 6. Using a labelled polymer reagent—cell- or tissue-bound antigen is detected by binding of an unlabelled primary antibody, then a polymeric reagent incorporating secondary antibody directed against the species in which the primary antibody was raised along with a high concentration of label.

2. Materials

2.1. Silane/APES Treatment of Microscope Slides (See Note 1)

1. Acetone.
2. Acetone/APES or silane solution: 2% v/v aminopropyltriethoxysilane (APES) or 3-(triethoxysilyl)-propylamine (silane) in acetone; make fresh on the day of use. Discard after treating a maximum of 1,000 slides.
3. Distilled/deionised water: discard and refresh after every five racks of slides have passed through.

2.2. Preparation of Cells Cultured on Coverslips

1. Cells cultured under standard conditions.
2. Foetal calf serum (FCS) free cell culture medium.
3. Alcohol or autoclave sterilised round glass coverslips (13 mm diameter, thickness 0).
4. Dental wax or Parafilm.

5. 0.1 M PIPES buffer, pH 6.9: stir 12.1 g PIPES [piperazine-*N*,*N*′-bis-(2-ethanesulfonic acid)] into 50 ml ultrapure water to give a cloudy solution. Add approximately 40 ml 1 M NaOH and the solution should clear. Check pH and adjust to 6.9, if necessary, using 1 M NaOH. Add ultrapure water to give a final volume of 400 ml.
6. 3% v/v Paraformaldehyde in 0.1 M PIPES buffer, pH 6.9: place 3 g paraformaldehyde in a 250-ml conical flask, add 30 ml ultrapure water, loosely stopper, and heat on a 60°C hotplate, in a fume cupboard, for about 30 min to give a cloudy solution. Add 1 M NaOH (the purest grade), with continual stirring, until the solution clears. Add ultrapure water to give a total volume of 50 ml, then add 50 ml 0.2 M PIPES buffer, pH 6.9 (see above). Divide into 10-ml aliquots and store frozen. Defrost in a warm water bath for use.
7. 0.1% v/v Triton X-100 or Saponin in 0.1 M PIPES buffer, pH 6.9.

2.3. Smears Prepared from Cells in Suspension

1. Cells in suspension (see Note 2).
2. Silane-treated glass microscope slides (see Subheading 2.1 and Note 1).
3. Aluminium foil or cling film/Saran Wrap.
4. Acetone.

2.4. Frozen (Cryostat) Sections

1. Chunk of fresh tissue approximately 0.5 cm^3 in size.
2. Cryostat embedding medium, for example “OCT” or similar.
3. Isopentane or hexane.
4. Liquid nitrogen.
5. Silane-coated clean glass microscope slides (see Subheading 2.1 and Note 1).
6. Acetone.
7. Aluminium foil or cling film/Saran Wrap.

2.5. Fixed, Paraffin Wax-Embedded Sections

1. Paraffin wax-embedded tissue blocks.
2. 20% v/v Ethanol or industrial methylated spirits (IMS) in distilled/deionised water.
3. Silane-treated glass microscope slides (see Subheading 2.1 and Note 1).
4. Xylene (see Note 3).
5. Absolute ethanol or IMS.
6. 70% v/v Ethanol or IMS in distilled/deionised water.
7. Deionised/distilled water.

2.6. Buffers for Blocking, Dilutions, and Washes

1. Washing buffer: Tris-buffered saline (TBS), pH 7.4–7.6: 60.57 g Tris, 87.0 g NaCl dissolved in 1 L distilled/deionised water. Adjust pH to 7.4–7.6 using concentrated HCl. Make up to total volume of 10 L using distilled/deionised water. This buffer is recommended for all washes, unless otherwise stated. Phosphate buffered saline (PBS), pH 7.4–7.6 can be substituted.
2. Blocking buffer: 5% v/v normal horse or goat serum in washing buffer. All immunocytochemical methods incorporate a step where cell and tissue preparations are incubated with a blocking buffer to reduce nonspecific binding of antibodies (see Note 4).
3. Dilution buffer for antibodies: 3% v/v normal horse or goat serum in blocking buffer. Antibodies are diluted to their working concentration in buffer containing a low percentage of normal serum. Again, this reduces the nonspecific binding of antibodies and minimises "dirty" background labelling (see Note 4).

2.7. Enzyme-Based Antigen Retrieval Methods (See Note 5)

1. Trypsin solution: 1 mg/ml crude, type II trypsin, from porcine pancreas (see Note 6) and 1 mg/ml calcium chloride in washing buffer (see Subheading 2.6) warmed to 37°C.
2. Protease solution: protease XXIV, bacterial, 7–14 U/mg in washing buffer (see Subheading 2.6) pre-warmed to 37°C.
3. Pepsin solution: pepsin from porcine stomach, 1:2,500, 600–1,000 U/mg in 0.01 M hydrochloric acid, pre-warmed to 37°C.
4. Neuraminidase solution: neuraminidase (sialidase), type V from *Clostridium perfringens* at a concentration of 0.1 U/mg in 0.1 M sodium acetate; pH to 5.5 with citric acid, containing 0.01% w/v calcium chloride, pre-warmed to 37°C.

2.8. Microwave Oven Heat-Mediated Antigen Retrieval Method (See Note 7)

1. Citrate buffer, pH 6.0: 2.1 g citric acid dissolved in 1 L distilled/deionised water pH to 6.0 using concentrated NaCl.
2. Distilled/deionised water.

2.9. Quenching Endogenous Enzyme

Methanol/hydrogen peroxide solution: 3% v/v hydrogen peroxide in methanol. Make up fresh every 2–3 days.

2.10. Examples of Some Immunocytochemical Labelling Techniques

2.10.1. Direct Method

For all these methods blocking buffer and washing buffer are required (Subheading 2.6). In addition: monoclonal or polyclonal antibody labelled with a fluorescent or enzyme label made up at optimum working dilution in dilution buffer (see Subheading 2.6).

2.10.2. Simple Indirect Method

1. Unlabelled monoclonal or polyclonal antibody made up at optimum working dilution in dilution buffer (see Subheading 2.6).

2. Fluorescent- or enzyme-labelled secondary antibody directed against the immunoglobulins of the species in which the primary antibody was raised, made up at optimum working dilution in dilution buffer (see Subheading 2.6).

2.10.3. Simple Avidin–Biotin Method

1. Biotin-labelled monoclonal or polyclonal primary antibody made up at optimum working dilution in dilution buffer (see Subheading 2.6).
2. Avidin or streptavidin labelled with a fluorescent or enzyme label made up at optimum working dilution in dilution buffer (see Subheading 2.6 and see Note 8).

2.10.4. Avidin–Biotin Complex Method

1. Biotin-labelled monoclonal or polyclonal primary antibody made up at optimum working dilution in dilution buffer (see Subheading 2.6).
2. Fluorescent- or enzyme-labelled avidin–biotin complex (ABC) made up according to manufacturer's instructions (see Note 9).

2.10.5. Peroxidase–Anti-peroxidase or Alkaline Phosphatase–Anti-alkaline Phosphatase Methods (See Note 10)

1. Unlabelled monoclonal or polyclonal antibody made up at optimum working dilution in dilution buffer (see Subheading 2.6).
2. Unlabelled "bridging" antibody directed against the species in which the primary antibody was raised, made up in excess concentration (see Note 11) in dilution buffer (see Subheading 2.6).
3. Peroxidase–anti-peroxidase (PAP) or alkaline phosphatase–anti-alkaline phosphatase (APAAP) raised in the same species as the primary antibody (e.g., when using a mouse monoclonal primary, use *mouse* PAP or APAAP, when using rabbit polyclonal primary antibody, use *rabbit* PAP or APAAP), made up at optimum working dilution in dilution buffer (see Subheading 2.6).

2.10.6. Simple Polymer Reagent Method

1. Unlabelled monoclonal or polyclonal antibody made up at optimum working dilution in dilution buffer (see Subheading 2.6).
2. Secondary antibody directed against the immunoglobulins of the species in which the primary antibody was raised, in polymer complex, with enzyme or fluorescent label, made up at optimum working dilution in dilution buffer according to manufacturer's guidelines (see Subheading 2.6).

2.11. Enzyme Development Methods

2.11.1. Diaminobenzidine for HRP Label

1. Washing buffer (see Subheading 2.6).
2. DAB-H_2O_2: 3,3-DAB tetrahydrochloride (DAB) 0.5 mg/ml in washing buffer (see Subheading 2.6). This substance is potentially carcinogenic (see Note 12). Add H_2O_2 to give a concentration of 0.03% v/v immediately before use.

2.11.2. Fast Red for AP Label

1. TBS pH 8.2–9.0: 6.57 g Tris, 8.7 g NaCl dissolved in a total volume of 1 L distilled/deionised water. Adjust pH to 8.2–9.0 using concentrated HCl.
2. Stock solution of naphthol phosphate: dissolve 20 mg naphthol AS-MX phosphate sodium salt in 500 μl *N,N*-dimethylformamide in a small glass vessel (see Note 13).
3. Stock solution fast red salt: dissolve 20 mg fast red salt in 1 ml TBS, pH 8.2–9.0.
4. Levamisole hydrochloride.

2.12. Counterstaining

1. Mayer's haematoxylin solution (see Note 14).
2. 1% v/v Ammonia in tap water.

2.13. Mounting

1. For fluorescently labelled preparations: an anti-fade mounting medium, for example, "Citifluor" or similar.
2. For AP/fast red labelled, or other preparations labelled with an alcohol-soluble chromogenic product: an aqueous mountant, for example, "Aquamount" or similar.
3. For HRP/DAB labelled, or other preparations labelled with an alcohol-insoluble chromogenic product: 70% v/v ethanol or IMS in distilled/deionised water; 95% v/v ethanol or IMS in distilled/deionised water; absolute alcohol; xylene (see Note 3); and an appropriate xylene-based mounting medium e.g. "Depex" or similar.

3. Methods

3.1. Silane Treatment of Microscope Slides (See Note 1)

1. Place slides in a slide carrier and immerse in acetone for 5 min.
2. Immerse in acetone/silane solution for 5 min.
3. Immerse in two consecutive baths of either acetone or distilled/deionised water for 5 min each.
4. Drain slides, dry either at room temperature or in a warm oven, and store in closed boxes at room temperature indefinitely.

3.2. Preparation of Cells Cultured on Coverslips

1. Wash cultured cells in fresh FCS free culture medium.
2. Aspirate and discard the medium.
3. Scrape cells from the flask using a rubber scraper, and re-suspend in fresh FCS-free culture medium.
4. Count cells, and sub-culture 1×10^5 cells into Petri dishes. Place sterile coverslips in Petri dishes.
5. Allow cells to proliferate for 24 h at 37°C under 5% CO_2.
6. Carefully remove coverslips using fine forceps or the edge of a scalpel blade.

7. Place coverslips, cell side up, onto a piece of dental wax or "Parafilm" for support, and cover each with 100 μl cold 3% v/v paraformaldehyde in 0.1 M PIPES buffer, pH 6.9, for 15 min.
8. Wash thoroughly in 0.1 M PIPES buffer, pH 6.9.
9. Permeabilize in 0.1% v/v Triton X-100 or 0.1% v/v Saponin in 0.1 M PIPES buffer, pH 6.9, for 10 min.
10. Wash thoroughly in 0.1 M PIPES buffer pH 6.9.

3.3. Smears Prepared from Cells in Suspension

1. Place a drop of cells in suspension approximately 5 mm from one end of a silane-treated glass microscope slide.
2. Place a second microscope slide on top of the first, allowing approximately 1 cm of glass to protrude at either end, and allowing the drop to spread between the two.
3. Drag one slide over the other in a rapid, smooth movement, spreading the cells in a thin smear over the surface of both slides.
4. Air-dry the slides for approximately 5 min. The slides may then be used at once, or wrapped individually in foil or cling film/Saran Wrap and stored in the freezer until required. If stored frozen, allow to thaw to room temperature before use.
5. When ready for use, fix by dipping in acetone for 1 min and air-dry.

3.4. Frozen (Cryostat) Sections

1. Using a sharp, clean blade, cut a solid tissue block of approximately 0.5 cm^3 (see Note 15).
2. Place the tissue on cryostat chuck thickly coated in cryostat embedding medium (see Note 16).
3. Using long-handled tongs, pick up the chuck and immerse chuck and tissue in isopentane or hexane pre-cooled in liquid nitrogen for approximately 1–2 min (see Note 17).
4. Place the frozen chuck in the cabinet of the cryostat and leave to equilibrate for approximately 30 min.
5. Using the cryostat, cut 5–10-μm thick sections and pick them up on clean, dry silane-treated microscope slides. Allow to air-dry for between 1 h and overnight.
6. Sections may then either be stored until required by wrapping individually or back-to-back in aluminium foil or cling film/Saran Wrap, sealing in polythene bags or boxes containing dessicant, and storing in the freezer, or may be used at once. If stored frozen, allow to thaw and equilibrate to room temperature before opening.
7. Immediately before use, dip slides in acetone for 1–10 min and air-dry for approximately 5 min.

3.5. Fixed, Paraffin Wax-Embedded Sections

1. Cool the paraffin wax-embedded tissue blocks on ice for approximately 15 min.
2. Cut 4–7 μm thick sections by microtome (see Note 18).
3. Float sections out on a pool of 20% v/v ethanol or IMS in distilled/deionised water on a clean glass plate supported by a suitable receptacle such as a glass beaker or jar (see Note 19).
4. Carefully transfer the sections, floating on the alcohol, onto the surface of a water bath heated to 40°C—they should puff out and become flat (see Note 20).
5. Separate out individual sections very gently, using the tips of fine, bent forceps.
6. Pick up individual sections on silane-treated microscope slides.
7. Allow slides to drain by up-ending them on a sheet of absorbent paper for 5–10 min.
8. Dry slides either in a 37°C incubator overnight or in a 60°C oven for 15–20 min (see Note 21). Slides may then be cooled, stacked or boxed, and stored at room temperature in a dust-tight container until required.
9. When required, soak slides in xylene for approximately 15 min to remove the paraffin wax.
10. Transfer through two changes of absolute ethanol or IMS, then through one change of 70% v/v ethanol or IMS in distilled/deionised water, then distilled/deionised water, agitating the slides vigorously for 1–2 min at each stage to equilibrate (see Note 22).

3.6. Buffers for Blocking, Dilutions, and Washes

These buffers are referred to in the following methods.

3.7. Enzyme-Based Antigen Retrieval Methods (See Note 5)

When using fixed paraffin wax-embedded tissues, methods to retrieve antigens damaged or sequestered by the harsh fixation and processing may be necessary, or may significantly enhance results. These methods are not appropriate for other types of cell or tissue preparation (see Note 23).

1. Trypsinisation: immerse slides in a bath of trypsin solution at 37°C, in an incubator or water bath, for 5–30 min, then wash in running tap water for 5 min.
2. Protease treatment: place slides face up in a suitable chamber (see Note 24) and apply a few drops of protease solution to cover the tissue preparation. Incubate at 37°C, for 5–30 min, and wash in running tap water for 5 min.

3. Pepsin treatment: place slides face up in a suitable chamber (see Note 24) and apply a few drops of pepsin solution to cover the tissue preparation, or immerse slides in a bath of pepsin solution. Incubate at 37°C, for 5–30 min, then wash in running tap water for 5 min.
4. Neuraminidase treatment: place slides face up in a suitable chamber (see Note 24) and apply a few drops of neuraminidase solution to cover the tissue preparation. Incubate at 37°C, for 5–30 min, then wash in running tap water for 5 min.

3.8. Microwave Oven Heat-Mediated Antigen Retrieval Method

An alternative to enzyme-mediated antigen retrieval is heat-mediated retrieval. The method given below is for heat-mediated treatment using a microwave oven, but other methods exist using, for example, pressure cooking or autoclaving (see Note 7).

1. Immerse slides in citrate buffer, pH 6.0, in any suitable microwave-safe container, such as a plastic sandwich box (see Note 25).
2. Place in a conventional microwave oven and heat on full power until the buffer boils.
3. Reduce the power to "simmer" or "defrost" for 5 min, so that the buffer simmers gently. Check the level of the buffer, and top up with hot distilled/deionised water if necessary. Heat on "simmer" or "defrost" for a further 5 min.
4. Allow slides to cool at room temperature for 30 min (see Note 26).
5. Wash in running tap water for 5 min.

3.9. Quenching Endogenous Enzyme

If HRP is to be employed as the label, then endogenous peroxidase must be quenched as follows. This step is most conventionally performed immediately prior to the addition of the primary antisera.

1. Immerse slides in methanol/hydrogen peroxide solution for 20 min.
2. Wash in running tap water for approximately 5 min.

If AP is to be employed as the label molecule it may be necessary to quench endogenous AP, although it is usually destroyed by processing to paraffin wax rendering this procedure unnecessary. If required, 1 mM levamisole is added to the final enzyme development medium.

3.10. Examples of Some Immunocytochemical Labelling Techniques

As described in Subheading 1, a number of basic immunocytochemical techniques are available which vary in their relative complexity and sensitivity. Illustrative examples are listed here which should give good results, but the researcher is urged to experiment and adapt these basic techniques to give optimum results in his/her experimental system. Other techniques also exist.

3.10.1. Direct Method (See Fig. 1)

1. Incubate slides with blocking buffer for 30 min.
2. Drain slides, and wipe around cell or tissue preparation using a clean, dry tissue.
3. Incubate slides with (fluorescent or enzyme) labelled monoclonal or polyclonal antibody in a humid chamber (see Note 24), for between 30 min and 2 h at room temperature, or at 4°C overnight.
4. Wash in three changes of washing buffer (see Note 27).
5. If fluorescently labelled antibody is employed, mount and view directly using a fluorescent microscope. If enzyme-labelled antibody is used, proceed to enzyme development as described in Subheading 3.11 onwards.

3.10.2. Simple Indirect Method (See Fig. 2)

1. Incubate slides with blocking buffer for 30 min.
2. Drain slides, and wipe around cell or tissue preparation using a clean, dry tissue.
3. Incubate slides with unlabelled monoclonal or polyclonal antibody in a humid chamber (see Note 24), for 30 min to 2 h at room temperature, or at 4°C overnight.
4. Wash in three changes of washing buffer (see Note 27).
5. Incubate slides with either fluorescent or enzyme-labelled secondary antibody directed against the immunoglobulins of the species in which the primary antibody was raised (see Note 28), in a humid chamber (see Note 24), for 1 h.
6. Wash in three changes of washing buffer (see Note 27).
7. If fluorescently labelled secondary antibody is employed, mount and view directly using a fluorescent microscope. If enzyme-labelled secondary antibody is used, proceed to enzyme development as described in Subheading 3.11 onwards.

3.10.3. Simple Avidin–Biotin Method (See Fig. 3)

1. Incubate slides with blocking buffer for 30 min.
2. Drain slides, and wipe around cell or tissue preparation using a clean, dry tissue.
3. Incubate slides with biotin-labelled monoclonal or polyclonal primary antibody in a humid chamber (see Note 24), for 30 min to 2 h at room temperature, or at 4°C overnight.
4. Wash in three changes of washing buffer (see Note 27).
5. Incubate with avidin or streptavidin labelled with fluorescent or enzyme label, in a humid chamber (see Note 24), for 30 min.
6. Wash in three changes of washing buffer (see Note 27).
7. If fluorescently labelled avidin or streptavidin is employed, mount and view directly using a fluorescent microscope. If

enzyme-labelled avidin or streptavidin is used, proceed to enzyme development as described in Subheading 3.11 onwards.

3.10.4. Avidin–Biotin Complex Method (See Fig. 4)

1. Incubate slides with blocking buffer for 30 min.
2. Drain slides, and wipe around cell or tissue preparation using a clean, dry tissue.
3. Incubate slides with biotin-labelled monoclonal or polyclonal primary antibody in a humid chamber (see Note 24), for 30 min to 2 h at room temperature, or at 4°C overnight.
4. Wash in three changes of washing buffer (see Note 27).
5. Incubate with fluorescent or enzyme-labelled ABC, in a humid chamber (see Note 24), for 30 min.
6. Wash in three changes of washing buffer (see Note 27).
7. If fluorescently labelled ABC is employed, mount and view directly using a fluorescent microscope. If enzyme-labelled ABC is used, proceed to enzyme development as described in Subheading 3.11 onwards.

3.10.5. Peroxidase–Anti-peroxidase or Alkaline Phosphatase–Anti-alkaline Phosphatase Methods (See Fig. 5)

1. Incubate slides with blocking buffer for 30 min.
2. Drain slides, and wipe around cell or tissue preparation using a clean, dry tissue.
3. Incubate slides with unlabelled monoclonal or polyclonal primary antibody in a humid chamber (see Note 24), for 30 min to 2 h at room temperature, or at 4°C overnight.
4. Wash in three changes of washing buffer (see Note 27).
5. Incubate with an unlabelled "bridging" antibody (see Note 11) for 1 h, in a humid chamber (see Note 24).
6. Wash in three changes of washing buffer (see Note 27).
7. Incubate with PAP or APAAP for 1 h, in a humid chamber (see Note 24).
8. Wash in three changes of buffer (see Note 27).
9. Proceed to enzyme development as described in Subheading 3.11 onwards.

3.10.6. Simple Polymer Reagent Method (See Fig. 6)

1. Incubate slides with blocking buffer for 30 min.
2. Drain slides, and wipe around cell or tissue preparation using a clean, dry tissue.
3. Incubate slides with unlabelled monoclonal or polyclonal antibody in a humid chamber (see Note 24), for 30 min to 2 h at room temperature, or at 4°C overnight.
4. Wash in three changes of washing buffer (see Note 27).

5. Incubate slides with polymer complex of secondary antibody directed against the immunoglobulins of the species in which the primary antibody was raised (see Note 28) and fluorescent or enzyme label, in a humid chamber (see Note 24) according to manufacturer's instructions.
6. Wash in three changes of washing buffer (see Note 27).
7. If fluorescently labelled secondary antibody is employed, mount and view directly using a fluorescent microscope. If enzyme-labelled secondary antibody is used, proceed to enzyme development as described in Subheading 3.11 onwards.

3.11. Enzyme Development Methods

3.11.1. Diaminobenzidine for HRP Label

1. Wash in three changes of washing buffer (see Note 27).
2. Incubate with DAB-H_2O_2 for 10 min (see Note 12).
3. Wash in running tap water for 5 min.
4. Proceed to counterstaining (Subheading 3.12) and mounting (Subheading 3.13).

3.11.2. Fast Red for AP Label

1. Wash slides briefly in TBS, pH 8.2–9.0.
2. Take 18.5 ml TBS pH 8.2–9.0, add 500 μl stock solution of naphthol phosphate and mix, then add Levamisole to give a 1 mM solution and mix, then add 1 ml fast red solution and mix. Filter and apply to slides immediately.
3. Immerse the slides in fast red solution for 5–30 min (see Note 29).
4. Wash in running tap water for 5 min.
5. Proceed to counterstaining (Subheading 3.12) and mounting (Subheading 3.13).

3.12. Counterstaining

1. Immerse in Mayer's haematoxylin solution for 3–5 min.
2. "Blue" by immersing in running tap water for 5 min, or dip briefly in 1% v/v ammonia in tap water then wash in tap water (see Note 30).
3. Proceed to mounting (Subheading 3.13).

3.13. Mounting

1. Preparations labelled using fluorescent tags should be mounted directly in an anti-fade fluorescent mountant such as "Citifluor" or similar.
2. Preparations labelled using alcohol-soluble chromagens such as fast red for AP, should be mounted directly in an aqueous mountant such as "Aquamount" or similar.
3. Preparations labelled using alcohol-insoluble chromagens such as DAB for HRP should be dehydrated by immersing, with agitation, for 1 min each in 70% v/v, 95% v/v ethanol or IMS, then two changes of absolute ethanol, cleared by immersion, with

a agitation, in two changes of xylene, then mounted in a xylene-based mountant such as "Depex" or similar (see Note 31).

3.14. Viewing, Interpretation, and Quantification of Labelling Results

Slides should be viewed by light or fluorescence microscope, as appropriate. Fluorescently labelled preparations should be stored in the dark and in a refrigerator until viewing. They should be viewed as soon as possible after labelling as fluorescence will fade over time, and photographic images should be made for permanent record. Enzyme labels should be permanent, and slides may therefore be stored for longer. Some aqueous mountants deteriorate over time. Good labelling is indicated by a strong specific label and low, or preferably non-existent, background and nonspecific labelling.

It is often helpful to score labelling on an arbitrary scale where the observer estimates the percentage of cells, for example cancer cells, labelled (10, 50, and 95%, etc.) and the intensity of labelling on a scale of–(no labelling at all), +/– very weak labelling, + (weak but definite labelling) to ++++ (extremely intense labelling) to give results ranging from completely negative to 100% ++++. Preferably, this should be carried out by at least two independent observers and the results compared.

Many attempts have been made to quantify immunocytochemistry results using automated, computer-based approaches, but the author is unaware of any truly satisfactory and reproducible system applicable to all applications.

3.15. Some Common Problems and Suggestions Likely for Solutions

3.15.1. High Nonspecific Background Labelling

This is possibly the most common problem and can be caused by a number of different factors. The most common is probably insufficient washing between steps (see Note 27). The second most likely cause is employment of too high a concentration of one or more of the reagents. All reagents (primary antibody, secondary antibody, avidin and biotin products, PAP or APAAP, etc.) should be titrated carefully to give optimum results (see Table 2 and subheading 1.4). When using HRP as a label, residual endogenous peroxidase may sometimes be a problem—check by incubating a slide that has been treated simply with the standard methanol-hydrogen peroxide solution (Subheading 2.9) with the chromogenic substrate DAB-H_2O_2 (Subheading 2.11)—there should be no brown labelling present; if there is, this indicates the presence of unquenched endogenous peroxidase. If this is the case, try freshly made methanol-hydrogen peroxide, increase the concentration of hydrogen peroxide or the incubation time in this step, or try an alternative label, for example, AP.

If the problematic high background labelling is absent in a negative control where primary antibody is omitted, this would indicate a cross reaction between the primary antibody and some cell or tissue component. This may be remedied by one or more of the following: increasing the concentration of blocking serum or protein in blocking buffer (see Subheading 2.6 and Note 4), incorporating more sodium chloride (up to 0.1 M) into blocking, dilution and washing buffers, or adding a little detergent (e.g. 0.05% v/v Tween 20) to washing buffers.

In avidin–biotin-based methods, endogenous biotin can sometimes cause confusing results. The endogenous biotin can be blocked by applying unconjugated avidin (which binds to tissue-bound biotin), then saturating with further free, excess, unlabelled biotin. Avidin may also sometimes attach to ionic cell/tissue sites: this may be most easily remedied by increasing the pH of washing, dilution, and blocking buffer to 9.0, or may be avoided by using the slightly more costly streptavidin products instead of avidin.

3.15.2. Weak or Absent Labelling

The positive control—a preparation known to express the antigen of interest—should be checked. If satisfactory labelling is achieved here, it would suggest that the antigen is present in only low levels, or absent, in the test slides. If low levels only are present, perhaps indicated by weak labelling, a more sensitive detection technique should be employed. If formalin-fixed, paraffin wax-embedded material is being examined, antigen retrieval methods should be tried. If positive controls show inadequate labelling, all of the reagents should be systematically checked for reactivity.

4. Notes

1. We use silane-treated slides for all cell and tissue preparations in immunocytochemistry. Alternative, commercial, brand-named preparations are also available, but tend to be more expensive. Silane, or equivalent, treatment is essential if a heat-mediated antigen retrieval method is to be employed subsequently. Slides should not be agitated in the baths of reagents, as air bubbles will prevent the silane solution reaching the glass surface and will result in patchy and inadequate treatment.
2. Any cells in suspension are suitable—for example, blood, cancer cells in ascites or in pleural effusions taken from patients or from animal models, cells derived from solid tissue tumours and released into suspension, or cultured cells in suspension.
3. Xylene is potentially hazardous and should be handled with care in a fume cupboard. Modern, safer chemical alternatives are commercially available, but it is our experience that they give slightly less satisfactory results.
4. Blocking buffer and buffer for dilution of antibodies: the use of normal (non-immune) animal serum and its incorporation into working solutions of antibodies effectively reduces nonspecific background labelling. Goat or horse serum is

recommended here for simplicity as it is unlikely that the detection methods employed will involve specific recognition of antibodies raised in these species. It is also appropriate to employ normal (non-immune) serum from the final antibody-producing species in any detection method—for example, in an indirect method where a labelled rabbit secondary antibody is employed to detect binding of an unlabelled mouse monoclonal antibody, non-immune rabbit serum would be entirely appropriate. Solutions of an "inert" protein or protein mixture such as bovine serum albumin, casein, or commercially available dried skimmed milk powder are also routinely used, and are cheaper than non-immune serum. As a guide, a solution of 1–5% w/v solution in buffer is appropriate.

5. It is not possible to predict if enzyme-based antigen retrieval methods will be effective, or which to choose; try them out and select conditions which work best for your application. Test a range of treatment times, for example, 0, 5, 10, 20, and 30 min. Trypsinisation is the most widely used antigen retrieval technique, probably because it is relatively cheap and can be extremely effective. The other enzymes tend to be more expensive (especially neuraminidase), but are more selective. Try heat-mediated antigen retrieval too. These methods should only be used in conjunction with fixed, paraffin wax-embedded tissue preparations mounted on silane- (or equivalent, see Note 1) treated glass slides.
6. Use crude, type II trypsin, from porcine pancreas. Impurities (e.g. chymotrypsin) enhance its effect. Do not use purer (and more expensive!) products.
7. It is not possible to predict if heat-mediated antigen retrieval methods will be effective in any particular case. This can only be determined by trial and error. Other heat-mediated antigen retrieval methods, such as autoclaving and pressure cooking also exist. Try enzyme-based antigen retrieval methods (Subheadings 2.7 and 3.7) too. These methods should only be used in conjunction with fixed, paraffin wax-embedded tissue preparations mounted on silane- (or equivalent, see Note 1) treated glass slides.
8. Avidin is a glycoprotein extracted from egg white, and it has four binding sites for the vitamin biotin. Streptavidin is a protein, similar in structure to avidin, and is derived from the bacterium *Streptomyces avidinii*. Avidin products are generally cheaper than streptavidin products. Streptavidin is said to give a cleaner result, but avidin may be perfectly acceptable for most applications.
9. The ABC is available commercially in a convenient dropper bottle kit format. Follow kit instructions, which usually require that avidin and labelled biotin are combined 30 min before use. Avidin and labelled biotin are mixed together in a ratio such

that three of the four possible biotin-binding sites are saturated, leaving one free to combine with the biotin label attached to the primary antibody.

10. The PAP and APAAP techniques were extremely popular some years ago owing to their sensitivity, resulting from the multi-layering. They are less commonly used today and have been superseded to some extent by the avidin–biotin, ABC and labelled polymer methods.
11. The bridging antibody forms a link or bridge between the primary antibody and the PAP or APAAP complex. So, for example, if a monoclonal mouse primary antibody is used, the bridging antibody will be antisera raised against mouse immunoglobulins; if a rabbit polyclonal primary antibody is used, the bridging antibody will be antisera directed against rabbit immunoglobulins. This antisera should be applied in excess. The bridging antibody needs to be present in excess so that only one of the two possible antigen-binding sites of each bridging antibody molecule are occupied by primary antibody, leaving the second binding site free to bind to the PAP or APAAP complex. As a rough guide, in our experience, most commercially available secondary antibodies will work best in this context when diluted at around 1/50.
12. DAB is potentially carcinogenic. It should be handled with care, using gloves. Avoid spillages and aerosols. Work in a fume cupboard. After use, soak all glassware, etc. in a dilute solution of bleach overnight before washing. Wash working surfaces with dilute bleach after use. Clean up spillages with excess water then wash with dilute bleach. We usually make a concentrated DAB stock solution at 5 mg/ml in distilled/deionised water and store at –20°C frozen in 1-ml aliquots in 10-ml plastic screw top tubes until required. This minimises the risk from weighing out the powder when required. DAB is also available in convenient tablet or dropper bottle kit form, which is minimises hazards, but is more expensive.
13. Use glass as plastic will dissolve in the dimethylformamide.
14. A number of different haematoxylin solutions are commercially available. Mayer's, a progressive stain, is convenient, but other types are equally effective.
15. When cutting tissue, use a very sharp blade and use single, firm, swift, downwards strokes. Avoid hacking and crushing which will result in poor morphology. Cut the tissue into a straight-edged, geometrical shape, most usually a cube, as this will make sectioning easier. Clearly, when handling human tissue, appropriate health and safety guidelines should be adhered to.

16. The cryostat embedding medium acts as a support for the tissue. Apply it generously to the chuck and immerse the tissue block into a pool of it. Align the tissue block with a straight edge parallel with the cutting edge of the chuck.
17. To obtain optimum morphological integrity, the tissue should be frozen as rapidly as possible, avoiding the formation of ice crystals as these are the most common cause of loss of morphology. The most effective method is to immerse the tissue in isopentane or hexane, pre-cooled in liquid nitrogen. These solvents conduct heat away from the tissue more rapidly that liquid nitrogen alone. Other methods of freezing the block—for example, by using a commercially available freezing spray, by blasting with CO_2 gas, or even simply placing it in the chamber of the cryostat until frozen—are less effective. Cool the isopentane or hexane in an appropriate vessel that will withstand immersion in liquid N_2.
18. Most tissues cut best when chilled. An ice cube or handful of crushed should be applied to the surface of the tissue block every few minutes during cutting. This is particularly important when working in a warm room.
19. Small creases or wrinkles in the section should begin to flatten out; the effect may be enhanced by gentle manipulation using, for example, a soft paintbrush and/or forceps.
20. If the wax begins to melt, it is too hot. If sections remain wrinkled, it may indicate that the water is slightly too cold.
21. Do not heat directly on a hotplate as this may damage some antigens.
22. Take careful note of the appearance of the slides during this process. When slides are transferred from one solvent to the next, they initially appear smeary as the two solvents begin to mix. Vigorous agitation ensures that the slides equilibrate effectively with the new solvent. When they are equilibrated, the surface smearing will disappear. During the re-hydration process, if white flecks or patches become visible around the sections, this indicates that the wax has not been adequately removed—return the sections to xylene for a further 10–15 min. A common cause of poor immunocytochemistry results is the inadequate removal of paraffin wax.
23. Make enzyme solutions fresh immediately before use. Use glassware, solutions, and so on that have been pre-warmed to 37°C before use. Initially try a range of digestion times, for example, 0, 5, 10, 20, and 30 min. Digestion times of >30 min are not recommended as visible damage to tissue morphology becomes apparent. This is particularly evident when using trypsin.

24. A flat platform on which to place the slides is used. This has a lid and creates a humid chamber. Humidity is important so that small volumes of solution placed onto the surface of the slides do not evaporate and therefore either dry out completely or become more concentrated during the incubation step. Drying of cell/tissue preparations will result in high nonspecific background labelling. For small numbers of slides it may be convenient to place them face up in a lidded Petri dish lined with a disc of damp filter paper. For larger numbers of slides, commercially available incubation chambers—usually fashioned in Perspex and containing raised ridges to support slides over troughs partially filled with water to maintain a humid atmosphere—are convenient. These tend to be expensive to buy. They can be made "in-house" if appropriate facilities exist. It is possible to make a perfectly functional incubation chamber very simply using a large sandwich box with supports for slides formed from, for example, glass or wooden rods or glass burettes, supported with "Plasticine" or Blu-tac. Again, a little water or dampened filter paper may be added to the base of the chamber to maintain humidity.
25. Slides can be placed in commercially available slide carriers, which typically hold up to about 12 or 25 slides and may be housed in, for example, appropriately sized plastic sandwich boxes, or upright in plastic Coplin jars. Space the slides out evenly in the buffer. Do not overcrowd slides as this will result in "hot spots" and uneven antigen retrieval. Do not use metal racks to hold slides when using the microwave-medicated antigen retrieval methods.
26. This cooling down period is part of the retrieval method and should not be skipped.
27. We recommend vigorous washing in three changes of washing buffer. Each wash should consist of vigorous "sloshing up and down" of slides in buffer for about 30 s to 1 min, then allowing slides to stand in the buffer for about 4 min. Insufficient washing, in particular omitting one or more changes of buffer can result in unacceptably high levels of dirty background labelling.
28. For example, if a monoclonal mouse primary antibody was used, incubate with labelled secondary antisera raised against mouse immunoglobulins; if a rabbit polyclonal primary antibody was used, incubate with a labelled secondary antibody directed against rabbit immunoglobulins.
29. Monitor the progress of colour development by periodic examination using a microscope. Stop development when specifically labelled structures show deep red and before nonspecific background labelling begins to occur.

30. Cell/tissue preparations will initially stain deep purple-red after immersion in Mayer's haematoxylin. The stain changes to deep blue when exposed to mildly alkaline conditions (known as "bluing"). This is most commonly achieved by washing in the slightly alkaline tap water that is available in most areas. If "bluing" is unsuccessful owing to unusually acidic tap water, dip slides briefly in 1% v/v ammonia in tap water then wash in tap water.
31. Take careful note of the appearance of the slides during this process. When slides are transferred from one solvent to the next, they initially appear smeary as the two solvents begin to mix. Vigorous agitation ensures that the slides equilibrate effectively with the new solvent. When they are equilibrated, the surface smearing will disappear. White clouding of the xylene (it appears "milky") indicates contamination with water. Slides should be passed back through graded alcohols (absolute ethanol, 95, and 70% v/v ethanol) to tap water, solvents should be discarded and replaced, and the process repeated.

Reference

1. Coons AH, Creech HJ, Jones RM (1941) Immunological properties of an antibody containing a fluorescent group. Proc Soc Exp Biol Med 47:200–202
2. Coons AH, Leduc EH, Connolly JM (1955) Studies on antibody production. I: a method for the histochemical demonstration of specific antibody and its application to a study of the hyperimmune rabbit. J Exp Med 102:49–60
3. Coons AH, Kaplan MH (1950) Localisation of antigen in tissue cells. J Exp Med 91:1–13
4. Nakane PK, Pierce GB Jr (1966) Enzyme-labelled antibodies: preparation and application for the localization of antigens. J Histochem Cytochem 14:929–931
5. Mason DY, Sammons RE (1978) Alkaline phosphatase and peroxidase for double immunoenzymic labelling of cellular constituents. J Clin Pathol 31:454–462
6. Guesdon JL, Ternynck T, Avrameas S (1979) The uses of avidin–biotin interaction in immunoenzymatic techniques. J Histochem Cytochem 27:1131–1139
7. Hsu SM, Raine L, Fanger H (1981) Use of avidin–biotin–peroxidase complex (ABC) in immunoperoxidase techniques: a comparison between ABC and unlabelled antibody (PAP) procdures. J Histochem Cytochem 29:577–580
8. Cattoretti G, Pileri S, Parravicini C, Becker MH, Poggi S, Bifulco C, Key G, D'Amato L, Sabattini E, Feudale E, Reynolds F, Gerdes J, Rilke F (1993) Antigen unmasking on formalin fixed, paraffin embedded tissue sections. J Pathol 171:83–98
9. Cattoretti G, Suurjmeijer AJH (1995) Antigen unmasking on formalin-fixed paraffin embedded tissues using microwaves: a review. Adv Anat Pathol 2:2–9
10. Norton AJ, Jordan S, Yeomans P (1994) Brief, high temperature heatdenaturation (pressure cooking): a simple and effective method of antigen retrieval for routinely processed tissues. J Pathol 173:371–379
11. Bankfalvi A, Navabi H, Bier B, Bocker W, Jasani B, Schmid KW (1994) Wet autoclave pre-treatment for antigen retrieval in diagnostic immunocytochemistry. J Pathol 174:223–228
12. Riggs JL, Seiwald RJ, Burkhalter JH, Downs CM, Metcalf T (1958) Isothiocyanate compounds as fluorescent labelling agents for immune serum. Am J Pathol 34:1081–1097
13. Titus JA, Haughland R, Sharrows SO, Segal DM (1982) Texas red, a hydrophilic, red-emitting fluorophore for use with fluorescein in dual parameter flow microfluorometric and

fluorescence microscopic studies. J Immunol Meth 50:193–204

14. Graham RC, Karnovsky MJ (1966) The early stages of absorption of injected horseradish peroxidase in the proximal tubules of mouse kidney: ultrastructural cytochemistry by a new technique. J Histochem Cytochem 14: 291–302
15. Graham RC, Ludholm U, Karnovsky MJ (1965) Cytochemical demonstration of peroxidase activity with 3-amino-9-ethylcarbazole. J Histochem Cytochem 13:150–152
16. Burstone MS (1961) Histochemical demonstration of phosphatases in frozen sections with naphthol AS-phosphates. J Histochem Cytochem 9:146–153

Further Reading

Polak JM, van Noorden S (1997) Introduction to immunocytochemistry, second edition. Royal microscopical society handbooks, number 37. Bios Scientific Publishers, Oxford, UK.

Chapter 2

Lectin Histochemistry to Detect Altered Glycosylation in Cells and Tissues

Susan A. Brooks and Debbie M.S. Hall

Abstract

Lectins are naturally occurring carbohydrate-binding molecules. A very wide range of purified lectins are commercially available which exhibit a diversity of carbohydrate-binding preferences. They can be used in the laboratory to detect carbohydrate structures on, or in, cells and tissues in much the same way that purified antibodies can be employed to detect cell- or tissue-bound antigens using immunocytochemistry. As lectins can distinguish subtle alterations in cellular glycosylation, they are helpful in exploring the glycosylation changes that attend both transformation to malignancy and tumour progression. In this chapter, methodologies are given for appropriate preparation of many types of cell and tissue preparations, including cells cultured on coverslips (which can be used for live-cell imaging), cell smears, and frozen (cryostat) and fixed, paraffin wax-embedded tissue sections. Heat- and enzyme-based carbohydrate retrieval methods are covered. Basic detection methods, which can be readily adapted to the researcher's needs, are given for direct (labelled lectin), simple indirect (labelled secondary antibody directed against the lectin), and avidin–biotin (biotinylated lectin) and avidin–biotin complex. The use of both the enzyme label, horseradish peroxidase, and fluorescent labels is considered.

Key words: Histochemistry, Lectin, Glycan, Frozen, FFPE tissue, Direct, Indirect

1. Introduction

1.1. Lectins and Their Applications

A lectin is a "protein or glycoprotein of non-immune origin, not an enzyme, that binds to carbohydrates and agglutinates cells" (1). Lectins are naturally occurring substances, which occur ubiquitously in nature. Purified preparations used in laboratory research are most commonly derived from plant or, sometimes, invertebrate sources. They can be exploited in the laboratory to detect and reveal carbohydrate structures in or on the surface of cells or, for example, as elements of glycoproteins separated and immobilised by SDS-PAGE and Western blotting, in very much the same way that antibodies

Miriam Dwek et al. (eds.), *Metastasis Research Protocols*, Methods in Molecular Biology, vol. 878,
DOI 10.1007/978-1-61779-854-2_2, © Springer Science+Business Media, LLC 2012

can be used to reveal specific antigens. Lectins can detect subtle alterations in cellular glycosylation. This is of interest in metastasis research as there is evidence that marked glycosylation changes can attend both transformation to malignancy and tumour progression.

1.2. Nomenclature of Lectins

Lectins are named after the source from which they are derived—sometimes the Latin name (for example, *Bandeirea simplicifolia* lectin or *Dolichos biflorus* lectin), sometimes the common name (for example, peanut lectin or wheatgerm lectin), or sometimes by a slightly obscure historical term (for example, Concanavalin A or Con A for the lectin from *Canavalia ensiformis*, the jack bean). The term lectin is used fairly interchangeably with the older term "agglutinin", as in peanut agglutinin (PNA) or *Helix pomatia* agglutinin (HPA). Lectins are often referred to by an abbreviation for their names, for example PNA or *Dolichos biflorus* agglutinin (DBA); obscurely, PHA which stands for phytohaemagglutinin is, for historical reasons, the abbreviation usually employed for the lectin derived from *Phaseolus vulgaris*. Many sources yield more than one lectin, termed isolectins, which may have quite different carbohydrate binding specificities (for example, the gorse, *Ulex europaeus*, yields two major isolectins *Ulex europaeus* agglutinin I or UEA-I, which has a strong binding preference for fucose, and *Ulex europaeus* agglutinin II or UEA-II, which has a strong binding preference for *N*-acetylglucosamine).

1.3. Monosaccharide Binding Specificity and Natural Binding Partners

The sugar binding specificity of a lectin is usually quoted in terms of the monosaccharide (or sometimes the slightly more complex structure, perhaps a disaccharide) that most effectively inhibits its binding to its naturally occurring complex—often undefined—binding partners. For example, the lectin from the Roman snail, *Helix pomatia* (HPA), is said to have a monosaccharide binding specificity for α-D-*N*-acetylgalactosamine, as does the lectin from the horse gram *Dolichos biflorus* (DBA). It is very important, however, to understand that the combining site of the lectin commonly encompasses more than the terminal monosaccharide (in this example, α-D-*N*-acetylgalactosamine) in an oligosaccharide chain, and that lectins with nominally identical monosaccharide binding specificities will actually recognise subtly different complex binding partners and will give quite different results in lectin histochemistry. In the example given above, HPA and DBA will give quite distinct labelling patterns in lectin histochemistry in spite of their apparent identical nominal binding specificities.

1.4. The Range of Lectins Available Commercially

Many hundreds of lectins are readily available, at modest cost, from commercial sources. Very few have well-defined binding characteristics, and little or nothing is known about some. In most cases, limited information, such as monosaccharide-binding preference, will be available. It is, therefore, often the case that the researcher may need to test several lectins before the tool most ideally suited to the task is selected.

Although the literature abounds with lectin histochemical studies, surprisingly little useful data have yet been revealed by their use. This may be in the most part owing to the conservative nature of many lectin histochemical studies, which often concentrate on a limited panel of much-used and well-characterised lectins, probably chosen for their nominal coverage of the major monosaccharides present in human tissues and often purchased as a convenient kit from one of the major suppliers. Progress seems more likely if researchers are more adventurous and use a broader spectrum of the lectins available.

1.5. The Range of Lectin Histochemistry Methods

Lectin histochemistry can be carried out on a range of different cell and tissue preparations, including, as described in this chapter, frozen or cryostat sections of fresh tumour tissue, formalin-fixed, paraffin wax-embedded surgical specimens, cells in suspension (either cultured cell lines or clinical samples, such as blood samples, ascites, or pleural effusions), and cultured cancer cell lines either grown on coverslips or pelleted, fixed, and processed to paraffin wax blocks. If cells or tissues are fixed and processed to paraffin wax, antigen retrieval methods (trypsinisation or heat-mediated retrieval methods) may assist the recovery of sequestered carbohydrates and improve labelling results significantly.

A very large number of histochemical labelling techniques are available for detection of lectin binding to cells and tissues. Each has advantages and disadvantages, and the simple and basic techniques described here should all work well and give good results. Lectin binding may be visualised by either a fluorescent (especially suitable for fresh cell smears or cells cultured on coverslips) or an enzyme label which yields a coloured product (the enzyme label in the methods given here is horseradish peroxidase, which yields a deep chocolate brown product with the chromogenic substrate diaminobenzidine (DAB)/H_2O_2, but alternative enzyme labels, such as alkaline phosphatase are also commonly used). Detection methods range from the very straightforward, but relatively insensitive, "direct method" to the more complex, multi-step, and more sensitive "avidin–biotin complex" (ABC) method. Other, even more sensitive, but further complex methods exist.

1.6. Some Examples of Glycosylation Changes, Detectable by Lectin Histochemistry, of Interest in Metastasis Research

There are a number of major alterations in cellular glycosylation, detectable by lectin histochemistry, that appear to be of interest in metastasis research. This field is reviewed elsewhere (2–4). Some examples, to be briefly discussed here, include altered sialylation, increased synthesis of complex β1-6-branched *N*-acetylglucosamine oligosaccharides (detected by the lectin PHA), and incomplete synthesis and/or synthesis of truncated structures. It seems likely that in the rapidly advancing field of glycobiology, as the functions of complex carbohydrates and their interactions with their receptors become better understood, their relevance and interactions in the complex processes of metastasis will increasingly be revealed.

1.6.1. Alterations in Sialylation

Sialic acids (neuraminic acids) are commonly found at the non-reducing termini of oligosaccharide chains, and disturbances in sialylation associated with malignancy and tumour progression have been reported. Increased synthesis of sialylated structures, for example the sialylated form of the Tn epitope (sialyl Tn, sTn), sialyl Lewis A (sLe^a), and sialyl Lewis X (sLe^x), have been associated with malignancy and metastatic competence, as has over-expression of sialyltransferase enzymes involved in synthesis of sialylated structures (5–8). Conversely, other studies have correlated failure in sialylation mechanisms with experimental metastatic capacity (9).

1.6.2. PHA Binding to Detect Complex β1-6-Branched Oligosaccharides

Increased synthesis of complex β1-6-branched oligosaccharides, detected by the lectin PHA, has been associated to increased metastatic potential of tumour cells in a number of studies (10–12).

1.6.3. Incomplete Synthesis/Synthesis of Truncated Structures

Incompletely synthesised or truncated carbohydrate structures, resulting from a failure in normal chain extension, are a feature of cancer cells. Attachment of a single *N*-acetylgalactosamine residue to the polypeptide backbone yields a simple structure, known as the Tn antigen. This is extended in normal cells, either by addition of sialic acid to yield sialyl Tn (sTn) or addition of other monosaccharides to build more complex structures. However, it is frequently unelaborated in cancer where its presence has been described as a marker of poor prognosis and an interesting potential target for immunotherapy (see Subheading 2 for review). *Helix pomatia* lectin (HPA) recognises Tn and other oligosaccharides terminating in α-D-*N*-acetylgalactosamine. Positive labelling of cancers by HPA using lectin histochemistry is associated with metastatic competence and poor patient survival in a range of human adenocarcinomas, including breast, colon, gastric, prostate, and oesophageal (13–16).

2. Materials

2.1. Silane/APES Treatment of Microscope Slides (See Note 1)

1. Acetone.
2. Acetone/silane: 7 ml 3-(triethoxysilyl)-propylamine (APES) in 400 ml acetone; make fresh on the day of use. Discard after treating a maximum of 1,000 slides.
3. Distilled water.

2.2. Cell and Tissue Preparations

2.2.1. Frozen or Cryostat Sections

1. Chunk of fresh tissue approximately 0.5 cm^3 in size.
2. OCT embedding medium.
3. Isopentane.
4. Liquid nitrogen.
5. APES-coated clean glass microscope slides (see Note 1).
6. Acetone.
7. Aluminium foil.

2.2.2. Fixed, Paraffin Wax-Embedded Sections

1. Paraffin wax-embedded tissue blocks (see Note 2).
2. 20% v/v ethanol (or industrial methylated spirits, IMS) in distilled water.
3. APES-coated glass microscope slides (see Note 1).
4. Xylene.
5. Absolute ethanol or IMS.
6. 70% v/v ethanol or IMS in distilled water.
7. Distilled water.

2.2.3. Carbohydrate Retrieval from Fixed, Paraffin Wax-Embedded Sections; Trypsinisation

1. Tris-buffered saline (TBS), pH 7.6: 60.57 g Tris, 87.0 g NaCl dissolved in 1 l distilled water. pH to 7.6 using concentrated HCl. Make up to a total volume of 10 l using distilled water.
2. Trypsin solution: 1 mg/ml crude, type II trypsin, from porcine pancreas (see Note 3) in TBS, pH 7.6.

2.2.4. Carbohydrate Retrieval from Sections Using Heat

1. Citrate buffer, pH 6.0: 2.1 g citric acid dissolved in 1 l distilled water. pH to 6.0 using concentrated NaCl.
2. Distilled water.

2.2.5. Cell Smears

1. Cells in suspension (see Note 4).
2. APES-coated glass microscope slides (see Note 1).
3. Aluminium foil.
4. Acetone.

2.3. Preparing Cultured Cells on Coverslips

1. Cells should be cultured under standard conditions to 70% confluence or as appropriate.
2. Cell culture medium, without foetal calf serum (FCS).
3. Alcohol- or autoclave-sterilised round glass coverslips (13 mm diameter, thickness 0).
4. Piperazine-*N*,*N'*-bis-[2-ethanesulfonic acid] (PIPES) buffer 0.2 M, pH 6.9: Stir 12.1 g PIPES into 50 ml ultrapure water to give a cloudy solution. Add approximately 40 ml 1 M NaOH and the solution should clear. Check pH and adjust to 6.9, if

necessary, using 1 M NaOH. Add ultrapure water to give a final volume of 200 ml.

5. PIPES buffer 0.1 M, pH 6.9: Prepare 0.2 M PIPES buffer, as above, but add ultrapure water to give a total volume of 400 ml, instead of 200 ml, or take a small volume, e.g. 50 ml of 0.2 M PIPES buffer, and dilute 1:1 with ultrapure water.
6. 3% v/v paraformaldehyde in 0.1 M PIPES buffer, pH 6.9: Place 3 g paraformaldehyde in a 250-ml conical flask, add 30 ml ultrapure water, loosely stopper, and heat on a 60°C hotplate, in a fume cupboard, for about 30 min to give a cloudy solution. Add 1 M NaOH (the purest grade), with continual stirring, until the solution clears. Add ultrapure water to give a total volume of 50 ml, and then add 50 ml 0.2 M PIPES buffer, pH 6.9 (see above). Divide into 10-ml aliquots and store frozen. Defrost in warm water bath for use.
7. 0.1% v/v Triton X-100 in 0.1 M PIPES buffer, pH 6.9.

2.4. Preparing Cultured Cells in Cell Pellets or Appropriate, for Paraffin Wax Embedding

1. Cells cultured to 70% confluence under standard conditions.
2. Cell medium, without FCS.
3. 4% v/v formol saline in distilled water, at 4°C: Dissolve 4.25 g NaCl in 500 ml of a 4% v/v formaldehyde solution in distilled water.
4. Lectin buffer: 60.57 g Tris, 87.0 g NaCl, 2.03 g $MgCl_2$, 1.11 g $CaCl_2$ dissolved in 1 l distilled water. pH to 7.6 using concentrated HCl. Make up to a total volume of 10 l using distilled water (see Note 5).
5. 4% w/v agarose in lectin buffer: Suspend 4 g low-gelling-temperature agarose in 100 ml lectin buffer and heat gently (in a water bath or microwave oven, with regular stirring) until the agarose dissolves (approximately 60°C). Cool to approximately 37–40°C (the agarose should thicken slightly as it begins to gel) for use.

2.5. Quenching Endogenous Peroxidase

1. Methanol/hydrogen peroxide solution: 3% v/v hydrogen peroxide in methanol. Make up fresh every 2–3 days.

2.6. Labelling Methods

2.6.1. Direct Fluorescent Label Method

1. Fluorescently labelled lectin, for example fluorescein isothiocyanate (FITC)- or tetramethylrhodamine (TRITC)-labelled lectin.
2. Lectin buffer: 60.57 g Tris, 87.0 g NaCl, 2.03 g $MgCl_2$, 1.11 g $CaCl_2$ dissolved in 1 l distilled water. pH to 7.6 using concentrated HCl. Make up to a total volume of 10 l using distilled water (see Note 5).

2.6.2. Direct Horseradish Peroxidase Label Method

1. Horseradish peroxidase-labelled lectin.
2. Lectin buffer: 60.57 g Tris, 87.0 g NaCl, 2.03 g $MgCl_2$, 1.11 g $CaCl_2$ dissolved in 1 l distilled water. pH to 7.6 using concentrated HCl. Make up to a total volume of 10 l using distilled water (see Note 5).
3. DAB-H_2O_2: 3,3-diaminobenzidine tetrahydrochloride, 0.5 mg/ml, in lectin buffer. Add H_2O_2 to give a concentration of 5% v/v immediately before use (see Notes 6–8).
4. Mayer's haematoxylin (see Note 9).

2.6.3. Biotin-Labelled Lectin: Simple Avidin–Biotin Method

1. Biotin-labelled lectin.
2. Lectin buffer: 60.57 g Tris, 87.0 g NaCl, 2.03 g $MgCl_2$, 1.11 g $CaCl_2$ dissolved in 1 l distilled water. pH to 7.6 using concentrated HCl. Make up to a total volume of 10 l using distilled water (see Note 5).
3. Streptavidin labelled with horseradish peroxidase.
4. DAB-H_2O_2: 3,3-diaminobenzidine tetrahydrochloride, 0.5 mg/ml, in lectin buffer. Add H_2O_2 to give a concentration of 5% v/v immediately before use (see Notes 6–8).
5. Mayer's haematoxylin (see Note 9).

2.6.4. Indirect Antibody Method

1. Native, unlabelled lectin.
2. Lectin buffer: 60.57 g Tris, 87.0 g NaCl, 2.03 g $MgCl_2$, 1.11 g $CaCl_2$ dissolved in 1 l distilled water. pH to 7.6 using concentrated HCl. Make up to a total volume of 10 l using distilled water (see Note 5).
3. Peroxidase-labelled rabbit polyclonal antibody directed against the lectin (see Note 10).
4. DAB-H_2O_2: 3,3-diaminobenzidine tetrahydrochloride, 0.5 mg/ml, in lectin buffer. Add H_2O_2 to give a concentration of 5% v/v immediately before use (see Notes 6–8).
5. Mayer's haematoxylin (see Note 9).

2.6.5. Avidin–Biotin Complex Method

1. Biotin-labelled lectin.
2. Lectin buffer: 60.57 g Tris, 87.0 g NaCl, 2.03 g $MgCl_2$, 1.11 g $CaCl_2$ dissolved in 1 l distilled water. pH to 7.6 using concentrated HCl. Make up to a total volume of 10 l using distilled water (see Note 5).
3. Horseradish peroxidase ABC (see Note 11).
4. DAB-H_2O_2: 3,3-diaminobenzidine tetrahydrochloride, 0.5 mg/ml, in lectin buffer. Add H_2O_2 to give a concentration of 5% v/v immediately before use (see Notes 6–8).
5. Mayer's haematoxylin (see Note 9).

2.7. Dehydration, Clearing, and Mounting Slides for Viewing by Microscopy

1. 70% v/v ethanol or IMS in distilled water.
2. 100% ethanol or IMS.
3. Xylene.
4. Xylene-based mounting medium, for example "Depex".
5. Coverslips.

3. Methods

3.1. APES Treatment of Microscope Slides (See Note 1)

1. Place up to a maximum of 1,000 clean glass microscope slides in slide carriers.
2. Place the first slide carrier in a trough containing enough acetone to cover the slides, and leave for 5 min.
3. Transfer, smoothly and without agitation, to a trough containing enough acetone/APES to cover the slides, and leave for 5 min.
4. Transfer, smoothly and without agitation, to a trough containing enough distilled water to cover the slides, and leave for 5 min. Discard the water and replace with fresh after five carriers of slides have passed through.
5. Transfer, smoothly and without agitation, to a second trough of distilled water, and leave for 5 min. Discard the water and replace with fresh after five carriers of slides have passed through.
6. Transfer to absorbent paper and allow to drain.
7. Allow to dry thoroughly—either at room temperature or more quickly in a suitable incubator, glass drying cabinet, or oven.
8. When completely dry, store in labelled boxes until required.

3.2. Different Types of Cellular and Tissue Preparations

Cells and tissues can be prepared in a number of different ways for lectin histochemistry. Some of the most common and more useful preparations are listed below.

3.2.1. Frozen or Cryostat Sections

1. Using a sharp, clean blade, cut a solid tissue block of approximately 0.5 cm^3 (see Note 12).
2. Place the tissue on an OCT-coated cryostat chuck (see Note 13).
3. Using long-handled tongs, pick up the chuck and immerse chuck and tissue in isopentane pre-cooled in liquid nitrogen for approximately 1–2 min (see Note 14).
4. Place the frozen chuck in the cabinet of the cryostat and leave to equilibrate for approximately 30 min.

5. Using the cryostat, cut 5-µm-thick sections and pick them up on APES-coated (see Note 1) clean glass microscope slides. Allow to air-dry for approximately 5 min.
6. Sections may then either be stored until required by wrapping individually in aluminium foil and storing in the freezer or may be used at once. If stored frozen, allow to thaw and equilibrate to room temperature before use.
7. Immediately before use, dip slides in acetone for 1 min and air-dry for approximately 5 min.

3.2.2. Fixed, Paraffin Wax-Embedded Sections

1. Cool wax-embedded tissue blocks on ice for approximately 15 min.
2. Cut 5-µm-thick sections by microtome (see Note 15).
3. Float sections out on a pool of 20% v/v ethanol in distilled water on a clean glass plate supported by a suitable receptacle, such as a glass beaker or a jar (see Note 16).
4. Carefully transfer the sections, floating on the alcohol, onto the surface of a water bath heated to 40°C—they should puff out and become flat (see Note 17).
5. Separate out individual sections very gently using the tips of fine, bent forceps.
6. Pick up individual sections on clean, APES-coated glass microscope slides (see Note 1).
7. Allow slides to drain by upending them on a sheet of absorbent paper for 5–10 min.
8. Dry slides either in a 37°C incubator overnight or on a hotplate at 60°C for 20 min. Slides may then be cooled, stacked, or boxed, and stored at room temperature in a dust-tight container until required.
9. When required, soak slides in xylene for approximately 15 min to remove the paraffin wax.
10. Transfer through two changes of absolute ethanol or IMS, then through one change of 70% v/v ethanol or IMS in distilled water, and then distilled water, agitating the slides vigorously for 1–2 min at each stage to equilibrate (see Note 18).

3.2.3. Trypsinisation for Carbohydrate Retrieval (See Note 19)

1. Immerse slides in trypsin solution at 37°C, in an incubator or water bath, for 5–30 min (see Notes 3 and 20).
2. Wash in running tap water for 5 min.

3.2.4. Heat-Mediated Carbohydrate Retrieval

1. Immerse slides in citrate buffer, pH 6.0, in a suitable microwave-safe container, such as a plastic sandwich box (see Note 21).
2. Place in a conventional microwave oven and heat on full power until the buffer boils.

3. Reduce the power to "simmer" or "defrost" for 5 min so that the buffer simmers gently. After 5 min, check the level of the buffer, and top up with hot distilled water if necessary. Heat on "simmer" or "defrost" for a further 5 min.
4. Allow slides to cool at room temperature for 30 min (see Note 22).
5. Wash in running tap water for 5 min.

3.2.5. Cell Smears

1. Place a drop of cells in suspension approximately 5 mm from one end of a clean APES-coated glass microscope slide (see Note 1).
2. Place a second clean APES-coated glass microscope slide on top of the first, allowing approximately 1 cm of glass to protrude at either end and allowing the drop to spread between the two slides.
3. Drag one slide over the other in a rapid, smooth movement, spreading the cells in a thin smear over the surface of both slides.
4. Air-dry the slides for approximately 5 min. They may then be used at once or wrapped individually in foil and stored in a freezer until required. If stored frozen, allow to thaw to room temperature before use.
5. When ready for use, fix by dipping in acetone for 1 min and air-dry.

3.2.6. Methods for Preparing Cultured Cells

Grown on Coverslips

1. Wash cultured cells, grown to 70% confluence, in fresh medium.
2. Aspirate and discard the medium.
3. Scrape cells from the flask using a rubber scraper and re-suspend in fresh medium.
4. Count cells, and subculture 1×10^5 cells into Petri dishes.
5. Place sterile coverslips in Petri dishes, and allow cells to proliferate for 24 h at 37°C under 5% CO_2.
6. Carefully remove coverslips using fine forceps or the edge of a scalpel blade.
7. Place coverslips, cell side up, onto a piece of dental wax or Parafilm for support, and cover each with 100 μl cold 3% v/v paraformaldehyde in 0.1 M PIPES buffer, pH 6.9, for 15 min.
8. Wash thoroughly in 0.1 M PIPES buffer, pH 6.9.
9. Permeabilize in 0.1% (v/v) Triton X-100 (diluted in 0.1 M PIPES buffer, pH 6.9) for 10 min (see Note 23).
10. Wash thoroughly in 0.1 M PIPES buffer, pH 6.9.

Fixed and Processed for Paraffin Wax Embedding

1. Wash cultured cells, grown to 70% confluence, in several changes of fresh medium without FCS.
2. Scrape cells from the flask using a rubber scraper.
3. Transfer to a 50-ml centrifuge tube and centrifuge at 1,100 × *g* for 5 min to pellet the cells.
4. Gently re-suspend the pellet (see Note 24) in fresh medium without FCS, and centrifuge at 1,100 × *g* for 5 min. Repeat.
5. Aspirate the supernatant and discard.
6. Gently re-suspend the cell pellet (see Note 24) in 4% v/v formol saline for 2 h at 4°C to fix the cells.
7. Centrifuge, as previously, aspirate, and discard the supernatant.
8. Wash the cell pellet in three changes of lectin buffer—centrifuge, aspirate, and discard the supernatant, add fresh buffer, and re-suspend the cells, as previously.
9. Gently re-suspend the cell pellet in a drop of warm (37°C) liquid agarose and allow to set at room temperature (see Note 25).
10. Process the cell/agarose pellets for paraffin wax embedding and then cut sections as detailed in Subheading 3.2.2. Carbohydrate retrieval methods (Subheading 3.2.3 and 3.2.4) may be appropriate.

3.3. Quenching Endogenous Peroxidase (See Note 26)

1. Immerse slides in methanol/hydrogen peroxide solution for 20 min.
2. Wash in running tap water for approximately 5 min.
3. Proceed to Subheading 3.4.6.

3.4. Examples of Some Histochemical Labelling Techniques (See Note 27)

3.4.1. Direct Fluorescent Label Method (See Note 28)

1. Incubate slides with fluorescently labelled lectin at a concentration of 10 μg/ml in lectin buffer, in a humid chamber, for 1 h (see Note 29).
2. Wash in three changes of lectin buffer (see Note 30).
3. Proceed to Subheading 3.4.6.

3.4.2. Direct Horseradish Peroxidase Label Method (See Note 31)

1. Incubate slides with horseradish peroxidase-labelled lectin at a concentration of 10 μg/ml in lectin buffer, in a humid chamber (e.g. a sandwich box lined with damp filter paper), for 1 h (see Note 29).
2. Wash in three changes of lectin buffer (see Note 30).
3. Incubate with DAB-H_2O_2 for 10 min (see Notes 6–8).
4. Wash in running tap water for 5 min.
5. Counterstain in Mayer's haematoxylin for 2 min (see Note 9).
6. "Blue" in running tap water for approximately 5 min (see Note 32).
7. Proceed to Subheading 3.4.6.

3.4.3. Biotin-Labelled Lectin: Simple Avidin–Biotin Method (See Note 33)

1. Incubate slides with biotin-labelled lectin at a concentration of 10 μg/ml in lectin buffer, in a humid chamber, for 1 h (see Note 29).
2. Wash in three changes of lectin buffer (see Note 30).
3. Incubate with streptavidin labelled with peroxidase at a concentration of 5 μg/ml in lectin buffer, in a humid chamber, for 30 min.
4. Wash in three changes of lectin buffer (see Note 30).
5. Incubate with DAB-H_2O_2 for 10 min (see Notes 6–8).
6. Wash in running tap water for 5 min.
7. Counterstain in Mayer's haematoxylin for 2 min (see Note 9).
8. "Blue" in running tap water for approximately 5 min (see Note 32).
9. Proceed to Subheading 3.4.6.

3.4.4. Indirect Antibody Method (See Note 10)

1. Incubate slides with unlabelled lectin at a concentration of 10 μg/ml in lectin buffer, in a humid chamber, for 1 h (see Note 29).
2. Wash in three changes of lectin buffer (see Note 30).
3. Incubate with horseradish peroxidase-labelled rabbit polyclonal antibody directed against the lectin, at a dilution of 1/100 in lectin buffer, in a humid chamber, for 1 h.
4. Wash in three changes of lectin buffer (see Note 30).
5. Incubate with DAB-H_2O_2 for 10 min (see Notes 6–8).
6. Wash in running tap water for 5 min.
7. Counterstain in Mayer's haematoxylin for 2 min (see Note 9).
8. "Blue" in running tap water for approximately 5 min (see Note 32).
9. Proceed to Subheading 3.4.6.

3.4.5. Avidin–Biotin Complex Method (See Note 34)

1. Incubate slides with biotin-labelled lectin at a concentration of 10 μg/ml in lectin buffer, in a humid chamber, for 1 h (see Note 29).
2. Wash in three changes of lectin buffer (see Note 30).
3. Incubate with horseradish peroxidase ABC, prepared according to the manufacturer's instructions (see Note 11) in a humid chamber, for 30 min.
4. Wash in three changes of lectin buffer (see Note 30).
5. Incubate with DAB-H_2O_2 for 10 min (see Notes 6–8).
6. Wash in running tap water for 5 min.
7. Counterstain in Mayer's haematoxylin for 2 min (see Note 9).
8. "Blue" in running tap water for approximately 5 min (see Note 32).
9. Proceed to Subheading 3.4.6.

3.4.6. Dehydration, Clearing, and Mounting Slides for Viewing by Microscopy (See Note 35)

1. Dehydrate by passing through 70% v/v ethanol or IMS spirit in distilled water, then two changes of 100% ethanol or IMS, and then clear in xylene. Agitate slides for 1–2 min at each stage to equilibrate.
2. Horseradish peroxidase-labelled slides should be mounted in a xylene-based mounting medium, for example Depex. Fluorescently labelled slides should be mounted in a commercially available fade-resistant mounting medium.

3.5. Positive and Negative Controls and Confirmation of Specificity of Labelling

1. Positive control: A cell or tissue preparation that is known to give positive labelling from a previous experiment makes the ideal positive control. If this is not possible, kidney—either human or animal—makes an excellent positive control for labelling with many lectins owing to its diverse glycosylation patterns.
2. Negative control: Omit the lectin. Labelling should be completely abolished.
3. Confirmation of specificity: Add 0.1–0.5 M monosaccharide for which the lectin shows the greatest affinity (for example, for *Ulex europaeus* isolectin I, add fucose; for *Helix pomatia* lectin, add *N*-acetylgalactosamine) to the lectin solution, at its working dilution, about 30 min prior to incubation with the cell or tissue preparation (see Note 36).

3.6. Interpretation of Labelling Results

Slides should be viewed using a light or fluorescence microscope. Good labelling is indicated by a strong specific label (either deep brown for horseradish peroxidase/DAB or fluorescent), low, or preferably non-existent, background, and non-specific labelling (see Note 37). Omission of the lectin should abolish labelling completely. The competitive inhibition experiment with an appropriate monosaccharide should abolish or at least dramatically diminish labelling.

It is often helpful to score labelling on an arbitrary scale, where the observer estimates the percentage of cells, for example cancer cells, labelled (10%, 50%, 95%, etc.) and the intensity of labelling on a scale of – (no labelling at all), +– (very weak labelling), and + (weak but definite labelling) to ++++ (extremely intense labelling) to give results ranging from completely negative to 100% ++++.

4. Notes

1. APES alters the electrical charge on the glass of the microscope slide so that cell and tissue preparations adhere much more firmly. We routinely APES treat all our microscope slides before use. It is absolutely essential if slides are to be subjected to

heat-mediated retrieval methods as the aggressiveness of the treatment will otherwise dislodge even the best cell/tissue preparation. It is desirable, but not essential, for other applications. Alternative, commercial, brand-named preparations are also available, but are more expensive. Other adhesives, for example glycerol gelatin or poly-L-lysine, will also improve adherence of cells/tissues to slides but are not as effective and are inadequate for preparations that are to be subjected to heat-mediated retrieval methods. Once acetone/APES solution is prepared, it should be used the same day and then discarded. 400 ml acetone/7 ml APES is sufficient to treat up to about 1,000 microscope slides; it is, therefore, a good idea to treat a batch of 1,000 slides, and then store them for use as required.

2. Paraffin wax-embedded tissue blocks are often the most convenient source of material for clinical studies as tumour specimens are routinely fixed in formalin and processed to paraffin wax for sectioning, labelling, and histopathological diagnosis in most clinical centres. Such blocks are often stored in hospital archives for years or decades, providing a hugely valuable resource for retrospective studies on glycosylation as related to the metastatic potential of tumours.

3. Crude, type II trypsin from porcine pancreas works best as the presence of impurities (e.g. chymotrypsin) assists its effect. Do not use purer (and more expensive!) products.

4. Any cells in suspension are suitable—for example, blood, cancer cells in ascites or in pleural effusions taken from patients or animal models, cells derived from solid tissue tumours and released into suspension, or cultured cells in suspension.

5. Lectin buffer is TBS with $CaCl_2$ and $MgCl_2$ added, and is good for all lectin histochemistry applications. We recommend it for all dilutions and washes. Many lectins are known to require Ca^{2+} and Mg^{2+} to stabilise their carbohydrate-binding site(s). The requirements of other lectins remain unknown. As a starting point, it is most inadvisable to use phosphate-buffered saline (PBS) for lectin histochemistry as the phosphate ions may bind with and sequester metal ions from the lectin-binding sites.

6. DAB is potentially carcinogenic and should be handled with care. Wear gloves. Avoid spillages and aerosols. Work in a fume cupboard. After use, soak all glassware, etc. in a dilute solution of bleach overnight before washing. Wash down worktops with dilute bleach after use. Clean up spillages with excess water, and then wash with dilute bleach.

7. DAB is available in dropper bottle kit form from major suppliers. This is more expensive than preparing solutions from powdered reagent, but is convenient and safer.

8. We usually make a concentrated DAB stock solution at 5 mg/ml in lectin buffer and freeze in a 1-ml aliquot in 10-ml plastic screw-top tubes until required. This minimises the risk associated with weighing out powder.
9. Mayer's haematoxylin: A number of different haematoxylin solutions are commercially available. We find Mayer's, which is a progressive stain, convenient, although other types are equally effective.
10. Peroxidase-labelled rabbit polyclonal antibodies against lectins: Only a very limited range of polyclonal antisera against lectins are available commercially, so lack of availability may limit the application of this method.
11. ABC is available from major suppliers, often as convenient dropper bottle kits. Avidin has four binding sites for biotin. Avidin and labelled biotin are mixed together in such a ratio that three of the possible four binding sites on avidin are saturated with labelled biotin, leaving one free to combine with the biotin label attached to the lectin. To use these, follow the manufacturer's instructions; these usually require that the avidin and labelled biotin solutions are mixed together in the appropriate ratio at least 30 min prior to use.
12. When cutting tissue, use a very sharp blade, for example a new razor blade or scalpel, and use single, firm, swift, downwards stokes. Do not hack at the tissue and avoid crushing it as this will result in poor morphology. Cut the tissue into a geometrical shape with straight edges, preferably a square, as this will make sectioning easier.
13. OCT acts as a support for the tissue block during sectioning. Embed the tissue block in a generous pool of it. Align the tissue block, in the OCT, with one of its straight edges parallel with the cutting edge of the chuck, as this will make sectioning easier.
14. It is possible to freeze the tissue block in other ways—for example, by immersing in liquid nitrogen, using a commercially available freezing spray, blasting with CO_2 gas, or even simply placing it in the chamber of the cryostat until frozen. However, to obtain optimum morphological integrity, the tissue should be frozen as rapidly as possible, avoiding the formation of ice crystals, and the best way is to immerse in isopentane pre-cooled in liquid nitrogen. Hexane pre-cooled in liquid nitrogen is also very good. Isopentane/hexane are far better conductors of heat than liquid nitrogen alone and, therefore, facilitate extremely rapid and effective freezing. Safety note: Take appropriate precautions when using liquid nitrogen and solvents.

15. Most tissues can be cut effectively when very cold. Ice (for example, an ice cube) should be applied to the surface of the tissue block every few minutes during cutting.
16. During this step, if small creases or wrinkles are observed in the sections, these will flatten out. Gentle manipulation using, for example, a soft paintbrush and/or forceps can aid this.
17. If sections remain wrinkled, it may indicate that the water is slightly too cold. If the wax begins to melt, it is too hot.
18. When slides are transferred from one solvent to the next, they initially appear "smeary" as the two solvents begin to mix. Vigorous agitation, i.e. "sloshing them up and down", ensures that the slides equilibrate effectively with the new solvent. When they are equilibrated, the surface smearing will disappear. If white flecks or patches become visible around the sections, this indicates that the wax has not been adequately removed—return the sections to xylene for a further 10–15 min. A common cause of poor labelling results is inadequate removal of paraffin wax.
19. Carbohydrate retrieval methods: In some cases, if the lectin histochemistry results on the fixed, paraffin wax-embedded tissues are disappointing, it may be because the carbohydrate moieties have been sequestered during tissue fixation and processing to paraffin wax. Results can often be improved by carbohydrate retrieval techniques, including trypsinisation or heat treatment. It is not possible to predict if carbohydrate retrieval methods will be effective or which to choose; try both and select which conditions work best for your application (other heat-mediated retrieval methods, such as heating using an autoclave or pressure cooking, may also be appropriate). N.B.: These methods are not appropriate for fresh cell/tissue preparations, such as frozen sections or cell smears.
20. Pre-warm all glassware, solutions, etc. to 37°C before use. Make up trypsin solution fresh immediately before use; trypsin will autolyse (self-digest) and lose activity over time. Initially, try a range of trypsinisation times, e.g. 0, 5, 10, 20, and 30 min. Incubation times of >30 min are not recommended as visible damage to tissue morphology becomes apparent.
21. Slides can conveniently be placed in commercially available slide carriers, which typically hold up to about 25 slides, or upright in Coplin jars. Do not overcrowd slides as this results in "hot spots" and uneven carbohydrate retrieval. When using microwave heat retrieval methods, do not use metal slide carriers!
22. The cooling down period is part of the retrieval method and should not be skipped.

23. If cell surface carbohydrates only are of interest, this and subsequent steps may be omitted.
24. To re-suspend cells, either vortex in short sharp bursts or tap the tube sharply. Do not suck the cells up and down in a pipette as this will damage them.
25. To suspend cells in liquid agarose at 37°C, pre-warm a glass pipette to 37°C and transfer cells to a small drop (approximately 0.5 ml) of agarose in a small plastic tube, such as a 1.5-ml Eppendorf tube. Tap the tube several times sharply to mix. Leave agarose to set at room temperature for approximately 30 min. The agarose/cell pellet can be removed by either cutting the plastic tube away or dislodging the pellet from the tube using a syringe needle or pin.
26. If horseradish peroxidase is used to visualise lectin binding, endogenous peroxidase enzyme within the cell or tissue preparation may contribute to non-specific and confusing labelling. This may be overcome by quenching endogenous enzyme. Endogenous peroxidase is usually quenched immediately prior to incubation with the lectin. This step is not required if fluorescent labels are used. Other enzyme labels, for example alkaline phosphatase, are also commonly used and, although these will not require this step, they may require their own blocking procedure.
27. A large number of different histochemical techniques can be used to detect lectin binding to cell and tissue preparations. The choice of technique is dependent on the type of label preferred by the experimenter—fluorescent or enzyme (coloured)—and the balance between quick, simple techniques which generally lack sensitivity and the more sensitive techniques which tend to be more time consuming, contain more steps, and require more reagents (and are, therefore, slightly more expensive). All methods given here should give good results. Other methods also exist. The methods described in chapter 1 by Brooks in this volume can be readily adapted for use with lectins instead of primary antibodies using the general guidelines given in this chapter.
28. Fluorescent labels can give exceptionally beautiful results, but the fluorescence is ephemeral and preparations should be viewed (and if a permanent record is required, photographed) as soon as possible after labelling. Slides can be stored in a refrigerator, in the dark, for up to about 6 months, but some preparations will deteriorate significantly during that time. The fluorescence-based method described is especially good for cells in suspension and cell smears. It may also be used for tissue sections, and use of a commercially available fluorescent nuclear counterstain is then recommended to aid interpretation of tissue morphology.

29. Lectins are usually purchased as a purified powdered product. We routinely solubilise the powder in lectin buffer to give a stock solution of 1 mg/ml, which is stable in a refrigerator for several months. Lectin stock solution can then be diluted to the optimum working dilution as required. It is best practice to prepare working solutions immediately prior to use and discard unused solution at the end of the experiment. A working concentration of 10 μg/ml is often the optimal concentration for most lectin histochemistry. It may be a good idea to try a range of concentrations, e.g. 2.5, 5, 10, 20, and 40 μg/ml, to assess which gives the best results.
30. We recommend vigorous washing in three changes of lectin buffer. Each wash should consist of vigorous "sloshing up and down" of slides in buffer for about 30 s–1 min, and then allowing slides to stand in the buffer for about 4 min. Insufficient washing, in particular omitting one or more changes of buffer, can result in unacceptably high levels and dirty background labelling.
31. This method has the advantage of being quick, simple, and straightforward. It lacks the sensitivity of some of the longer, multi-step, and avidin–biotin methods. We have found that horseradish peroxidase-labelled lectins can sometimes give subtly different binding results to native (unlabelled) or biotin-labelled lectins, possibly owing to stearic hindrance of lectin binding site by the relatively large peroxidase molecule (17).
32. Mayer's haematoxylin is a port-wine red colour, and cell/tissue preparations will stain purple/red after immersion in it. The haematoxylin changes to a deep blue when exposed to mildly alkaline conditions (known as "bluing"). The mild alkalinity of tap water is usually sufficient. If bluing is unsuccessful owing to unusually acidic tap water, soak slides instead in tap or distilled water to which a few drops of ammonia or sodium hydroxide have been added.
33. This is a really useful method and is highly recommended for most applications; it is also quick, straightforward, and sensitive, generally giving good results. The biotin label appears to be small enough not to interfere with the combining site of the lectin and therefore does not appear, in our hands, to alter its binding characteristics (17) (see Note 31).
34. This is a very sensitive method that gives good, clean results. It should be employed when the simpler methods, for example the simple avidin–biotin method, give weak labelling.
35. After labelling, slides are in an aqueous medium. Before they can be mounted in a xylene-based mounting medium, they must therefore be dehydrated through graded alcohols and cleared in xylene. Aqueous mounting media are also commercially

available which, historically, tended to give inferior results to xylene-based mounting media. Some of the modern ones are, however, very good and may be used instead without the need for dehydration and clearing.

36. For a few lectins, this may not be possible as they are only inhibited by complex sugars, not monosaccharides (consult manufacturer's literature). If this is the case, inhibition with a heavily glycosylated glycoprotein, e.g. fetuin or asialofetuin, should give effective inhibition, as should incubation of the lectin in the presence of 5% v/v normal (human or animal) serum.

37. If dirty, non-specific background labelling is observed, this may indicate inadequate washing between stages in the labelling protocol (see Note 30). If careful attention to washing does not satisfactorily limit background labelling, incorporation of a blocking agent into lectin, antibody, and streptavidin–horseradish peroxidase solutions may be indicated. The simplest and most effective blocking agent is to include 3% w/v bovine serum albumin. Normal serum, often used as a blocking agent in immunocytochemistry, is not appropriate as the many heavily glycosylated molecules present will effectively inhibit specific labelling.

References

1. Goldstein IJ, Hughes RC, Monsigny M, Osawa T, Sharon N (1980) What should be called a lectin? Nature 285:66
2. Brooks SA, Carter TM, Royle L, Harvey DJ, Fry SA, Kinch C, Dwek RA, Rudd PM (2008) Altered glycosylation of proteins in cancer: what is the potential for new anti-tumour strategies. Anticancer Agents Med Chem 8:2–21
3. Dwek MV, Brooks SA (2004) Harnessing changes in cellular glycosylation in new cancer treatment strategies. Curr Cancer Drug Targets 4:425–442
4. Hakomori S (2001) Tumor-associated carbohydrate antigens defining tumor malignancy: basis for development of anti-cancer vaccines. Adv Exp Med Biol 491:369–402
5. Maramatsu T (1993) Carbohydrate signals in metastasis and prognosis of human carcinomas. Glycobiology 3:294–296
6. Kanangi R (1997) Carbohydrate mediated cell adhesion involved in hematogenous metastasis of cancer. Glycoconj J 14:577–584
7. Renkonen R, Mattila P, Marja-Leena M, Rabina J, Toppila S, Renkonen J, Hirvas L, Niittymaki J, Turunen JP, Renkonen O, Paavonen T (1997) *In vitro* experimental studies of sialyl Lewis x and sialyl Lewis a on endothelial and carcinoma cells: crucial glycans on selectin ligands. Glycoconj J 14: 593–600
8. Nakano T, Matsui T, Ota T (1996) Benzyl-alpha-GalNAc inhibits sialylation of O-glycosidic sugar chains and enhances experimental metastatic capacity in B16BL6 melanoma cells. Anticancer Res 16:3577–3584
9. Yogeeswaren G, Salk PL (1981) Metastatic potential is positively correlated with cell surface sialylation of cultured murine tumor cells. Science 212:1514–1516
10. Dennis JW, Laferte S, Waghorne C, Breitman ML, Kerbel RS (1987) Beta 1-6 branching Asn-linked oligosaccharides is directly associated with metastasis. Science 236:582–585
11. Fernandes B, Sagman U, Auger M, Demetrio M, Dennis JW (1991) Beta 1-6 branched oligosaccharides as a marker of tumor progression in human breast and colon neoplasia. Cancer Res 51:718–723
12. Korczak B, Goss P, Fernandes B, Baker M, Dennis JW (1984) Branching N-linked oligosaccharides in breast cancer. Adv Exp Med Biol 353:95–104

13. Brooks SA, Leathem AJC (1991) Prediction of lymph node involvement in breast cancer by detection of altered glycosylation in the primary tumour. Lancet 338:71–74
14. Schumacher U, Higgs D, Loizidou M, Pickering R, Leathem A, Taylor I (1994) *Helix pomatia* agglutinin binding is a useful prognostic indicator in colorectal carcinoma. Cancer 74:3104–3107
15. Kakeji Y, Maehara Y, Tsujitani S, Baba H, Ohno S, Watanabe A, Sugimachi K (1994) *Helix pomatia* agglutinin binding activity and lymph node metastasis in patients with gastric cancer. Semin Surg Oncol 10:130–134
16. Yoshida Y, Okamura T, Shirakusa T (1994) An immunohistochemical study of *Helix pomatia* agglutinin binding on carcinomas of the esophagus. Surg Gynecol Obstet 177: 299–302
17. Brooks SA, Lymboura M, Schumacher U, Leathem AJC (1996) *Helix pomatia* lectin binding in breast cancer - methodology makes a difference. J Histochem Cytochem 44: 519–524

Further Reading

Brooks SA, Leathem AJC, Schumacher U (1997) Lectin histochemistry – a concise practical handbook. BIOS, Oxford, UK

Carter TM, Brooks SA (2006) Detection of aberrant glycosylation in breast cancer using lectin histochemistry. Methods Mol Med 120:201–216

Chapter 3

Histopathological Assessment of Metastasis

Derek E. Roskell and Ian D. Buley

Abstract

In spite of advances in the fields of immunohistochemistry and molecular biology, in clinical practice much of the assessment of metastases still relies on light microscopy using conventional histological stains. This is not so much a reflection of a reluctance by histopathologists to adopt new techniques, but more an indication that for most malignancies an enormous amount of useful prognostic data can be gained from relatively unsophisticated assessment of tissues, and that many of the strongest studies of prognostic factors in malignancy predate the era of molecular diagnostics. Although it is undoubtedly true that newer techniques have added prognostic information in the assessment of many tumors, and many, such as the measurement of estrogen receptor status in breast cancer, could be considered routine, a skilled assessment of the morphology of the tissues still provides the fundamental basis of assessing prognosis in the vast majority of cases.

Key words: Pathological, Specimen, Immunochemistry, Primary cancer, Secondary cancer, Metastasis

1. Introduction

In spite of recent advances in the fields of immunohistochemistry and molecular biology, in clinical practice much of the assessment of metastases still relies on light microscopy using conventional histological stains. This is not so much a reflection of reluctance by histopathologists to adopt new techniques, but more an indication that for most malignancies an enormous amount of useful prognostic data can be gained from relatively unsophisticated assessment of tissues. Many of the strongest studies of prognostic factors in malignancy predate the era of molecular diagnostics. While it is undoubtedly true that newer techniques have added prognostic information to the assessment of many tumors, many, such as the measurement of estrogen receptor and HER2 status in breast cancer cells, are considered routine. A skilled assessment of the morphology of tissues will provide the fundamental basis for assessing prognosis in the vast majority of cases.

Miriam Dwek et al. (eds.), *Metastasis Research Protocols*, Methods in Molecular Biology, vol. 878,
DOI 10.1007/978-1-61779-854-2_3,

1.1. Metastatic Potential of Primary Tumors: Typing and Subtyping

For some malignant tumors the presence of metastatic disease may be obvious by clinical examination at the time of presentation. Therapeutic interventions that remove the whole primary tumor (in the absence of overt metastatic disease) do not always, and in many cancers seldom, result in long-term cure. Probably the most important factor in determining the metastatic potential of a primary tumor is the correct identification of the tumor type. Some tumors, such as basal cell carcinoma of the skin, are very unlikely to metastasize in any circumstance, while others, such as small cell carcinoma of the lung, metastasize in almost every case. For these tumors, simply making the diagnosis usually offers an adequate assessment of metastatic potential.

Detailed morphological subtyping of tumors can also be informative. The World Health Organisation (WHO) Histological Classification of Tumors of the Breast lists more than 30 different subtypes of carcinoma and many of these particular histological appearances have implications for the incidence of metastatic spread. However, the majority of common malignancies fall into a group in which the incidence of metastasis varies considerably, and a more detailed assessment must be undertaken in order to establish the likely prognosis on a case-by-case basis.

2. Methods

2.1. Stage and Grade

Histopathological assessment of the excised primary tumor to predict the likelihood of recurrence or metastasis involves assessing two aspects of tumor growth: the stage and the grade. Although these are often confused and used interchangeably, these simple terms are quite different. The stage simply refers to how far the tumor has spread within the tissue of origin or beyond, while the grade refers to the perceived aggressiveness of the cancer cells, regardless of how far they have invaded the host tissue.

2.1.1. Stage

Staging a malignancy by assessment of the primary site has its limitations, as this will not detect the presence of tumor in tissues that have been left in the patient. However, there are clues that can point to the likelihood of tumor cells having escaped to form metastatic deposits. The factors to be considered include the proximity of the surgical margin, the size of the tumor, the presence of lymphatic or vascular invasion, and the invasion of the tumor through structures known to provide a physical barrier keeping the tumor from areas rich in lymphatics. For example, invasion of a colonic carcinoma into and through the muscle of the bowel wall confers a worse prognosis, as does the invasion of a malignant melanoma of the skin into the deeper layers of the dermis which contain the lymphatics. These are aspects of staging that can be assessed from the resected tissue specimen. Numerous clinical and pathological staging proformas have been developed, the most well known of which is the TNM (tumor, nodes, metastasis) system (1). In this system numerical values are given to aspects of the tumor which then gives a measure to its stage in terms of the primary tumor (size, the extent to which it has spread into surrounding tissues), lymph nodes (number(s) and site(s) of affected nodes), and the presence or absence of distant metastases. A summary of the TNM staging system as applied to colonic carcinoma is shown in Table 1.

Table 1
Summary of TNM6 staging of colonic carcinoma

Primary tumor T

Tis	Carcinoma in situ or invasion of lamina propria only
T1	Invasion of submucosa
T2	Invasion of external muscle (muscularis propria)
T3	Invasion through muscularis propria into subserosa, or pericolic tissue
T4	Invasion of other organs or structures or through visceral peritoneum

Lymph node N

N0	No lymph node metastasis
N1	Metastasis in 1–3 pericolic nodes
N2	Metastasis in four or more pericolic nodes

Distant metastasis M

MX	Cannot be assessed
M0	No distant metastasis
M1	Distant metastasis

Stage grouping

0	Tis	N0	M0
I	T1 T2	N0 N0	M0 M0
IIA IIB	T3 T4	N0 N0	M0 M0
IIIA IIIB IIIC	T1,T2 T3,T4 Any T	N1 N1 N2	M0 M0 M0
IV	Any T	Any N	M1

For an example of scoring using the TNM system see Note 1. The score for the individual TNM components can be used to place the tumor into a reliable prognostic group, in this case, stage III out of IV. As would be expected, most of the weight in this numerical "stage grouping" system is given to the scores for distant (M) and lymph node (N) metastasis.

2.1.2. Grade of the Tumor

The morphological grading of tumors relies on assessing their rate of growth and their degree of differentiation. Growth rates may be assessed by counting mitotic figures in the histological sections, this is usually expressed as the number of mitoses per ten standard high-power fields on the microscope. An alternative method involves immunohistochemical staining in order to identify the proportion of tumor cells which express proteins present in cells that are actively dividing (see Note 3). Other surrogate markers of growth rate also contribute to the assessment of grade. Necrosis is thought to be a reflection of a fast-growing tumor outgrowing its blood supply, so the presence of necrosis usually correlates with a higher-grade tumor.

The degree of differentiation of a malignancy refers to how much the tumor resembles normal tissue. As such, a well-differentiated tumor is assigned a lower grade than a poorly differentiated one (see Note 2). An anaplastic tumor is so poorly differentiated that no recognizable feature of the tissue is present. Pleomorphism among the nuclei of tumor cells reflects the degree of difference within a population of tumor cells. In a normal tissue the nuclei largely resemble each other, while the severe pleomorphism observed in a tumor may be a reflection of genetic diversity and mutations in the tumor cells. The more pleomorphism is present, the greater the likelihood there is of a subset of the tumor cells that will develop the genetic capability to metastasize.

The number of mitoses, degree of differentiation, and nuclear pleomorphism are all used to place a malignant tumor into a grade. For many of the more common malignancies, such as breast carcinoma, these features are scored numerically and combined to give an assessment of grade that is as reproducible and standardized as possible (Table 2).

Other factors that are useful in the grading of some tumors include the degree of inflammatory response, which implies a better prognosis, for example, in colonic cancer and melanoma.

The pattern of the invading margin can be significant in head and neck squamous carcinoma and in colonic carcinoma where a worse prognosis may be associated with an infiltrative, rather than a compressive, expansile margin.

In many tumors, differentiation towards a particular tissue type may be as significant as lack of differentiation (anaplasia) in establishing a poor prognosis. An example of this is differentiation towards trophoblastic tissue. The normal function of the trophoblast is to form the part of the placenta that invades the wall of the

Table 2
Histological grading of breast carcinoma

Tubule formation (score 1–3)	Nuclear pleomorphism (score 1–3)	Mitotic count (using nomogram related to microscope field diameter), score 1–3
1: Majority of tumor >75%	1: Mild	1: Low
2: Moderate 10–75%	2: Moderate	2: Moderate
3: Little or none <10%	3: Severe	3: High
Overall scores	Grade 1 Grade 2 Grade 3	3–5 Points 6–7 Points 8–9 Points

uterus in order to gain nutrients from the mother's blood. In normal pregnancy trophoblastic cells can be found circulating in the maternal bloodstream. Trophoblastic differentiation in an ovarian or testicular germ cell tumor can confer an ability for highly aggressive invasion and blood-borne spread.

2.2. Metastasis to Lymph Nodes

The presence, or absence, of metastasis in the lymph nodes forms part of the staging of malignancies. Prognosis may be affected not just by the presence of lymph node metastasis but also by the number of nodes involved, their site, the size of the deposits, and the extension of the tumor through the capsule of a node. Detailed assessment of lymph nodes is therefore important in establishing the prognosis and the possible need for further treatment, such as chemotherapy.

The assessment of lymph nodes in an excised surgical specimen is relatively straightforward, although there are some diagnostic problems which the pathologist may face.

2.2.1. Morphology of Lymph Node Metastasis, Micrometastases, and "Isolated Tumor Cells"

Metastatic tumor arrives at a lymph node via the afferent lymphatics which empty into the subcapsular sinus, and it is here that the majority of "early" metastases are first seen. The pattern of growth usually resembles the primary tumor quite closely, and very well-differentiated tumors may appear histologically benign but for the fact that they have spread to a lymph node. Diagnosis is difficult, however, when poorly cohesive individual cells spread into the substance of the node and epithelial structures are not produced. This is frequently the case with lobular carcinoma of the breast wherein malignant cells may resemble macrophages and lymphoid cells. In such cases, histochemical stains, in this case for mucin, or immunohistochemical stains, may reveal a far greater metastatic burden than would otherwise be suspected (see Note 4). Equally, it is important to recognize that the proliferation of epithelioid cells in the subcapsular sinus do not always imply metastasis. A frequent finding in lymph nodes, including those draining sites of malignancy, is sinus histiocytosis, a benign proliferation of macrophages that are sometimes mistaken for malignancy. The diffuse expansion of the sinuses throughout the node suggest immune reactivity, however, in cases where there is uncertainty, immunohistochemistry using antibodies directed against macrophage-specific markers, for example, CD68 (see Note 5) or cytokeratins (see Note 4) assist in resolving diagnostic difficulties.

A pathological curiosity which may be mistaken for metastasis is the presence of ectopic normal tissue within a lymph node structure. Thus, normal salivary gland tissue may be mistaken for metastatic carcinoma in a neck node, and nests of melanocytic naevus cells can be found both within lymph nodes and lymphatic vessels, giving rise to ambiguous diagnosis of malignant melanoma. Malignancies of these cell types may present clinically as lymph node metastasis with an occult primary. It is important that the benign nature of these ectopic tissues is recognized in order to prevent damaging and unnecessary investigations or therapy.

The detection of small or occult metastases and how they correlate with prognosis is an area of some controversy. As conventional histological sections are just four thousandths of a millimeter thick, a single section will miss a proportion of small deposits. The more sections examined and the more refined the techniques used to detect tumor cells, the greater the number of metastases that will be detected. As most prognostic data has been collected using simple histological methods, the prognostic and therapeutic significance of very small metastases is not clear (see Note 5).

The staging of malignancies of lymphoid origin, lymphomas, can present particular difficulties when assessing lymph nodes for the presence of disease, as often the malignant cells resemble a phase of normal differentiation of lymphoid cells. Distinguishing malignancy depends upon assessing the growth pattern(s) of the cells within the node. However, in more subtle cases this may involve establishing the presence of immunohistochemical markers, either of a particular stage in lymphoid differentiation, or through assessment of genetic damage, for instance through chromosomal translocation.

2.3. Diagnosis of Metastasis to Other Sites

Much of what has already been discussed regarding lymph node metastasis applies to metastasis in other sites. Particular problems may be encountered, however, in distinguishing metastasis in an organ from a primary malignancy at that site. This is obviously only a difficulty if the metastasis is of a type that could occur there as a primary tumor, for example, the problem encountered when finding a deposit of adenocarcinoma in bone is in determining the primary site of the tumor, as primary adenocarcinoma does not arise in bone. Conversely, an adenocarcinoma in the ovary could be either a primary cancer or a metastasis from elsewhere.

A carcinoma in the lung, liver, or other organ may be a primary tumor if there is an absence of an overt cancer in alternative primary site(s). Similarly, the presence of a single tumor deposit is often indicative of cancer, as in many cases multiple deposits imply metastasis within the organ. One should bear in mind, however, that primary carcinomas, for example, a primary liver carcinoma, may seed metastases within the same organ, rather unsurprisingly as tumors frequently demonstrate tissue tropism in the distribution of metastasis.

Certain cell types prefer to grow in a particular environment, and malignant liver cells settle preferentially and establish metastatic deposits in the liver. Another situation in which multiple primary tumors arise is in particular with "cancer syndromes." These individuals are at high risk of developing certain malignancies and frequently suffer multiple primary tumors in one or more organs at the same time. Probably the best simple morphological indicator of primary, rather than metastatic, disease is the presence of an in situ carcinoma adjacent to the invasive tumor. In situ carcinoma or severe dysplasia is the preinvasive stage of "malignancy" and in many tumors this is an important stage in the development of clinical disease. Specific antigens or other molecular markers may also give clues as to whether a tumor is primary to that site, or is a secondary malignancy that has metastasized from elsewhere.

2.3.1. Establishing the Site of an Unknown Primary Tumor

An enlarged lymph node may be the first presentation of malignancy. In such cases the major distinction to be made is between a reactive enlargement, a primary tumor of the lymph node (almost always a lymphoma) and a metastatic deposit. Fine needle aspiration (FNA) cytology is a rapid and minimally invasive investigation that allows such a distinction to be made in many cases. In clinical practice, a standard procedure following an FNA diagnosis of probable lymphoma would be to excise the node and submit it to detailed histological and immunohistochemical analysis in order to place the lymphoma into a precise diagnostic category. A cytological diagnosis of metastatic malignancy, however, is not generally followed by excision of the node, unless the enlarged node is causing, or threatens, local complications. Instead, a detailed consideration of the clinical history and examination, followed by investigations, particularly radiological imaging, directed at identifying the primary site is undertaken. Sometimes the pattern of metastatic disease strongly suggests a particular primary site, reflecting tissue tropism. Small cell carcinoma of the lung, for example, is a frequent cause of metastasis to the adrenal gland, and may present at this site before a primary lung cancer is detected. Nevertheless, in a proportion of cases the pathologist is faced with an excised lymph node, core biopsy or FNA cytology containing a metastasis from an unknown primary. Even when considering only a few cells from an FNA, some suggestion of the likely primary sites can usually be made.

A few malignancies are immediately recognizable by their characteristic morphology in tissue sections. These are, however, the exceptions, and more often the pathologist is able to place the tumor into a category by its type of differentiation, for example, squamous or adenocarcinoma, which will serve to narrow down the likely primary site(s). Within these groups there are additional features, such as calcified psammoma bodies which are associated with papillary carcinomas of thyroid and ovary, or clear cell morphology associated with renal carcinoma. The identification of these features narrows the broad category of adenocarcinoma to a more focused area. It should be noted, however, that few morphological features are absolutely specific.

The use of tissue markers detected either by immunohistochemistry or in raised levels in the blood of the patient can aid in determining the primary site of the tumor. Typically tumors are stained with antibodies directed against markers found in different cell types this helps to refine the morphological diagnosis (see Note 6).

2.3.2. Histopathology and the Clinical Detection of Metastasis

Clinical or radiological detection (X-rays and other scans) of new lesions in a patient with a history of malignancy are likely to indicate metastatic disease. However, as the diagnosis of metastasis has considerable implications for the future treatment and life expectancy of the patient, it is important that a conclusive tissue diagnosis should be made wherever possible. Enlarged lymph nodes in a patient with malignancy do not wholly imply metastasis, the presence of inflammation associated with a primary tumor, even in a benign tumor, may lead to reactive enlargement of regional lymph nodes. This is particularly likely to occur following biopsy or surgery of the primary tumor, as the tissue damage from these procedures can lead to lymphadenopathy.

Cytology is a diagnostically useful and minimally invasive means of diagnosing metastasis at most sites, often radiological imaging (ultrasound or CT scan) is used as guidance for FNA if the lesion is in a deep location where it cannot be palpated. Pleural effusions, urine, sputum, and other fluids are easily collected and can also be examined for the presence of malignant cells, although multiple samples may be necessary to enable unequivocal diagnosis. Core biopsy is a widely used alternative to FNA, here a small cylinder of intact tissue is removed of approximately 1 mm diameter. Core biopsies of this type have the advantage of the morphology of the tissue remaining intact with more material being available for immunostaining; however, the disadvantage is an increased risk of hemorrhage or perforation of organs by the larger needle that is used in this procedure. Occasionally the difficult location of a putative metastatic deposit (e.g. in the brain) and overwhelmingly supportive radiological evidence reduces the requirement for a tissue diagnosis, but there are often cases in which diagnosis is ambiguous without histological assessment of the tissues. Examples of this include multiple liver deposits that may be observed on ultrasound scanning, or a "hot spot" observed on radioisotopic bone scans, or shadowing on a chest X-ray.

2.3.3. Intra-operative Diagnosis of Metastasis

Occasionally, immediate diagnosis of metastasis is required if an unexpected deposit is found during surgery. The progress of the operation and the type of surgery performed may depend on whether or not the lesion found, distant from the main tumor, is a metastasis or whether the surgical resection margin is free of tumor. In such cases FNA is an option, but the usual diagnostic method involves removal of a small piece of tissue and histology on the frozen section.

Conventional processing of tissues for histology involve fixation in formalin followed by chemical processing to embed the dehydrated tissue in a paraffin wax block, from which sections are cut with a microtome, de-waxed, and stained. This processing takes considerable time, particularly for larger pieces of tissue, and even the most rapid processing takes several hours. The use of frozen sections avoids the fixation, processing, and wax-embedding stages, shortening the process to a few minutes. The fresh tissue is rendered solid and can be cut into thin slices by freezing in liquid nitrogen and cutting in a cryostat (see Note 7) (see the Chapter 1 "Immunochemistry" by Brooks, in this volume). The sections are then dried and may be stained immediately. While the technique is relatively fast, the major disadvantages are that the morphology of the tissues is poorly displayed compared to paraffin sections, and that only relatively small pieces of tissue can be successfully cut in this way. In many cases metastatic tumor can be reliably identified from a frozen section, but more subtle examples may be missed and there is an increased risk of false positive diagnosis.

Apart from the occasional need for immediate diagnosis, surgical access at the time of primary tumor resection offers opportunities for assessing the spread of tumor and therefore may contribute to accurate staging of the disease. Needle core biopsy of local or distant lymph nodes or other lesions may be possible if they can be visualized by scanning using diagnostic imaging techniques, or felt, and isotonic fluid washings from, for example, the peritoneal cavity, can be examined for the presence of malignant cells.

2.4. When Metastasis is Not Malignant

The discussion so far has assumed that a tumor that has spread to lymph nodes or a distant site is by definition malignant. While this is almost universally correct, there are some interesting exceptions that may cause diagnostic difficulty. These fall into two categories:

benign tumors that spread and tumors that were once malignant but no longer pose a threat to the patient.

The first group: benign tumors that spread, is largely made up of tumors that grow inside blood vessels. These may break away from the main tumor mass and be carried to a distant site but are not classified as metastases per se (see Note 8). The second, unusual, group is made up largely of tumors originating from germ cells or from precursor tissues of developing organs in children (so-called "blastomas"). These primitive, poorly differentiated malignancies are different from conventional poorly differentiated malignancies. In this case, proliferation of the primitive cells from which organs they originated results in a tumor with little visible differentiation, but the cells may retain their normal capacity to mature into fully differentiated adult tissues. The apparent maturation of these tumors may be seen following chemotherapy, which kills the primitive cells but has little effect on the differentiated ones, so that subsequent biopsy of a metastatic lesion may demonstrate only the mature, fully differentiated tissue which, though it undoubtedly represents metastasis, is no longer a threat to the patient.

3. Notes

1. Under the TNM staging system a colonic carcinoma that has invaded the submucosa and has spread to four pericolic nodes with no evidence of distant metastasis would be staged (classified) as T1, N2, and M0.
2. The less the tissue resembles normal tissue, the more likely it is to have lost the factors that allow organization and adhesion of the cells, and these changes correlate with metastatic potential.
3. An example of a protein that is produced when cells are actively dividing is Ki-67. Further details can be found in Chapter 1, by Brooks in this volume.
4. Antibodies directed against cytokeratins may be used to identify epithelial cells of the breast that have metastasized to the lymph nodes.
5. In breast cancer management, micrometastases are defined as between 200 μm and 2 mm in size. Metastases below 200 μm are defined as "isolated tumor cells" (ITCs). These ITCs are currently regarded as node negative in TNM staging whereas micrometastases are regarded as positive with their own specific "Nm" designation. Antibodies directed against cytokeratins may be used to identify the epithelial cells of the breast that have metastasized to the lymph nodes. For detection of macrophages an antibody directed against CD68 may be used.

6. Antigenic markers vary in their specificity and sensitivity for different tumor types and the site(s) of origin of the tumor. Some have high sensitivity and specificity, for example, the prostate specific antigen for prostate cancer, whereas others such as the cytokeratin subtypes give broader guidance for the identification of epithelial cells. Using several antibodies maximizes diagnostic accuracy.
7. A cryostat is essentially a microtome in a refrigerated cabinet. Further details regarding the preparation of tissues and cryostat sectioning can be found in Chapter 1 on immunochemistry by Brooks, in this volume.
8. An example of a benign tumor that can travel away from the main tumor mass is an atrial myxoma, which grows inside the heart. This passive embolization of a tumor is not equivalent to the active invasion of blood vessels by a malignancy, and while tumor emboli can cause disease by obstructing blood vessels, and they can continue to grow within vessels at the distant site, they do not break out of the blood vessel and invade surrounding tissues. Thus the detection of such a tumor away from its primary site does not constitute metastasis in the malignant sense and does not have the prognostic or therapeutic implications of malignancy.

Reference

1. Sobin LH, Wittekind CL (eds) (2002) TNM classification of malignant tumors, 6th edn. New York, Wiley-Liss

Further Reading

Bishop PW. An immunohistochemical vade mecum. http://www.e-immunohistochemistry.info

Kumar V, Abbas AK, Fausto N (eds) (2004) Robbins and cotran pathologic basis of disease, 7th edn. Saunders, Philadelphia, USA

Rosai J (ed) (2004) Ackerman's surgical pathology, 9th edn. Mosby, St Louis, USA

Chapter 4

Immunocytochemical Detection and Characterisation of Individual Micrometastatic Tumour Cells

Stephan Braun and Klaus Pantel

Abstract

Detection of micrometastases in the bone marrow (BM) of cancer patients may be a helpful method for early detection of relapse. The diverse methodology employed across different laboratories, however, renders comparison of results difficult. This chapter describes a robust reliable system for the immunochemical detection of individual micrometastatic tumour cells from the BM aspirates of patients with solid epithelial tumours, based on the binding of antibody to cytokeratin proteins.

Key words: Bone marrow aspirate, Micrometastases, Epithelial membrane antigen, Cytokeratin mucin, Cytospin, Silver stain

1. Introduction

The early haematogenous spread of cancer cells is regarded as a major cause for the later development of metastatic disease in patients with completely resected solid tumours. Such metastases account for the majority of cancer-related deaths in industrialised nations. Current procedures for tumour staging often fail to detect micrometastases, and accordingly, immunocytochemical assays with monoclonal antibodies directed against epithelial antigens enable the detection of individual micrometastatic carcinoma cells in secondary organs. The development of monoclonal antibodies directed against epithelial cell proteins has opened a diagnostic window to allow the detection of disseminated carcinoma cells. Several studies have shown that the bone marrow

Miriam Dwek et al. (eds.), *Metastasis Research Protocols*, Methods in Molecular Biology, vol. 878, DOI 10.1007/978-1-61779-854-2_4, © Springer Science+Business Media, LLC 2012

(BM) is an important, clinically relevant indicator organ for early disseminated cancer cells which may derive from tumour tissues of epithelial organs such as breast, lung, colon, rectum, prostate, kidney, ovary, oesophagus, and pancreas (for review see ref. 1). The use of various detection antibodies and marker antigens has contributed to the wide range of heterogeneity in the methodology applied to the identification of metastatic cancer cells.

Evaluation of the antibodies used in many of the studies (particularly with respect to specificity and sensitivity) has revealed considerable discrepancies amongst the antibodies used and this has somewhat hampered the development of reliable and reproducible assays. Many studies aimed at the identification of metastatic epithelial cancer cells have been performed with monoclonal antibodies directed against either cytokeratin (2–5), a major constituent of the cytoskeleton in epithelial cells (15), or against membrane-bound mucins, such as epithelial membrane antigen (EMA), human milk fat globule (HMFGM), and tumour-associated glycoprotein-12 (6–8). Analysing BM samples from non-carcinoma patients, however, we and others (9–13) have demonstrated that, in contrast to antibodies directed against cytokeratin polypeptides, the use of antibodies directed against epithelial mucins result in a considerable number of false positive results due to the cross reactivity of these antibodies with autochthonous BM cells. As a result, cytokeratin has emerged as a potential "standard marker" for the detection of disseminated epithelial tumour cells in BM (13).

A priority at this time must therefore be the development of standardised protocols for the detection of disseminated tumour cells of epithelial origin. A pooled analysis of almost 5000 patients with primary breast cancer patients demonstrated the independent prognostic significance of disseminated tumor cells in BM (14). In this chapter we provide a standardised protocol for the detection of minimal residual tumour cells in BM aspirates collected from cancer patients. The methods described can detect from 10^{-5} to 10^{-6} nucleated cells. Epithelial tumour cells are identified using a broad-spectrum monoclonal antibody (A45-B/B3) directed against cytokeratin. The system uses the sensitive alkaline phosphatase anti-alkaline phosphatase staining technique for detection of antibody binding. Moreover, to investigate the malignant nature of such cells, we designed a double-marker assay to allow for the simultaneous detection and characterisation of individual micrometastatic carcinoma cells using a protocol that combines immunogold-alkaline phosphatase double labelling of two antigens. This methodology allows the localisation of two antigens within the same cellular compartment. The proposed protocols may help to improve current tumour staging and assist in identification of relevant therapeutic targets with potential consequences for adjuvant anti-cancer therapy (13).

2. Materials

2.1. Bone Marrow Preparation

1. 10× Concentrated phosphate-buffered saline (PBS) stock solution is prepared using 1.5 M NaCl, 80 mM Na_2HPO_4, 20 mM KH_2PO_4, pH 7.4.
2. Hank's salt solution, store at 4°C.
3. Percoll density gradient: prepare using 100% Percoll stock solution (take 100 ml Percoll and add to 9 ml Hank's salt solution), dilute with sterile NaCl to give 0.9–50% Percoll (1.065 g/ml). Alternatively Ficoll can be used (1.077 g/ml). The Percoll or Ficoll is stored at 4°C.
4. Erythrocyte lysis buffer comprising 10 mM $KHCO_3$, 150 mM NH_4Cl, 0.1 mM EDTA, pH 7.4 (see Note 1).

2.2. Immunocytochemistry (See Note 2)

1. PBS-AB serum is used for preparation of all the antibody stock solutions. For this 10% v/v of AB serum (Biotest, Germany) is added to PBS (see Note 3).
2. BSA-C serum is used to block non-specific binding of the antibody and is prepared by adding 0.1% v/v acetylated bovine serum albumin (BSA) (Aurion-BSA-C, Biotrend, Köln, Germany) to the PBS-AB serum prepared above.
3. For single labelling protocols dilute the monoclonal antibody A45-B/B3 (Baxter, Germany) with PBS-AB serum to give a final concentration of 1–2 μg/ml (see Note 4).
4. For double labelling, dilute the alkaline phosphatase-conjugated monoclonal antibody A45-B/B3 F_{ab} fragment (Baxter) in PBS-AB serum to a final concentration of 1–2 μg/ml.
5. For single labelling, dilute the rabbit anti-mouse detection antibody (see Note 5) 1 in 20 in the PBS-AB serum. Dilute the alkaline phosphatase anti-alkaline phosphatase (APAAP) complex (e.g., D651, Dakopatts, Germany) 1 in 100 in PBS-AB serum.
6. For double labelling prepare the permeabilization solution comprising PBS supplemented with 0.1% v/v Triton X-100 and fixation solution comprising PBS with 10% w/v paraformaldehyde.
7. For double labelling, dilute the immunogold-conjugated, terminal F_c fragment-specific goat anti-mouse antibody (GE Healthcare, Germany) 1:40 in PBS-AB serum.
8. Prepare the immunogold post fixation buffer comprising 2% v/v glutaraldehyde in PBS.
9. New Fuchsin Development Solutions are required during each staining run. We prepare sufficient for four Hellendahl jars, with a final volume 253.3 ml.

   Levamisole solution: Add 62.5 ml, 0.2 M Tris to 187.5 ml water with 90 mg Levamisole.

New fuchsin solution: 50 mg $NaNO_2$ should be dissolved in 1.25 ml water and is added to 0.5 ml new fuchsin solution (prepared in a 5% w/v solution in 2 M HCl) the solutions should be shaken well, until air bubbles appear.

Naphthol phosphate solution: 125 mg Naphthol AS-BI phosphate sodium salt is dissolved in 1.5 ml *N,N*-dimethylformamide (DMF) (see Note 2).

The freshly prepared Levamisole and new fuchsin solutions are mixed and stirred well before the naphthol solution is added. Immediately prior to use the mixture is filtered into Hellendahl jars.

10. Silver enhancement solution: a Silver Intense Kit (GE Healthcare) is used as recommended by the manufacturer.
11. Cover slips are sealed using Kaiser's glycerine-gelatin using a melting temperature of 50°C.

3. Methods

3.1. Bone Marrow Preparation

1. Transfer a maximum of 10 ml BM aspirate (see Note 1) in 50 ml tubes; add Hank's salt solution (4°C) to a final volume of 20 ml.
2. Centrifuge for 10 min at 180 × *g* at room temperature; meanwhile prepare 8 ml of 50% Percoll (4°C) in a fresh 15-ml tube. Alternatively, prepare 20 ml of Ficoll in a fresh 50-ml tube.
3. Remove the (serous) supernatant without disturbing the BM cell pellet. Add 1 ml 10% Percoll to 5 ml pellet and transfer solution carefully onto the Percoll gradient. Alternatively add PBS to the BM pellet to give a final volume of 10 ml and place the solution carefully onto the Ficoll gradient. Avoid mixing the cells and Percoll/Ficoll prior to centrifugation. Centrifuge for 20 min at 1,000 × *g* at room temperature, then centrifuge again for a further 30 min at 1,230 × *g* at room temperature.
4. Remove the interphase containing the mononucleated cells (MNC) and transfer to a fresh 50-ml tube, add PBS (4°C) to give a final volume of 50 ml, and centrifuge for 10 min at 540 × *g* at 4°C.
5. Remove the supernatent (wash) solution then resuspend the pellet and lyse the erythrocytes by adding 2-10 ml lysis buffer (4°C) and incubate for 5 min at room temperature (see Note 8).
6. Add PBS to a final volume of 50 ml, and centrifuge for 10 min at 540 × *g* at 4°C.
7. Remove the supernatent (the wash solution) and resuspend cells in a final volume of 2–5 ml PBS (4°C). Count cells and adjust cell concentration to 1×10^6 MNC/ml (see Note 1).

8. Prepare cytospin slides for cytocentrifugation. Add a total of 0.5 ml cellular suspension (= 0.5×10^6 MNC) per spot in the cell chamber.
9. Centrifuge the cytospin preparations for 3 min at $120 \times g$ at room temperature. Let the slides dry for 12 h and either perform the immunostaining step immediately or store the slides at –20 or –80°C for use later (see Note 9).

3.2. Immuno-cytochemistry

1. Thaw the frozen slides of BM preparations (and the appropriate positive and negative control cell line preparations) at 37°C for at least 15 min.
2. Prepare appropriate dilutions of antibody stock solution in PBS-AB serum. Place thawed slides in a humid chamber for the subsequent incubation steps; the slides must not dry out during the entire process of immunostaining.

3.2.1. Single Labelling with APAAP Immunostaining

1. Blocking step: incubate slides with PBS-AB serum for 20 min, pour off the blocking solution (do not wash slides).
2. Detection antibody: incubate the slides with the A45-B3 antibody for 45 min, and wash thoroughly in PBS for 10 min. Repeat the PBS wash step three further times (see Note 2).
3. Bridging antibody: incubate with Z259 antibody for 30 min and wash thoroughly in PBS for 10 min with three changes of buffer (see Note 3).
4. APAAP complex—for example, D651: incubate antibody complexes for 30 min and wash thoroughly in PBS for 10 min with three changes of buffer (see Note 3).
5. New fuchsin development of alkaline phosphatase: incubate slides with freshly prepared, mixed, and filtered alkaline phosphatase-development solutions for 20 min (see Note 7). Finally, wash the slides in PBS for 10 min with three changes of buffer.
6. Seal the cells with cover slips using Kaiser's glycerol: gelatin solution.

3.2.2. Double Labelling

1. Immerse the slides in PBS Triton X-100 solution for 10 min at 4°C and then incubate in PBS with 1.0% w/v paraformaldehyde in PBS for 20 min at 4°C. Wash the slides in PBS for 10 min using three changes of buffer.
2. Blocking step: incubate slides in PBS with 10% v/v AB serum and 0.1% v/v BSA-C for 20 min, pour off the blocking solution.
3. Co-incubation with detection antibodies: mix both primary antibodies: alkaline phosphatase-conjugated monoclonal antibody A45-B/B3 F_{ab} fragment (for cytokeratin labelling) and a second unconjugated whole immunoglobulin (for labelling of target antigen) at the respective dilutions within the same tube.

Incubate the slide with the antibody mixture for 45 min, and wash thoroughly in PBS for 10 min with three changes of buffer.

4. Immunogold labelling: incubate with the F_c-terminus-specific, gold-conjugated goat anti-mouse antibody for 45 min. Wash thoroughly in PBS for 10 min with three changes of buffer followed by a further three washes with water.
5. Post-fixation of immunogold colloids: immerse the slides in the 2% v/v glutaraldehyde solution for exactly 5 min, and wash thoroughly in water for 10 min. Repeat the water-wash step a further three times.
6. Silver development of labelled immunogold: prepare the silver enhancement solution fresh according to the manufacturer's recommendations. Drop the development solution onto the cells after cooling the solution to 4°C. Monitor the silver development at regular intervals by viewing under the microscope (see Note 6). Stop the reaction (as soon as brownish to black silver granules become visible) by placing the slides in water. Wash the slides thoroughly in water for 10 min with three changes of water. Immerse the slides in PBS for 5 min (see Note 3).
7. New fuchsin development of alkaline phosphatase: incubate the slides in the freshly prepared, mixed, and filtered alkaline phosphatase-development solutions for 20 min. Wash the slides in PBS for 10 min with three changes of buffer, and mount with coverslips sealed using Kaiser's glycerol: gelatine solution.

It is important to incorporate a slide containing a positive control known to express the antigens of interest in order to be able to judge that negative immunostaining is due to lack of antigen expression rather than simply methodological failure. We also recommend using a set of slides that will be negative for the staining, by applying an antibody that does not cross react with human proteins. In this setting, we screen the identical number of cells (e.g. 2.0×10^6 per patient) as in the specific immunostaining (see Note 2).

To set up this assay as a routine method for the detection of tumour cells in BM of cancer patients, we recommend the incorporation of BM aspirates of non-carcinoma patients in order to substantiate the specificity of the approach (see Note 2).

4. Notes

1. Low numbers of MNC recovered from the Percoll/Ficoll interface could be due to a poor quality BM aspirate, for example, due to significant contamination with peripheral blood. This may also lead to an underestimation of the frequency of tumour cells detected by immunocytochemical screening (1). Good quality aspirates of 3–5 ml of BM contain more than 10^7 MNC.

It is recommended that a laminar air-flow working bench is used and gloves are worn when handling BM aspirates.

2. AB serum is an antibody-free serum from humans of blood group AB. It is good practice to use sterile disposable plasticware for the preparation of reagents, especially if BM MNCs are foreseen for cell culture or subsequent RNA/DNA analyses. For dilution of antibodies, remove immunoglobulin solutions with sterile pipette tips.
3. Single labelling: failure to produce a positive signal with the cytokeratin staining on the positive control specimens could be due to the use of damaged control specimens, for example: specimens that have been repeatedly freeze-thawed following cytospin preparation, too high dilution of the antibodies (especially the primary antibody A45-B/B3), or problems with the preparation of the developing solutions (precipitation of naphthol AS-BI phosphate in DMF due to wet glass ware). Since non-specific labelling of BM cells by the antibody A45-B/B3 occurs in less than 2% of non-carcinoma control patients (13), if such non-specific staining of the negative control specimens occurs, this may be predominantly caused by too high a concentration of antibodies. So far it remains unclear as to why, in a few cases, irrelevant antibodies with non-human (e.g., murine) specificity react with human BM cells on isotype control specimens of patient BM samples. To maintain the specificity in the interpretation of staining signals, we strongly recommend excluding these rare cases (usually below 1% of all cases in our series) as these give rise to indeterminate results.
4. Detection antibodies are available from a wide range of suppliers. We find that Z259 (Dakopatts) works well in our hands.
5. APAAP detection systems are available from a wide range of suppliers. We find that D651 (Dakopatts) works well in our hands.
6. Double labelling: for unambiguous interpretation of silver lactate precipitates with specifically bound immunogold-conjugates, we strongly recommend to use an epi-polarization filter unit (13). The use of this device enables recognition of the brown-to-black granulations other than the silver precipitate. Interpretations based on light microscopy alone may lead to the registration of non-specific signal and/or underestimation of the actual staining signal as this may be too weak when using non-polarized light microscopy. Special care must be taken to control the time- and temperature-dependent silver precipitation; we recommend that the positive control slide is continuously monitored under the microscope and as soon as the first slightly brown signal becomes visible the reaction is stopped. Development exceeding 30 min may lead to weak new fuchsin signals owing to chemical interaction with the alkaline phosphatase. For preparation of new fuchsin development solutions clean glassware should be used.

7. Particular care should be taken handling naphthol and DMF; a fume cupboard should be used where necessary and care should be taken during the disposal of naphthol-containing waste (always refer to the manufacturers safety data sheet).
8. The degree of erythrocyte contamination can influence the success of the staining protocol. Only include this step if the assessment of the erythrocyte pellet (the red pellet) renders this step necessary.
9. We find slides can be stored at –20°C for 4–8 weeks or at –80°C for 1–2 years without any change in the immunochemical staining results.

References

1. Pantel K, Brakenhoff RH (2004) Dissecting the metastatic cascade. Nat Rev Cancer 4: 448–456
2. Braun S, Pantel K, Muller P, Janni W, Hepp F, Kentenich CR, Gastroph S, Wischnik A, Dimpfl T, Kindermann G, Riethmuller G, Schlimok G (2000) Cytokeratin-positive cells in the bone marrow and survival of patients with stage I, II, or III breast cancer. N Engl J Med 342:525–533
3. Cote RJ, Rosen PP, Lesser ML, Old LJ, Osborne MP (1991) Prediction of early relapse in patients with operable breast cancer by detection of occult bone marrow micrometastases. J Clin Oncol 9:1749–1756
4. Lindemann F, Schlimok G, Dirschedl P, Witte J, Riethmüller G (1992) Prognostic significance of micrometastatic tumour cells in bone marrow of colorectal cancer patients. Lancet 340:685–689
5. Pantel K, Brakenhoff RH, Brandt B (2008) Detection, clinical relevance and specific biological properties of disseminating tumour cells. Nat Rev Cancer 8: 329–340
6. Mathieu MC, Friedman S, Bosq J, Caillou B, Spielmann M, Travagli JP, Contesso G (1990) Immunohistochemical staining of bone marrow biopsies for detection of occult metastasis in breast cancer. Breast Cancer Res Treat 15:21–26
7. Diel IJ, Kaufmann M, Costa SD, Holle R, von Minckwitz G, Solomayer EF, Kaul S, Bastert G (1996) Micrometastatic breast cancer cells in bone marrow at primary surgery: prognostic value in comparison with nodal status. J Natl Cancer Inst 88:1652–1664
8. Harbeck N, Untch M, Pache L, Eiermann W (1994) Tumour cell detection in the bone marrow of breast cancer patients at primary therapy: results of a 3-year median follow-up. Br J Cancer 69:566–571
9. Taylor-Papadimitriou J, Burchell J, Moss F, Beverley P (1985) Monoclonal antibodies to epithelial membrane antigen and human milk fat globule mucin define epitopes expressed on other molecules. Lancet 1:458
10. Delsol G, Gatter KC, Stein H, Erber WN, Pulford KA, Zinne K, Mason DY (1984) Human lymphoid cells express epithelial membrane antigen. Lancet 2:1124–1128
11. Heyderman E, Macartney JC (1985) Epithelial membrane antigen and lymphoid cells. Lancet 1:109.
12. Pantel K, Schlimok G, Angstwurm M, Weckermann D, Schmaus W, Gath H, Passlick B, Izbicki JR, Riethmüller G (1994) Methodological analysis of immunocytochemical screening for disseminated epithelial tumor cells in bone marrow. J Hematother 3:165–173
13. Fehm T, Braun S, Muller V, Janni W, Gebauer G, Marth C, Schindlbeck C, Wallwiener D, Borgen E, Naume B, Pantel K, Solomayer E (2006) A concept for the standardized detection of disseminated tumor cells in bone marrow from patients with primary breast cancer and its clinical implementation. Cancer 107: 885–892
14. Braun S, Vogl FD, Naume B, Janni W, Osborne MP, Coombes RC, Schlimok G, Diel IJ, Gerber B, Gebauer G, Pierga JY, Marth C, Oruzio D, Wiedswang G, Solomayer EF, Kundt G, Strobl B, Fehm T, Wong GY, Bliss J, Vincent-Salomon A, Pantel K (2005) A pooled analysis of bone marrow micrometastasis in breast cancer. N Engl J Med 353:793–802
15. Pantel K, Izbicki J, Passlick B, Angstwurm M, Häussinger K, Thetter O, Riethmüller G (1996) Frequency and prognostic significance of isolated tumour cells in bone marrow of patients with non-small cell lung cancer without overt metastases. Lancet 347:649–653

Chapter 5

The Polymerase Chain Reaction

Hazel M. Welch

Abstract

The polymerase chain reaction (PCR) has had a significant impact on all aspects of the molecular biosciences, from cancer research to forensic science. The sensitivity and specificity inherent in the technique allow minute quantities of genetic material to be detected while the unique properties of thermostable DNA polymerase ensure that abundant copies are reliably reproduced to levels that can be visualized and/or used for further applications. This chapter describes applications of PCR and PCR-RT to investigate primary cancer and metastatic disease at both the DNA and mRNA expression levels.

Key words: Polymerase chain reaction, PCR-RT, Quantitative real-time PCR, Quantitative real-time RT-PCR

1. Introduction

Of all the techniques employed in the molecular biosciences, the polymerase chain reaction (PCR) is probably the most widely used; it certainly has had the greatest impact. Prior to the development of PCR, obtaining sufficient genetic material for mapping, sequencing, mutation screening, or for use as a probe would take at least a week of full-time work; this can now be achieved within a few hours. The importance of PCR was recognized in 1993 when its inventor K. Mullis received the Nobel Prize. PCR has also become a household term, forming the cornerstone of many popular television and cinema dramas.

DNA amplification by PCR was first described in 1985. Sequence-specific oligonucleotide primers flanking codon 6 of the β-globin gene (the region mutated in sickle cell anaemia) were designed and used in a series of repeated reactions. (1). Subsequently, the use of a thermostable DNA polymerase from *Thermus aquaticus* (*Taq*) was developed for use in these reactions

Miriam Dwek et al. (eds.), *Metastasis Research Protocols*, Methods in Molecular Biology, vol. 878, DOI 10.1007/978-1-61779-854-2_5, © Springer Science+Business Media, LLC 2012

(or cycles); *Taq* polymerase retains significant enzymatic activity following extensive incubation at 95°C. Consequently, it was no longer necessary to add fresh enzyme at the start of each cycle. In addition, sensitivity, specificity, yield, and the length of target amplified, were significantly improved (2). The technique immediately leant itself to automation and considerable development by the biotechnology industries.

In basic PCR, the system is relatively simple; this is due to (a) the availability of sequence data and Web-based tools for designing primers coupled with (b) quality products from biotechnology companies making the technique readily accessible. In its simplest form, the requirements are template DNA, a pair of oligonucleotide primers, one sense (or forward) and the other antisense (or reverse), flanking the region (or target) of interest (Fig. 1), a thermostable

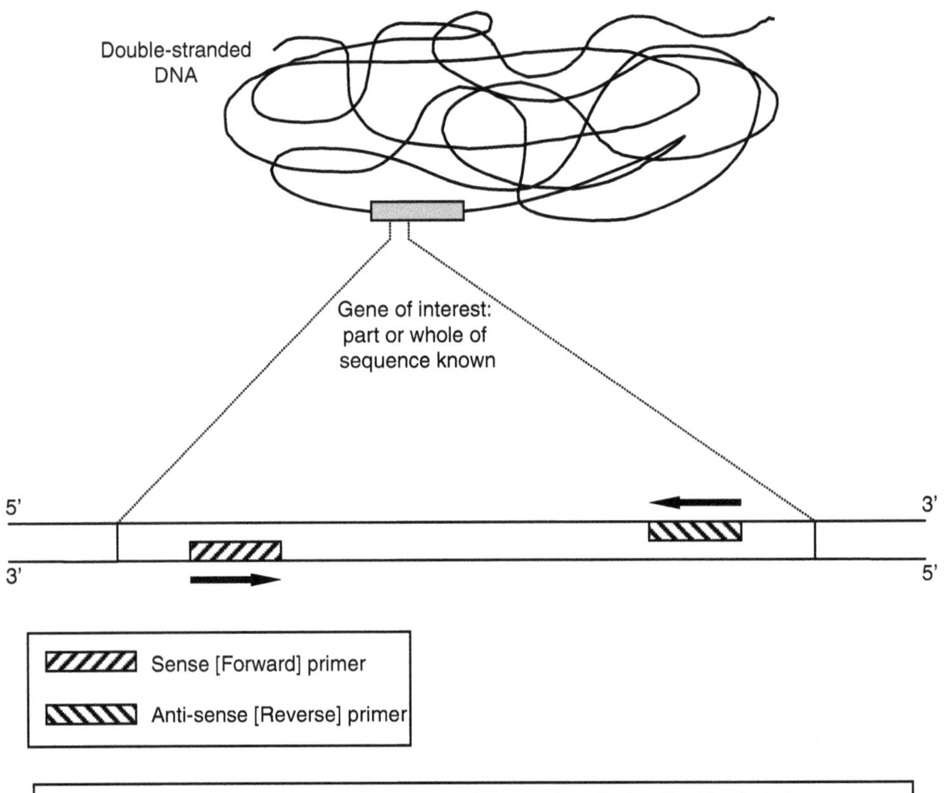

Fig. 1. Simplified organization of a gene of interest and orientation of sense and antisense primers.

DNA polymerase, deoxynucleotide triphosphates (dNTPs), and an appropriate buffer solution (containing Mg^{2+}). The entire reaction takes place in a thermal cycler and consists of repeated cycles comprising three basic steps. Step 1 is denaturation (or melting) of the double-stranded DNA template at >90°C; this is followed by step 2, rapid cooling to allow annealing of the primers to the separated DNA strands, and step 3, rapid heating to 72°C for primer extension (copying of the target region) by the DNA polymerase (Fig. 2). The product of each cycle becomes the substrate for the next, hence "chain reaction" or amplification. Each cycle is repeated 30–40 times and the amplified products (amplicons or fragments) are then (usually) visualized by gel electrophoresis. Assuming 100% efficiency, the copy number is doubled during each cycle so that after ten cycles the amplicon is multiplied by a factor of $\sim 10^3$ and after 20 cycles by a factor of $>10^6$, all in just a few hours.

This chapter focuses on the applications of PCR and reverse transcriptase PCR (RT-PCR) to metastasis research. This may involve mutation analysis, determination of gene copy number, identification of circulating tumour cells (3), chromosome translocations, promoters silenced by hypermethylation, or an examination of gene expression, most of which can be performed by conventional RT-PCR.

The latest equipment includes quantification/"real-time" technologies (quantitative real-time PCR, Q-PCR, or quantitative real-time RT-PCR, Q-RT-PCR), where the amount of product (equivalent to the number of copies) is measured after each cycle; but whichever method is used, the optimum cycling parameters are best established using a conventional thermal cycler. The full range of applications for PCR and RT-PCR is vast and cannot be covered in detail here. The biotechnology and specialist companies are excellent sources for additional protocols, literature and troubleshooting; and worth pursuing. However, the design of (a) each assay, (b) primers, and (c) techniques to exclude contamination is universal and of paramount importance; thus, emphasis will be placed on establishing techniques that are appropriate for most applications.

1.1. Gene and Sequence Analysis Using PCR

PCR can be applied to the specific detection and amplification of a single (copy) gene or related genes that are part of a large family; the distinction resides with primer design. For the former, it is essential that the primers are specific (i.e. recognize unique sequences, perhaps including codes for Methionine (Met) or Tryptophan (Trp). Primers designed to identify related family members should recognize conserved sequences and/or incorporate alternative bases, especially for amino acids that have multiple codons, i.e. make use of the degeneracy of the genetic code (although degeneracy at the 3′ end of the primer should be avoided). Similarly, if gene sequence data is limited (or only partial amino acid sequence is available), a mixture of primers

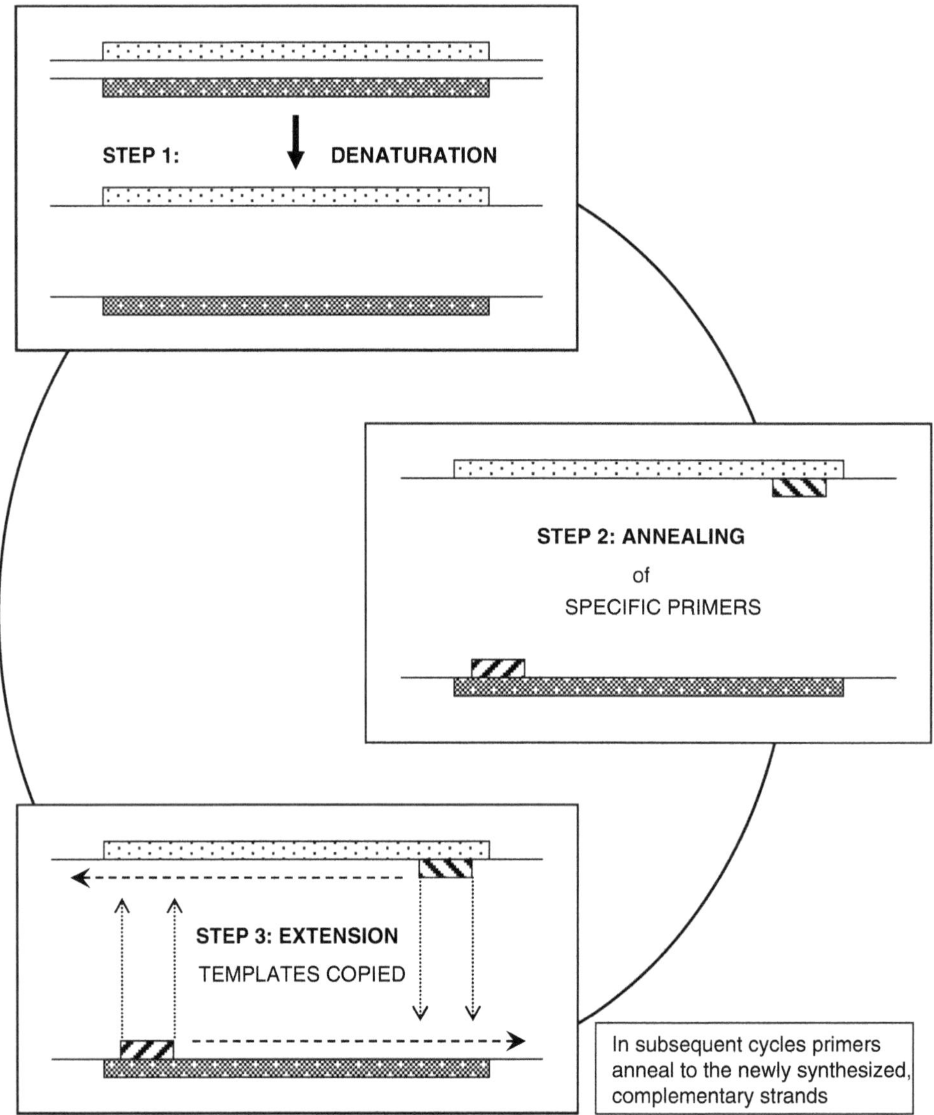

Fig. 2. Polymerase chain reaction: During the first cycle, both strands of the target are copied and are not limited in length at the 3′ ends; however, templates for subsequent cycles are limited by the first primer. Thus, at the end of the second and subsequent cycles, the newly synthesized fragments are defined by the primers: the number of these fragments increases exponentially with each cycle.

containing alternative bases at specific sites can be designed. The basis of the differential display technique is the detection of random fragments of expressed sequences and this allows different populations to be compared: here, a subset of mRNAs is targeted by an "anchored" oligo-dT primer while a random decamer is used for the sense primer (4).

The choice of polymerase is the second most important consideration. Reactions that require high-fidelity amplification (i.e. a minimum number of mismatches) are best performed using an enzyme with proofreading ($3' \rightarrow 5'$ exonuclease) activity (e.g. *Pfu*). Applications typically include expression cloning; mutation, or single-nucleotide polymorphism (SNP) analyses. However, when long regions are required for cloning purposes, amplifying a fragment longer than 5 kb becomes problematical and a combination of two enzymes, one with proofreading ability and the other without (e.g. *Taq*), is advised (5). For most general applications, *Taq* polymerase is suitable: it is a robust enzyme, produces good yields, and, although it does not have $3' \rightarrow 5'$ proofreading activity, does have $5' \rightarrow 3'$ exonuclease activity.

Conventional PCR has also been adapted for methylation-specific PCR (MSP) (see below) and quantitative and semi-quantitative analyses (e.g. gene copy number and expression levels). The latter adaptations are generally superseded by Q-PCR: here, the quantity of product is measured at the end of each amplification cycle (6). This system has a number of advantages but is also not without drawbacks (7). The simplest protocols incorporate a fluorescent dye that intercalates with double-stranded DNA (e.g. SYBR green). There are many variations and the method of choice is influenced by the equipment/supplier (including whether capillaries, tubes, or 96-well plates are used), but the principle of measuring the level of fluorescence as the reaction progresses is common to all procedures.

1.2. Gene Expression Analysis by Methylation-Specific PCR

MSP is a specialized form of PCR and is described in detail in Chapter 15 by Lehmann, Albat, and Kreipe in this volume. Gene silencing is associated with hypermethylation of CpG dimers (to give 5′-methylcytosine) (8) notably exemplified in breast tumours by *14-3-3σ* silencing (9). Different sets of primers distinguish methylated from non-methylated DNA. The protocol requires additional treatment of the sample: prior to PCR, DNA is treated with sodium bisulphite (10) which converts non-methylated cytosine residues to uracil; uracil is recognized by primers containing thymine. In contrast, methylated cytosine is protected from the sodium bisulphite and remains unmodified; thus cytosine residues are recognized by primers containing guanine. Ideally, MSP should be carried out in conjunction with RT-PCR because tumour samples frequently contain both non-methylated as well as methylated DNA sequences.

1.3. Gene Expression Analysis by RT-PCR and Quantitative Real-Time RT-PCR

RT-PCR and Q-RT-PCR are essentially PCR but with an additional (prior) reverse transcriptase step to convert mRNA into cDNA (complementary DNA), which is then used as the substrate for the reaction.

Different reverse transcriptase enzymes are available commercially: for example, AMV RT (from avian myeloblastosis virus) and M-MLV RT (from Moloney-murine leukaemia virus);

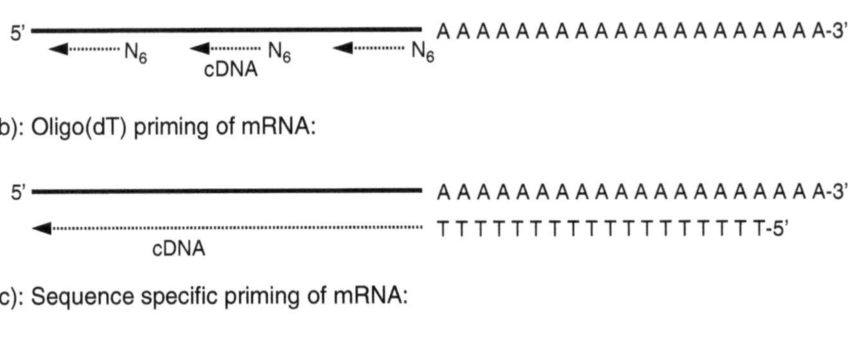

Fig. 3. First-strand synthesis of cDNA: Alternative priming for reverse transcriptase step.

it will be helpful to check company literature for the most appropriate enzyme for the given application. Reverse transcriptases are RNA-dependent DNA polymerases and require a "primed" template.

There are two alternative routes for RT-PCR: either conventional two-step, where cDNA is prepared in bulk and a sample used for each PCR reaction, or one-step, where both the reverse transcriptase and PCR reactions occur consecutively in the same tube. If two-step RT-PCR (or Q-RT-PCR) is considered, there are three alternative methods for priming mRNA for reverse transcription: (a) a random hexamer, (b) an oligo-dT (18–20 bases), or (c) the antisense (reverse) primer (Fig. 3). One-step RT-PCR utilizes the antisense (reverse) primer for reverse transcription; the advantages of this technique are reduced time, a considerably reduced risk of contamination, and, usually, increased specificity and yield. The advantages of two-step RT-PCR are that a number of reactions can be performed from a single population of cDNA and that the cDNA is stable when frozen. Repeated freezing and thawing introduce nicks and rapid deterioration of cDNA, with a subsequent decline in PCR efficiency and yield. Aliquots of cDNA should be stored in sterile tubes at –70°C to circumvent this problem. It is worthwhile comparing both techniques to determine which is most suitable for your application.

1.4. Type of Assay

Two of the most important aspects of PCR are specificity and sensitivity. Specificity of the reaction is determined, primarily, by the primers. These should recognize only the target of interest (this is sometimes difficult when the gene of interest is part of a family or alternative transcripts are expressed); specificity of primer-template recognition is also influenced by the annealing temperature. The sensitivity (amplification) ensures detection of a target that may be represented <1 copy per cell; however, without due care, the abundant products of a PCR reaction can easily lead to widespread contamination and false-positive results. Other important factors are the quality of the template, the reaction conditions, as well as the choice of enzyme. Enzyme choice is influenced by the type of assay to be performed, for example whether high fidelity, hot-start, or high throughput; the enzyme (and template) will also influence the reaction conditions. Thus, it is necessary to peruse products from different companies—ask for a free sample. My experience is that they are very willing to oblige!

1.5. Primer Location

There are a number of options available when considering which primers to use: sequences may be published already and verified by sequencing; be available commercially (although the sequences are unlikely to be known); be listed in one of a number of online resources (see Appendix); or you may choose to design them yourself. For the latter, knowledge of part of the whole gene sequence will help enormously. The Ensembl site is easy to navigate [http://ensembl.org/index.html] (see Example 2 below), although others are equally good.

For gene analysis, the most important consideration is that the primers recognize a unique sequence. If your gene is part of a family and alignment programmes are not easily available, then search other family members for homology to your primers. While viewing the related gene, bring up the "Find" box: "Ctrl" (for PC) or "Apple" (for Macs) + "F". Key in part of the primer sequence, do not key in all of it (five to seven bases should be sufficient to identify any homologous sequences), and be prepared to try two to three different sequences; flanking sequences can be checked by eye.

For expression analysis, detecting a unique sequence remains the priority with the additional requirement that only mRNA sequences are amplified. Thus, design primers that flank an intron (any contaminating DNA sequences will be immediately identifiable by size) and/or design one of the primers to cross the junction of two exons. For conventional PCR, the amplified fragment should be between 250 and 500 bp while for Q-PCR and Q-RT-PCR the amplified fragment should be much smaller, between 80 and 150 bp.

1.6. Primer Design

Whatever the decision regarding the location of your primers, the same principles apply when designing them; primers should be 18–30 bases in length and have 40–60% GC content or at least equivalent amounts so that they have similar T_m (melting temperatures) and non-complementary 3′ ends (to prevent primer–dimer formation; see Example 1 below). The primers should avoid three G or C residues at the 3′ end (to prevent interference with the enzyme) and should not contain runs of inter- or intra-complementary bases. Importantly: (a) when designing the antisense (reverse) primer, it should be complementary to the mRNA (sense) sequence, and (b) when recording or ordering primers (both sense and antisense), always follow the convention and write in 5′→3′ orientation. Technical information regarding the T_m and GC content can be checked once a draft of the primers has been planned. The company of your choice will have suitable online programs that will provide technical information; see Example 3.

Example 1
Primers that will form dimers:

(i) xxx.........GGATAT-3′ (alone)

(ii) xxx.........ACAAGT-3′ with xxx.........ACTTGT-3′

If further applications of the PCR product are planned (e.g. cloning), then the 5′ ends of the primers may contain restriction endonuclease sites or adaptors. Note: *Taq* polymerase adds an extra dNTP (usually A) to the 3′ terminus, while *Pfu* generates blunt-ended fragments. Alternatively, specific modifications can be added such as a probe for Q-PCR (see company literature).

Example 2
Steps to bring up the cDNA sequence of Epidermal Growth Factor Receptor (EGFR) and locate the primers: (a) sense CAGCCA-CCCATATGTACCATC and (b) antisense AACTTTGGGCG-ACTATCTGC that define an 80-bp fragment:

Ensembl: http://www.ensembl.org/index.html.

Search: *Human for EGFR*

Select: *Gene*

Select: *Protein ID (ENSP00000275493)*

Select: *cDNA* (from 'Transcript-Based Displays')

Scroll down the page to view cDNA (with amino acids) sequence

Press: *Ctrl + F keys* to bring-up Find box

Key in: CAGCCA (starts at 2980)

Example 3

Technical details of primers from Sigma-Genosys: http://sigma-genosys.com. www.sigmaaldrich.com/life-science/custom-oligos.html

Select: *Custom DNA oligos*

Select: *Standard synthesis & purification*

Select: *Custom Oligos ORDER NOW*

Select: *DNA oligos – modified & non-modified* BOOKMARK PAGE

Select: 2 (or more) items

Synthesis scale: 0.025 μM (this should be sufficient)

Purification: Desalt

Format: Dry

Concentration in water: None

Number of tubes: One

Key in your sequences

Select: *Technical data*

The primer details will be processed immediately and the following information provided: *Tm*, primer length, %GC, the degree of secondary folding and potential to form a primer dimer. Keying in a primer pair enables you to adjust the sequences so that they (a) have a similar Tm, (b) will not form secondary structures, or (c) form a primer dimer (alone; to check them as a pair, you will have to reverse the antisense order and compare the 3′ ends by eye).

2. Preparation and Materials

1. A clean laboratory coat and gloves are essential. Be prepared to change gloves regularly and discard them directly into an autoclave bag (or equivalent; the inside of protective gloves is an extremely rich source of nucleases). Adopt aseptic techniques, especially avoid touching your hair or glasses.
2. Designate separate areas for (a) isolating nucleic acids and setting up reactions and (b) processing PCR products. PCR products are the main source of contaminating material; it is essential that you keep them away from your set-up area.
3. Keep the work surface clean (use bleach and 70% *v/v* ethanol or industrial methylated spirits, IMSs, to rinse off the surface). Particles of dust are contaminated with pathogens, so these should not be allowed to build up.

4. Designate (clean) racks and ice buckets for molecular biology use.
5. Keep a designated set of pipettors for handling nucleic acids and setting up the reactions. Clean and calibrate them regularly.
6. Always use (good quality) filter tips (aerosol resistant) when handling nucleic acids and setting up the reactions.
7. Purchase RNase-free and DNase-free reaction tubes and consumables wherever possible. Autoclave them (in small amounts); always tip out onto a clean surface—do not reach in with your hand. Designate them for molecular biology use.
8. Purchase molecular biology-grade (PCR grade) water in 1.25-ml aliquots (e.g. from Promega DW0991). Designate these for use with nucleic acids; keep the water provided with each kit for setting up the reactions.
9. Purchase additional molecular biology-grade water (e.g. from Promega P1195) for other uses, such as preparing primers. Handle in a laminar flow cabinet to keep sterile. Always prepare a smaller aliquot for use; never use directly from the bottle.
10. Purchase an RNase inhibitor (e.g. RNasin; Promega).
11. Use thin-walled tubes (0.25 or 0.5 μl) for the reactions; these will ensure rapid transmission of temperature changes and contribute to a good yield, although conventional tubes are perfectly adequate.

3. Methods

3.1. Before You Start

1. Successful PCR is achieved following careful preparation; any slight differences in pipetting early in the proceedings are amplified during the process; thus, wherever possible, avoid repeatedly pipetting small volumes. Assays should be scaled up so that "master mixes" are prepared. These will significantly decrease the pipetting variations.
2. A number of methods are available for isolating DNA from solid tissues (first, carefully grind under DNase- and RNase-free conditions), blood, or cultured cells. Probably, the simplest and readily applicable are the guanidine thiocyanate-containing solvents [e.g. TRizol™ (Invitrogen) and TRI Reagent (Sigma-Aldrich)]. They have the added advantage of allowing both DNA and RNA to be isolated simultaneously (and protein if desired). Alternatively, should disposal of organic solvents be a problem, commercial kits employing the use of Ni^{+}/silica-based spin columns are easy and versatile to use (see Note 1).

3. Solubilise primers and prepare dilutions prior to setting up the reactions (and certainly not when any template is being prepared or in the vicinity) (see Note 2).
4. Three types of reaction will be considered: basic PCR, cDNA synthesis, and one-step RT-PCR.
5. Only use molecular-grade (PCR grade) water that is DNase and RNase free.
6. Thaw items carefully and then immediately place them on ice. Leave the enzyme in the freezer until required. Make sure that the solutions are evenly mixed prior to use (see Note 3).

3.2. Setting Up a Reaction (See Note 4)

1. The basic PCR reaction is universal and contains the following components:

   Suggested 100 μl reaction to start:

	Volume	Final concentration
DNA template	Variable	≤1 μg
Sense primer (10 μM)	10 μl	0.1 μM
Antisense primer (10 μM)	10 μl	0.1 μM
Buffer (usually 10× concentration)	10 μl	×1
dNTPs mix (10 mM each)	2 μl	200 μM each
Enzyme (e.g. 500 U/μl)	0.5 μl	2.5 U/μl
H_2O	To 100 μl	Variable

   The optimum template, primer, and Mg^{2+} concentrations and annealing temperature (the cycling parameters) should be determined empirically for each primer pair.
2. Regarding the DNA template: This can be of any size from high-molecular-weight (HMW) genomic DNA to a plasmid, recombinant, or even a PCR product (see Note 4).
3. It is highly unlikely that only one reaction will be set up; there should be at least one negative control to confirm that the reagents are not contaminated. Draw up a table including volumes for a single reaction, then multiple reactions, and then a final column that includes an extra 10% (to allow for pipetting variations).
4. Create Master Mixes.
5. The core Master Mix should contain H_2O, buffer, dNTPs, and enzyme. It is important to ensure the order in which reagents are added: (a) add H_2O first; (b) buffer next; and (c) dNTPs next; as an option, include the primer pair. Maintain the mixtures on ice. The template(s) can be added directly to the reaction tubes (see Note 4). When everything else has been set up, the enzyme can be added to the master mix last

and carefully mixed. Using a fresh tip every time, add an appropriate measure to each reaction tube (in the above example to reach a final volume of 100 μl). Always prepare at least one negative control; this should include everything, except the template (use H_2O to make up the volume).

3.3. Establishing Cycling Parameters

1. The first variable to consider is the annealing temperature. The starting point is usually 5°C lower than the T_m. In some thermal cyclers, it is possible to programme a gradient to quickly determine the optimum temperature; otherwise, this will need to be performed by trial and error. Spurious fragments may disappear with a rise of one degree; similarly, stronger bands may appear with a lowering of 1°C.
2. Include the amplification of a reference gene (also often called housekeeping gene, an example of this is GAPDH or β-actin); this will test your technique as well as the quality of the template, cycling conditions, and the reagents.
3. The primer and dNTP concentrations are not usually limiting factors, and should always be in excess (as above). However, too high a concentration of primers will interfere with the efficiency of the enzyme reaction and encourage incorrect primer annealing and aberrant amplifications.
4. The Mg^{2+} concentration is important (the enzyme reaction is a bi-substrate reaction and requires Mg^{2+} in the active site); too high a Mg^{2+} concentration will cause interference. However, Mg^{2+} also affects primer-template-enzyme stability and the Mg^{2+} concentration may need to be titrated.
5. The number of cycles required is influenced by the copy number of the starting template; usually, between 30 and 40 cycles will suffice.

3.4. Thermal Cycling (See Note 5)

The type of machine, enzyme, template, and T_m of the primers influence the cycling programme, although a machine that is able to rapidly change (ramping time) from one temperature to the next is essential. The T_m of the primer pair(s) determines the annealing temperature. The length of time at each step is determined by the user, usually 30–90 s. For a basic three-step cycling, the suggested settings to start are as follows:

Denaturation	94°C	1 min
Annealing	50–68°C	0.5 min
Extension	72°C	1 min
Number of cycles	35	
Final extension	72°C	10 min

Reaction products can be stored at room temperature overnight or at 4°C (or –20°C) for longer periods.

3.5. Reverse Transcriptase Reaction

The conditions best suited for each propriety enzyme are included with each purchase, but can be adapted for your own purposes. The reaction described below is standard for use with an oligo-dT_{18} primer, but can be adapted for alternative primers. To prepare more cDNA, either scale-up one reaction or set up a number of reactions and then pool them (± aliquots for freezing); doubling the amount of template will not double the yield!

1. For generating cDNA from 1 μg total RNA, first prepare a 6.5-μl volume containing the following:

	Volume	Final concentration
1 μg total RNA template	Variable	0.1 μg/μl
Olig-dT_{18} (0.5 μg/μl)	0.5 μl	25 ng/μl
dNTPs (10 mM)	0.5 μl	0.5 mM
H_2O to 6.5 μl	Variable	

Heat denature at 70°C for 10 min, chill immediately on ice, briefly centrifuge to collect condensation, and return to ice. To the above, add the following:

	Volume	Final concentration
First-strand buffer (×5 concentration)	2 μl	1×
*Dithiothreitol (100 mM)	1 μl	10 mM
RNasin (40 U/μl)	0.25 μl	1 U/μl
Enzyme (200 U/μl)	0.5 μl	10 U/μl

Incubate at 42°C for 60 min. Assess the cDNA product quantitatively and qualitatively using UV spectrophotometry, or by gel electrophoresis.

3.5.1. One-Step RT-PCR

There are many formulations for the OneStep RT-PCR (Qiagen™). The Qiagen protocol is given here and describes a 50 μl reaction; however, 15 μl yields sufficient product for analysis by gel electrophoresis and, therefore, the volumes can be adjusted as appropriate.

The same rules apply as for PCR: Always use fresh tips; add the enzyme to the master mix last; and keep everything on ice until transferring to the thermal cycler.

The table below describes a 50mL reaction; use this as a guide to plan a single, then multiple reactions.'

	Volume	Final concentration
RNA template	Variable	≤250 ng
RNasin (40 U/μl)	1.25 μl	1 U/μl
Sense primer (10 μM)	3 μl	0.6 μM
Antisense primer (10 μM)	3 μl	0.6 μM
Buffer (×5 concentration)	10 μl	1×
dNTPs mix (10 mM each)	2 μl	400 μM each
Enzyme (e.g. 500 U/μl)	2 μl	
H_2O	To 50 μl	Variable

Plan your assays to avoid pipetting small volumes. Include a negative control using H_2O to replace the template. If comparing a number of different templates for the expression of more than one gene, then set up at least two, if not three, mixes and include template with RNasin and H_2O.

One-Step RT-PCR (Qiagen) differs from basic PCR in that:

(a) The thermal cycler (and lid) must be preheated to 50°C prior to loading the samples.

(b) The first stage of the (RT-PCR) reaction includes an incubation step at 50°C for 30 min for reverse transcription.

(c) This is followed by an incubation step at 95°C for 15 min (to activate HotStar Taq polymerase and destroy the reverse transcriptases).

(d) A three-step cycling protocol follows as for basic PCR using an annealing temperature indicated by the primers (see above).

4. Notes

1. (a) There is no substitute for careful preparation of template, ideally to ensure uncontaminated, undegraded, non-fragmented HMW nucleic acid, whether it is DNA or RNA. Contamination with proteins (e.g. histones or other chromatin-associated proteins), inorganic ions (e.g. Na^+), compounds (e.g. EDTA and Na_2), organic molecules, or nucleic acids, as well as changes in pH, will influence the enzyme activity and primer annealing, hence the efficiency of the reaction and overall yield. However, when the target sequence is relatively short (<100 bp) or even present at

Table 1
Expression data and abundance of mRNA: Cells, especially cancer cells, vary enormously in their nucleic acid content. The following table gives a rough guide to the abundance of different RNA molecules, based on the RNA content of a typical human cell: 10–30 pg

Number of cells	Yield of tRNA	Yield of mRNA	Abundant transcripts/cell	Rare transcripts/cell
10^6	10–15 μg	750 ng	10^{10}	10^7
10^5	1–1.5 μg	75 ng	10^9	10^6
10^4	100–150 ng	7.5 ng	10^8	10^5
10^3	10–15 ng	750 pg	10^7	10^4
10^2	1–1.5 ng	75 pg	10^6	10^3
10	100–150 pg	7.5 pg	10^5	10^2
1	10–15 pg	~0.75 pg	~10^4	~10

Between 1 and 5% of this is mRNA (~0.5 pg) while the rest is composed of rRNA (80–85%) and tRNA (15–20%). Note that "rare" or low-abundance messages may include >10,000 different mRNA species and account for 45% of the mRNA population

very low copy number (<2 copies per cell), PCR is also achievable following relatively crude preparations. The limitations are that the template will not be stable and, therefore, will be subject to degradation; specificity may be affected and the yield will probably be low; refer to Table 1.

(b) To establish the cycling parameters: template extracted from a cancer cell line is ideal; be certain you have a positive control as well as a negative control, and then move on to extraction from tumour tissues.

(c) Good-quality nucleic acid is stable, but repeated freezing and thawing will degrade it. Prepare more than one aliquot to avoid frequent freezing and thawing of the same sample.

(d) Good-quality nucleic acid preparations should be concentrated and, therefore, are difficult to use for assessing the optical density, especially if >0.5 μg/μl (and the volume for reading in a UV spectrophotometer is 1 μl). Prepare different dilutions and only accept a reading when it is consistent.

2. (a) Follow the manufacturer's instructions for preparation; use molecular-grade water. Keep the stock solution at a high concentration (usually 100 μM).

(b) Prepare a "working" ×10 concentration solution (usually 10 μM) and prepare small aliquots and store frozen (if possible, number them 1/5, 2/5–5/5, etc.; this way, you will know when to start preparing more).

3. (a) *Taq* polymerases are extremely active, even at room temperature. It is essential that you keep them in the freezer until required and, when in use, keep them on ice. If appropriate, also prepare small aliquots to reduce the risk of denaturation and contamination.

   (b) Enzymes are provided in a 50% glycerol solution to prevent degradation by repeated freezing and thawing; because of the viscosity it is easy to transfer excess to the Master Mix. Always place your tip near the top of the solution.

4. Fresh tips should be used throughout the different procedures.

   (a) Early protocols gave instructions for 100 or 50 μl reactions and I have used these volumes for ease of comparison/calculations. However, if your next step is analysis by gel electrophoresis, then there is really no need for a large volume, assuming that you load between 15 and 20 μl per lane (freezers around the world must be full of surplus PCR products). We routinely set up 15 μl reactions and examine the entire product, hence no wastage.

   (b) If your product yield is low, do not assume that additional template is required: copy number is the important factor. Thus, if a gene has already been amplified and is part of a gene family (or has pseudogenes), an excess of template at the start will "mop-up" the primers (imagine iron filings on a magnet) and prevent the PCR from reaching an exponential phase. This is especially important when using a recombinant or amplicon as the template. When examining HMW DNA, start with a concentration of 100 ng/100 μl reaction, while 1 ng of a 50 kb λ DNA should be sufficient; titrate the template if necessary.

   (c) Alternatively, if the target sequence is rare (perhaps a translocation point), be prepared to set up a re-amplification reaction. Take an aliquot of your first product (even if nothing is visible on a gel) and re-amplify it using either the same primers or a second pair of "nested primers" (i.e. a set designed to lie within the target of the first pair).

   (d) If one template is being examined for a number of genes, then this can be added to the master mix and the different primer pairs dispensed to the reaction tubes (or vice versa). Adding template or primers to the master mix increases the risk of contaminating the reagents, although if the correct order is adhered to only the enzyme is at risk; thus take extra care to use a fresh tip. This adaptation has the advantage of reducing pipetting variations. Alternatively, consider preparing different master or core mixes, for example (1) H_2O + template (±RNasin) + dNTPs in one mix with (2)

H_2O + buffer + enzyme in another mix, and pipetting the primers directly in the reaction tubes. When you are confident of your technique and assured that the primers are working well, adapt the assays to reduce the risk of contamination and increase efficiency. It is commonplace to prepare three core mixes and pipette 5 μl from each to obtain the final 15 μl reaction volume when examining a number of templates for the expression levels of different genes.

5. Thermal Cycling.
   (a) Hot Start PCR requires an initial denaturation step (see manufacturer's recommendations). Ensure sufficient time for melting the HMW DNA at 94°C; otherwise, efficiency will be severely affected.
   (b) Allow enough time for the primers to anneal to their complementary sequences, but not long enough for the *Taq* polymerase to start amplifying spurious sequences.
   (c) Stringency is increased with higher temperatures, although primers with high percentage of GC residues require a higher melting point; however, avoid a very high annealing temperature because it may be too close to the optimum temperature of the polymerase (70–74°C).
   (d) A final extension of 10 min at 72°C is often included to ensure that amplified fragments are extended fully and are complete in order that they are of the size defined by the primers.

5. Appendix

Useful Web sites:

http://www.ncbi.nlm.nih.gov.

http://www.ncbi.nlm.nih.gov/tools/primer-blast/index.cgi?LINK_LOC=BlastHome.

www.promega.com.

www.qiagen.com.

www.sigma.com.

www.invitrogen.com.

http://www.ensembl.org/index.html.

http://sigma-genosys.com.

www.sigmaaldrich.com/life-science/custom-oligos.html.

https://www.roche-applied-science.com.

http://www.proteinatlas.org/index.php.

http://www.ncri-onix.org.uk/portal/#S1.

References

1. Saiki RK, Scharf S, Faloona F, Mullis KB, Horn GT, Erlich HA, Arnhein N (1985) Enzymatic amplification of b-globin genomic sequences and restriction site analysis for diagnosis of sickle cell anaemia. Science 230:1350–1354
2. Saiki RK, Gelfand DH, Stoffel S, Scharf SJ, Higuchi R, Horn GT, Mullis KB, Erlich HA (1988) Primer-directed enzymatic amplification of DNA with a thermostable DNA polymerase. Science 239:487–491
3. Fehm T, Hoffmann O, Aktas B, Becker S, Solomayer EF, Wallwiener D, Kimmig R, Kasimir-Bauer S (2009) Detection and characterization of circulating tumor cells in blood of primary breast cancer patients by RT-PCR and comparison to status of bone marrow disseminated cells. Breast Cancer Res 11:4(R59)
4. Liang P, Pardee AB (1992) Differential display of eukaryotic messenger RNA by means of the polymerase chain reaction. Science 257:967–971
5. Cheng S, Fockler C, Barnes WM, Higuchi R (1994) Effective amplification of long targets from cloned inserts and human genomic DNA. Proc Natl Acad Sci USA 91:5695–5699
6. Kubista M, Andrade JM, Bengtsson M, Forootan A, Jonak J, Lind K, Sindelka R, Sjoback R, Sjogreen B, Strombom L, Stahlberg A, Zoric N (2006) The real-time polymerase chain reaction. Mol Aspect Med 27:95–125
7. Bustin SA, Nolan T (2004) Pitfalls of the quantitative real-time reverse-transcriptase polymerase chain reaction. J Biomol Tech 15:155–166
8. Kulis M, Esteller M (2010) DNA methylation in cancer. Adv Genet 70:27–56
9. Ferguson AT, Evron E, Umbricht CB, Pandita TK, Chan TA, Hermekin H, Marks JR, Lambers AR, Futreal PA, Stampfer MR, Sukumar S (2000) High frequency of methylation at the 14-3-3σ locus leads to gene silencing in breast cancer. Proc Natl Acad Sci USA 97:6049–6054
10. Herman JG, Graff JR, Myohanen S, Nelkin BD, Baylin SB (1996) Methylation-specific PCR: a novel PCR assay for methylation status of CpG islands (DNA methylation/tumor suppressor genes/pl6/p15). Proc Natl Acad Sci USA 93:9821–9826

Chapter 6

Sodium Dodecyl Sulphate–Polyacrylamide Denaturing Gel Electrophoresis and Western Blotting Techniques

Christine Blancher and Rob Mc Cormick

Abstract

The study of proteins, their expression and post-translational modification, is a key process in molecular biology. Immunoblotting is a well-established and powerful tool for the study of proteins, which continues to evolve as new reagents and apparatus are developed. This chapter describes in detail the process by which proteins are extracted from cells, quantified, fractionated using poly-acrylamide gel electrophoresis, transferred to a membrane, and assessed by immunoblotting. Variations in experimental technique, and new technologies available to the researcher, are also discussed.

Key words: Western blotting, Sodium dodecyl sulphate–polyacrylamide denaturing gel electrophoresis, Immunoblotting, Protein extraction, Polyacrylamide gel electrophoresis

1. Introduction

The goal of Western-blotting, or more correctly immunoblotting, is to identify with a specific antibody a particular antigen in a midst of a complex mixture of proteins fractionated in a polyacrylamide gel and immobilised onto a membrane. Immunoblotting can be used to determine a number of important characteristics of proteins: the presence and semi-quantitation of an antigen, the relative molecular weight of the polypeptide chain, and the efficiency of extraction of the antigen.

Immunoblotting occurs in six stages: extraction and quantification of protein samples, resolution of the protein sample in sodium dodecyl sulphate–polyacrylamide denaturing gel electrophoresis (SDS–PAGE), transfer of the separated polypeptides to a membrane support, blocking non-specific binding sites on the membrane, addition of antibodies, and finally, a detection step.

Miriam Dwek et al. (eds.), *Metastasis Research Protocols*, Methods in Molecular Biology, vol. 878, DOI 10.1007/978-1-61779-854-2_6, © Springer Science+Business Media, LLC 2012

Sample preparation is important to obtain accurate separation of the proteins on the basis of molecular weight. Different procedures might be required to prepare the sample depending on whether the protein is extracellular, cytoplasmic, or membrane associated. Although there are exceptions, many nuclear and cytoplasmic proteins can be solubilised by lysis buffers that contain the non-ionic detergent Nonidet P-40 (NP-40) with either no salt at all or relatively high concentrations of salt (e.g., 0.5 M NaCl). However, the efficiency of extraction is often greatly affected by the pH of the buffer and the presence or absence of chelating agents (e.g., ethylenediaminetetraacetic acid, EDTA). Extraction of membrane-bound and hydrophobic proteins is less affected by the ionic strength of the lysis buffer, and is often facilitated by the addition of ionic and non-ionic detergents. Many methods of solubilisation, particularly those that involve the mechanical disruption of cells, release intracellular proteases that can digest the target protein. The susceptibility to attack, of different proteins by proteases, varies widely, with cell-surface and secreted proteins generally being more resistant than intracellular proteins. It is therefore important to maintain the cell extracts on ice to minimise proteolytic activity. In addition, a cocktail of protease inhibitors, which are available individually or as ready-made tablets, are commonly included in lysis buffers. The treatments used should disrupt the secondary or tertiary protein structure, thus preventing alterations in protein migration through the acrylamide matrix. Therefore, sample preparation should solubilise and denature proteins, dissociate polypeptides, and reduce disulphide bonds. This is achieved through a combination of SDS, a reducing agent (usually β-mercaptoethanol), and heat.

When comparing samples, it is important to load equivalent amounts of total protein onto the gel. Total protein concentration can be measured using a number of methods. The Bradford method is relatively accurate for most proteins except small basic polypeptides such as ribonuclease or lysozyme. It is hampered by detergent concentrations over approximately 0.2% (e.g., Triton X-100, SDS, and NP-40). Commercially available colorimetric assay systems are available (e.g., the Pierce—BCA protein assay reagent, or the Bio-Rad—DC protein assay) they are easier to use and relatively cheap.

The proteins diluted in a sample buffer are separated by gel electrophoresis. Most often, the electrophoresis system of choice will be a Tris/glycine discontinuous SDS–PAGE (1). The SDS coats the proteins, providing them with a negative charge proportional to their length. When the coated sample is run on an SDS polyacrylamide gel, the proteins separate by charge and by the sieving effect of the gel matrix. Sharp banding of the proteins is achieved using the discontinuous gel system which has stacking and separating gel layers that differ in, either salt concentration, pH, and acrylamide concentration, or a combination of these. To assess the relative molecular weight of the electrophoresed protein(s) a pre-stained molecular weight marker is run in the outer lane of the gel for comparison.

Once separated, proteins may be transferred (blotted) onto a second matrix that binds in a non-specific manner. Generally a nitrocellulose or polyvinylidene difluoride (PVDF) membrane is used for this step. Transfer is achieved by placing the membrane in direct contact with the gel and then placing this sandwich in an electric field to drive the proteins from the gel onto the membrane (2). Electrophoretic transfer can be undertaken by complete immersion of a gel-membrane sandwich in a conducting buffer (wet transfer), by placing the gel-membrane sandwich between gel pads (dry transfer), or by absorbent filter paper soaked in transfer buffer (semi-dry transfer). For wet transfer, the sandwich is placed in a buffer tank with platinum wire electrodes. For dry and semi-dry transfer, the gel-membrane sandwich is placed between two plates equipped with copper saturated calomel (SCE), stainless steel or platinum electrodes. Apparatus for both types of transfer are available commercially.

Before the blot can be processed for protein detection, it is essential to block the membrane to prevent non-specific absorption of the immunological reagents to the surface of the membrane. The most commonly used blocking solution that is compatible with nearly all detection systems is non-fat dried milk (3) this is often prepared in a solution containing the detergent Tween 20 in order to reduce excessive background staining (4). However, commercial blocking buffers are also available to optimise the detection of difficult antigen–antibody interactions.

Procedures for detection of a Western blot vary widely. One common variation is direct, compared with indirect, detection systems. With the direct detection method, the primary antibody is labelled with an enzyme or fluorescent dye. This detection method is not widely used as most researchers prefer the indirect detection method; wherein an unlabelled primary antibody specific to the target protein is first incubated with the membrane in the presence of blocking solution. The membrane is then washed and incubated with a labelled secondary antibody which is directed against the primary antibody. Labels may include biotin, fluorescent probes such as fluorescein or rhodamine, and enzyme conjugates such as horseradish peroxidase (HRP) or alkaline phosphatase (AP). After further washing, detection is achieved using the appropriate apparatus, for example, film with chemiluminescent labels, or a UV source plus CCD camera for fluorescent probes.

A vacuum-based method for blocking and antibody detection on the membrane has recently been developed. The SNAP i.d. system from Millipore draws reagents through the membrane using a vacuum. Because the reagents are forced through, rather than relying on diffusion, this method offers considerable time saving. Blocking, primary and secondary antibody application, plus washing can be achieved in less than 30 min, obviating the need for sometimes lengthy (e.g., overnight) incubations with primary antibodies.

2. Materials

2.1. Extraction and Quantification of Protein Samples

Lysis buffers that are commonly used to prepare extracts of mammalian cells for immunodetection are shown below. In the absence of any information about the target protein, we recommend trying the triple- and single-detergent lysis buffer before attempting more specialised methods of extraction. A large variety of protease inhibitor cocktails are also available commercially which inhibit a broad spectrum of proteases and phosphatases and should be chosen according to tissue sample type and protein studied.

1. Triple-detergent lysis buffer: 50 mM Tris, pH 8.0; 150 mM NaCl; 0.1% w/v SDS; 1% v/v Nonidet P-40 (NP-40); 0.5% w/v sodium deoxycholate and protease inhibitor cocktail.
2. Single-detergent lysis buffer: 50 mM Tris, pH 8.0; 150 mM NaCl; 1% w/v Triton X-100 or 1% v/v NP-40 and protease inhibitor cocktail.
3. High-salt lysis buffer: 50 mM HEPES, pH 7.0; 500 mM NaCl; 1% v/v NP40 and protease inhibitor cocktail.
4. High stringency lysis buffer: 6 M urea; 10 mM Tris, pH 6.8; 5 mM DTT; 1% w/v SDS; 10% v/v glycerol and protease inhibitor cocktail.
5. No-salt mild lysis buffer 50 mM HEPES, pH 7.0; 1% v/v NP-40 and protease inhibitor cocktail.
6. Ice cold phosphate buffer saline (PBS): weigh 11.5 g anhydrous di-sodium hydrogen orthophosphate (80 mM), 2.96 g sodium dihydrogen orthophosphate (20 mM), and 5.84 g sodium chloride (100 mM), transfer to a beaker and make up to 1 L with distilled water. Adjust the pH to 7.5. Alternatively, use ready prepared PBS solution or tablets dissolved in deionised/distilled water.
7. Preparing a standard curve for protein quantitation: prepare a series of known protein samples for a standard curve using a protein as similar in its properties to your sample as possible (e.g., if you are measuring antibody concentrations, then use purified antibody as the protein sample for the standard curve). If your sample is unknown, use an antibody as the standard protein. Bovine serum albumin (BSA) gives a value about two-fold higher than its weight for Bradford dye-binding assays (5) but it is fine for Lowry-based assays (6). The dilution of the

protein standard should be from 0.2 mg/ml to approximately 1.5 mg/ml, and the standards should always be prepared in the same buffer as the protein sample.

8. Bradford dye concentrate: dissolve 100 mg Coomassie Brilliant Blue R250 in 50 ml of 95% v/v ethanol. Add 100 ml concentrated phosphoric acid. Add distilled water to a final volume of 200 ml. The dye is stable at 4°C for at least 6 months. This dye concentrate is also available commercially. Alternatively use a colorimetric protein assay kit, for example, the Pierce BCA Kit, or the BioRad DC Kit.
9. "Rubber policeman" for scraping cell monolayers (Sarsted, Greiner).
10. Either a sonicator or homogeniser with immersible tip, or 23-G hypodermic needle, in order to shear chromosomal DNA to a lower viscosity.
11. Plastic or glass cuvettes with 1 cm path length matched to laboratory spectrophotometer. If analysing multiple samples it is easier to perform microplate assays using 96-well microtitre plates and a microplate reader.
12. Spectrophotometer capable of measuring the absorbance at wavelengths of 200–600 nm.

2.2. Resolution of the Protein Sample in SDS–PAGE

Three formats of gels are commonly available: mini gels (~8 cm×8 cm), midi gels (13 cm×9 cm) and large gels (16 cm×16 cm). For most of the analytical and some preparative fractionation techniques mini gels are the first choice.

1. The Bio-Rad Mini-PROTEAN Tetra cell vertical electrophoresis apparatus and casting equipment or the Invitrogen, Novex, X-Cell SureLock Mini-Cell vertical electrophoresis systems are commonly used and combine fast separation with an easy set-up (see Note 1). These systems all allow between one and four pre-cast or hand-cast mini gels of variable thickness to be run at the same time. The number of wells in the gel can be adjusted according to the comb that is used. A wide range of pre-cast gels are commercially available in both gradient and fixed concentrations of acrylamide, including different gel formulations and adapted running buffers. For each hand-cast gel an inner glass plate and outer glass plate, two spacers and a Teflon comb of the same thickness (0.75–1.5 mm) will be required.
2. Power pack.
3. Gel casting solutions: 29% w/v acrylamide/1% w/v bisacrylamide stock solution. It is preferable to purchase prepared acrylamide/bisacrylamide as it circumvents the need to weigh powdered acrylamide which is a neurotoxin. 10% w/v SDS stock solution prepared in deionised water and stored at room

temperature. Stacking gel buffer: 0.5 M Tris pH 6.8 prepared by dissolving 6.05 g Tris base in 40 ml distilled water, adjust to pH 6.8 by the slow addition of 1 M HCl and make up to 100 ml with distilled water. Resolving gel buffer: 1.5 M Tris pH 8.8 prepared by dissolving 90.9 g Tris base in 350 ml water, adjust to pH 8.8 by slow addition of 1 M HCl and make up to 500 ml with distilled water.

4. Catalysts: 10% w/v ammonium persulphate (APS) made fresh, stored for maximum of 1 week at 4°C or stored in aliquots of 1 ml at −20°C for several months. *N,N,N′,N′*-tetramethyl ethylene diamine (TEMED) used as supplied by the manufacturer.
5. Water-saturated butanol solution: prepare by combining 100 ml *n*-butanol and 5 ml distilled water in a bottle and shaking well. Store at room temperature.
6. SDS sample buffer: make up 10 ml, six times concentrated, sample buffer by mixing 6 ml 0.5 M Tris, pH 6.8; 1.2 g SDS; 3 ml glycerol; 600 μl β-mercaptoethanol; and 6 mg Bromophenol blue. SDS sample buffer is also available commercially from a number of suppliers.
7. Molecular weight markers: these are convenient for monitoring electrophoresis and transfer, and they are available prestained with blue-dye or multicoloured to provide easier band identification.
8. Tris–glycine-SDS running buffer: the formulation for the running buffer is 25 mM Tris, pH 8.3; 192 mM glycine (electrophoresis grade); 0.1% w/v SDS. Prepare a 5× concentrated solution by dissolving 15.1 g Tris base, 72 g glycine, 5 g SDS, make up to 1 L with deionised water. Dilute with deionised water for 1× working solution as required. This buffer is available commercially as well as ready to use Tris-Hepes-, Tris-Acetate-, MOPS-, and MES-SDS running buffer specifically adapted for precast gels (e.g., from BioRad, Pierce and Invitrogen).
9. Safety note: acrylamide and bisacrylamide are potent neurotoxins and are absorbed through the skin. Their effect is cumulative. Polyacrylamide is considered to be less toxic, but should still be treated with care as it may contain small quantities of non-polymerised material. The use of prepared acrylamide–bisacrylamide solution is recommended as it circumvents the requirement to weigh acrylamide powder.

2.3. Staining SDS–Polyacrylamide Gels

Total protein stains are commonly used for qualitative and quantitative analysis of proteins separated on gels or to visualise the proteins remaining on the gel after transfer. The most common method is to stain with Coomassie dye.

1. Coomassie Brilliant Blue solution: dissolve 0.25 g of Coomassie Brilliant Blue R250 in 90 ml methanol: H_2O (1:1 v/v) and add 10 ml of glacial acetic acid. Filter the solution through a Whatman No. 1 filter to remove any particulate matter.
2. Destain solution: mix 90 ml methanol: H_2O (1:1 v/v) and 10 ml of glacial acetic acid. Alternatively, ready-made Coomassie-based gel stain solutions are available which do not require methanol or acetic acid fixative and destaining (e.g., Piercenet, Invitrogen, and BioRad).
3. Safety note: methanol is highly flammable and toxic, with danger of very serious irreversible effects following inhalation, skin contact or swallowing.

2.4. Transfer of the Separated Polypeptides to a Membrane Support

1. Wet transfer apparatus: systems include the BioRad mini Trans-blot cell (buffer tank, two gel holder cassettes, four fibre pads, and mini Trans-blot module) or the Invitrogen XCell II Blot module (buffer tank, four blotting pads, and mini-gel blot module). This apparatus is commonly used and allows the transfer of 1–2 mini-gels in 1–2 h or overnight using 200–450 ml of transfer buffer.
2. Semi-dry transfer apparatus: systems include the BioRad Trans-Blot SD and Sigma Uniblot blotting system. These apparatus are ideal for rapid (15–60 min), high-intensity transfer and are best suited for proteins of mid-range molecular weight 10–100 kDa. They require very little transfer buffer but adjustments may need to be made to ensure optimal transfer conditions.
3. Dry transfer apparatus: systems include the Invitrogen iBlot Dry blotting System—a buffer-free system equipped with its own integrated power supply, it requires preassembled iBlot transfer stacks including cathode and anode gel pads and copper electrodes, nitrocellulose membrane, disposable sponge, and filter paper. The transfer of proteins is achieved in a very short period of time (6–10 min).
4. Power pack.
5. Sheets of absorbent filter paper (Whatman 3MM or equivalent) or BioRad extra-thick blotting paper (semi-dry transfer) cut to the gel size.
6. Transfer membrane: nitrocellulose membrane or PVDF. PVDF is more suitable for small molecular weight proteins.
7. Towbin transfer buffer (2) (for wet or semi-dry transfer systems): 25 mM Tris, 192 mM glycine, and 10–20% v/v methanol pH 8.3. For 2 L, dissolve 6.04 g Tris base and 28.8 g glycine in 1.6 L of distilled water, add 100–200 ml methanol, make up to 2 L with distilled water. It is not necessary to add acid or base to adjust the pH. Store at 4°C. SDS may be added to this buffer up to a final concentration of 0.1% w/v. Prepare

1 L by mixing 200 ml of 5× concentrated Tris–Glycine-SDS running buffer with distilled water and adding 100–200 ml of methanol (see Notes 2 and 3). The opposing effects of methanol and SDS in the blotting process can be exploited in semi-dry transfer because the buffer reservoirs (in this case, the filter paper on both sides of the gel) are independent. In a discontinuous system, methanol should be included in the buffer on the membrane side (anode) of the blot assembly and SDS used on the gel side (cathode), taking advantage of the positive effects of each component.

8. Discontinuous buffer system for semi-dry blotting: a new discontinuous buffer system using a Tris-CAPS (3-(cyclohexylamino)-1-propanesulfonic acid) buffer provides excellent results in semi-dry blotting. Prepare the Tris-CAPS buffer with 60 mM Tris base and 40 mM CAPS, pH 9.6. For 1 L of 5× concentrated stock solution dissolve 36.4 g Tris base, 43.3 g CAPS in distilled water. Prepare 100 ml of 1× transfer buffer for the anode (lower electrode) by mixing 20 ml of 5× concentrated Tris-CAPS with 15 ml methanol and 65 ml distilled water. Prepare 100 ml of 1× transfer buffer for the cathode (upper electrode) by mixing 20 ml of 5× Tris-CAPS with 1 ml 10% w/v SDS and 79 ml distilled water.

2.5. Total Protein Stain for Membranes

1. Ponceau S solution: for many years the red Ponceau S stain has been the best option for staining the membrane following Western blotting and before immunoblotting steps, despite some shortcomings of the system (low sensitivity, weak binding, detection limit: 250 ng protein). To prepare the solution use 0.1% w/v Ponceau S in 5% v/v acetic acid in water. Ponceau S is available in a ready-to-use solution from a number of suppliers.
2. Other reversible protein stains: the MemCode reversible protein stain from Piercenet is suited to both nitrocellulose and PVDF membranes. Similarly, the Invitrogen NOVEX reversible membrane protein stain kit works well too. Both of these stains are more sensitive than Ponceau S (detection limit: 10–25 ng protein), the dyes that they utilise have high avidity and the blue colour of the protein is easily photographed.

2.6. Blocking Non-specific Binding Sites on the Membrane

1. Blocking buffers: the blocking buffer should improve the sensitivity of the assay by reducing background interference. No single blocking buffer is compatible with every system. For this reason, a variety of blocking buffers in both Tris-buffered saline (TBS) and PBS are available (e.g., Pierce: Blotto, BSA, Casein, SEA BLOCK, Super block and Starting Block and so on). The choice of blocking buffer for a given blot depends on the antigen and the type of enzyme conjugate used. We use the Blotto

blocking solution prepared with 5% w/v non-fat dry milk (e.g., Marvel) and 0.05% v/v Tween 20, in PBS or TBS. For 1 L TBS use 2.4 g Tris base, 8 g NaCl and adjust to pH 7.6 with HCl. Store TBS at 4°C only for 1 or 2 days.

2. Platform shaker.

2.7. Addition of Antibodies

The choice of a primary antibody for a Western blot will depend on the antigen to be detected and which antibodies are available (7). A huge number of primary antibodies are available commercially and can be identified quickly by searching sites such as www.antibodyresource.com.

1. Primary antibody dilution buffer: PBS or TBS with 0.1% v/v Tween 20 and 5% w/v blocking agent (e.g., 5% w/v non-fat dry milk for monoclonal antibodies or 5% w/v BSA for polyclonal antibodies).
2. Choice of antibody: both monoclonal and polyclonal primary antibodies work well in Western blotting. Polyclonal antibodies are less expensive and less time consuming to produce and they often have a high affinity for the antigen. Monoclonal antibodies are valued for their specificity, purity, and consistency and usually result in lower background and less non-specific binding. Crude antibody preparations such as serum or ascites fluid are sometimes used for Western blotting, but the impurities present may increase background. The appropriate concentration of a primary antibody should be determined empirically in pilot experiments. The recommended test dilutions for polyclonal antibodies: 1:100–1:5,000; supernatants from cultured hybridoma cell: undiluted to 1:100; ascitic fluid from mice bearing hybrid myelomas; 1:1,000–1:10,000.
3. Washing steps are necessary to remove any unbound reagents and to reduce background thereby increasing the signal-to-noise ratio. Insufficient washing often results in high background, while excessive washing may result in decreased sensitivity caused by elution of the antibody and/or antigen from the blot. A recommended washing buffer is 0.1% v/v Tween 20 in PBS or TBS. Commercial washing buffers are also available (e.g. from Pierce).
4. The choice of secondary antibody depends upon the species of animal in which the primary antibody was raised (the host species). A wide variety of labelled secondary antibodies are available for use in Western blotting. The labels include biotin, fluorescein, rhodamine, DyLight Dyes, HRP, and AP. Antibody solutions for Western blotting are typically diluted from 1:100 to 1:500,000 from a 1 mg/ml stock solution. The optimal dilution of a given antibody with a particular detection system must be determined experimentally.

5. Heat-sealable plastic bags (e.g., Sears Seal-A-Meal or equivalent) and heating bag sealer.
6. Plastic trays for washing steps and secondary antibody incubation, usually pipette tip boxes or box lids are perfect for this.

2.8. Detection

Many different labels can be conjugated to antibodies. Radioisotopes were used extensively in the past, but they are expensive, have a short shelf-life, offer no improvement in signal-to-noise ratio and require special handling. Alternative labels are biotin, fluorophores, and enzymes. Using fluorophores results in fewer steps but special equipment is required to view the fluorescence. Also, a photograph must be taken for a permanent record of the results. Enzymatic labels are most commonly used and consistently produce excellent results. AP and HRP (or POD) are the two enzymes that are used extensively. An array of chromogenic, fluorogenic, and chemiluminescent substrates is available for use with either enzyme. Recently, a new detection technology has been developed using Quantum Dot particles (Intvitrogen) which can be either conjugated to secondary antibodies or streptavidin and is to be used in conjunction with biotin-labelled secondary antibodies. Advantages of the Quantum dot system include stability, the intensity of the signal, and multiple colours when exited using a UV source, this system enables the simultaneous detection of multiple proteins on a single membrane (8).

Chemiluminescent substrates used in conjunction with HRP secondary antibody conjugate generate a light signal that can be captured on a photographic film. These substrates have steadily gained in popularity because they offer several advantages over other detection methods: they are inexpensive, sensitive, fast, and stable. Several suppliers (e.g., Sigma, Pierce, GE Healthcare and Invitrogen) offer very good chemiluminescent detection systems with new improved sensitivity that allows detection and quantification of very low protein signals (e.g., Sigma, Pierce, GE Healthcare, and Invitrogen).

An X-ray film cassette, a roll of Saran Wrap (or similar), a timer, and autoradiography film are all recommended for chemiluminescence detection (e.g., Hyperfilm ECL, GE Healthcare) alternatively a quantitative Western blot imaging system may be used which incorporates a CCD camera (e.g., Alpha Innotech).

3. Methods

3.1. Extraction and Quantification of Protein Samples

3.1.1. Lysis of Cultured Mammalian Cells

Cells should be harvested as quickly as possible and all reagents kept on iced water. Wash cells twice with cold PBS. Using 2 ml PBS for a 35-mm dish and 4 ml PBS for a 100-mm dish, harvest cells into a 5-ml round bottom tube by scraping the plate with a rubber policeman. Alternatively, cells growing in suspension should be concentrated by centrifugation (1,000 × *g* for 5 min) then transferred to a 5-ml round bottom tube before washing twice with PBS. Pellet the cells by centrifuging at 3,000×g for 10 min at 4°C. Tip off the PBS overlying the pellet and carefully aspirate the last drops. Add 400–500 ml lysis buffer and vortex to homogenise the cells (see Note 4). Shear the chromosomal DNA by sonication or homogenisation (e.g. using an Ika, Ultra-turrax T8 disperser) for 30 s to 2 min depending on the power output of the apparatus. This should be sufficient to reduce the viscous lysate to manageable levels. The shearing process may also be undertaken by passing the sample three or four times through a 23-G hypodermic needle (see Note 5).

3.1.2. Protein Quantification

Several methods can be employed to determine the concentration of proteins, the method of choice depends upon the constituents of the storage buffers and the purity of the protein.

UV detection: this is the quickest of all methods for quantifying protein solutions. The absorbance at 280 nm is due primarily to the presence of the amino acids tyrosine and tryptophan. This method is used for pure protein solutions in buffers that did not absorb light at a wavelength of 280 nm. Take an aliquot of the protein sample and the standards and read the absorbance at 280 nm against the blank (lysis buffer). Calculate the protein concentration given that the absorbance at $A280$ nm for a 1 mg/ml solution of IgG is 1.35, IgM is 1.2; BSA is 0.7). A very rough approximation for other proteins is one absorbance unit is equal to 1 mg/ml. If the protein solution is contaminated with nucleic acids, read the absorbance versus a suitable control at 280 and 260 nm, and calculate the approximate concentration using the equation: Protein concentration (in mg/ml) = $(1.55 \times A280) - (0.76 \times A260)$.

Bradford assay: dilute the concentrated dye-binding solution 1 in 5 with distilled water. Filter if any precipitate develops. Add 5 ml of diluted dye-binding solution to each protein sample, standard, and blank (lysis buffer). Allow the colour to develop for at least 5 min but not longer than 30 min. The red dye will turn blue as it binds to the protein. Read the absorbance at 595 nm. The standard curve will be linear between about 20 and 150 μg/ml. Several other colorimetric assays are available commercially (e.g., Pierce BCA protein reagent and Bio-Rad, DC protein Assay) and are a modification of the Lowry method (6, 9). They offer the advantage of being faster (around 30 min) and easier to perform than the Lowry method, compatible with most ionic and non-ionic detergents, linear for working range from 20 to 2,000 μg/ml, and adaptable to microtitre plates.

3.2. Resolution of the Protein Sample in SDS Polyacrylamide Denaturing Gel Electrophoresis

1. Wash the glass plates and wipe with alcohol prior to use.
2. Make a sandwich of one large and one small glass plate separated by a spacer at both sides and lock into the casting stand with the smaller plate foremost. Ensure that the spacers and plates are all level.
3. Either pouring directly from a universal tube or using a Pasteur pipette, transfer around 2 ml of water between the plates of the sandwich. This will identify any leaks in the apparatus before

pouring the gel. If the system is leak-free, pour out the water and begin preparing the gel.

4. The gel is poured in two layers. A separating (resolving) gel is prepared first. This gel should fill all but the top 1 cm of the apparatus and is followed by a stacking gel. The stacking gel is the same for all gels but different final percentages of acrylamide are used to separate proteins of different molecular weights in the separating gel (see Table 1). As an approximate guide use a 5% gel for 60–200-kDa proteins (high-range molecular weight markers), 10% gel for 16–70-kDa proteins (mid-range molecular weight markers) and 15% gel for 12–45-kDa proteins (low-range molecular weight markers).
5. Add the 10% w/v APS and the TEMED to the mixture and swirl well immediately prior to pouring the gel as it will cause the polymerisation reactions to start, this then proceeds rapidly. Leave a small amount of gel in the universal tube after pouring; this can be used as a guide to when the main gel has set (see Notes 6 and 7).
6. Pour a thin (2–5 mm) layer of water-saturated butanol solution on top of the separating gel. This ensures a clean interface between the separating and stacking gels.
7. Once the separating gel has set (this usually occurs within 20 min at room temperature), pour off the water-saturated butanol solution, prepare and pour the stacking gel. Again pour the stacking gel immediately after adding the 10% w/v APS and TEMED. Pour the stacking gel to the top of the sandwich and insert the Teflon comb immediately. It is important when pouring the gels that you avoid the ingress of air bubbles. If air bubbles appear they can usually be teased to the edge of the gel and burst with a hypodermic needle. The stacking gel will polymerise within 40 min, again this can be checked using the residual liquid left in the universal tube.
8. When comparing samples separated by SDS–PAGE it is helpful to aim to load a constant quantity and volume of protein in all the wells. The amount loaded will vary depending on the supply of protein but typically 30–100 μg is a reasonable quantity of protein. It is advisable to check with the manufacturer for the maximum loading volumes of large gels. Dilute the protein samples in lysis buffer to the required concentration, in order to load as small a volume as possible, and then mix with the 6× SDS sample buffer to get the final solution.
9. Denature the proteins by heating at 95°C for 5 min, then leave to cool to room temperature prior to loading the gel.
10. Whilst heating the samples, remove the gel sandwich from the casting frame and attach to the electrophoresis apparatus.

Table 1
Solutions for preparing resolving and stacking gels for sodium dodecyl sulphate–polyacrylamide denaturing gel electrophoresis; modified from Sambrook et al. [10]

Components volumes (ml) enough to prepare two mini gels

Resolving gel final volume, 20 ml					
Acrylamide concentration (%)	6	8	10	12	15
H_2O	10.6	9.3	7.9	6.6	4.6
30% Acrylamide solution	4.0	5.3	6.7	8.0	10.0
1.5 M Tris (pH 8.8)	5.0	5.0	5.0	5.0	5.0
10% SDS	0.2	0.2	0.2	0.2	0.2
10% APS	0.2	0.2	0.2	0.2	0.2
TEMED	0.016	0.012	0.008	0.008	0.008
5% Stacking gel final volume, 10 ml					
H_2O	5.6				
30% Acrylamide solution	1.7				
0.5 M Tris (pH 6.8)	2.5				
10% SDS	0.1				
10% APS	0.1				
TEMED	0.01				

11. Fill the tank and electrophoresis apparatus with the Tris–glycine electrophoresis buffer.
12. Carefully remove the Teflon comb from the sandwich and, using a 1-ml Gilson pipette, wash the wells with electrophoresis buffer. This removes any non-polymerised gel that could interfere with loading.
13. Load one well at either edge of the gel with the pre-stained molecular weight markers, then continue to load the samples. Record the position of samples in relation to the marker.
14. Attach the electrophoresis apparatus to the electricity power supply (the positive electrode anode should be connected to the bottom buffer tank). Apply a voltage of 8 V/cm height of the gel. After the dye front has moved into the resolving gel, increase the voltage to 15 V/cm height and run the gel until the bromophenol blue marker dye reaches the bottom of the resolving gel (about 1 or 2 h for a small gel). Then turn off and disconnect the power supply.

15. Remove the sandwich plates from the electrophoresis apparatus and place them on a paper towel. To disassemble, slide a spacer between the glass plates and rotate to separate the plates.
16. The gel can now be stained with Coomassie Brilliant Blue or used for Western blotting.

3.3. Staining SDS–Polyacrylamide Gels with Coomassie Brilliant Blue

1. Immerse the gel in at least five volumes of staining solution and place on a slowly rotating platform for a minimum of 4 h at room temperature.
2. Remove the stain and save it for future use. Destain the gel by soaking it in destaining solution on a slowly rocking platform for 4–8 h, changing the destaining solution three or four times.
3. The more thoroughly the gel is destained, the greater the sensitivity of the method. Destaining for 24 h usually allows as little as 0.1 μg of protein to be detected in a single band. Using the following alternatives can result in faster staining or destaining: staining and destaining at higher temperatures (45°C); including a few grams of an anion-exchange resin or a piece of sponge in the destaining buffer, this absorbs stain as it leaches from the gel; destaining electrophoretically in apparatuses that are available commercially (see Note 8).
4. After destaining, the gels may be stored indefinitely in water, in a sealed plastic bag without any diminution in the intensity of staining.

3.4. Semi-dry Electrophoretic Transfer of the Separated Polypeptides to a Membrane Support

Semi-dry electrophoretic transfer (11): transfer of proteins from a gel to a membrane using a semi-dry method gives even and rapid transfer and does not require a large power source. It can also be adapted to handle stacks of gel-membrane sandwiches (up to six). A continuous buffer system using a Tris–glycine-SDS-based transfer buffer works very well with this type of apparatus. However, if further optimisation is needed, the use of a discontinuous buffer system with TRIS-CAPS buffer provides excellent results.

1. Rinse the electrode plates of the semi-dry apparatus with distilled water and wipe off any beads of liquid that adhere to them with Kim Wipes or similar absorbent tissue.
2. Wearing gloves, cut six sheets of absorbent paper (Whatman 3MM or equivalent) and one sheet of membrane (PVDF or nitrocellulose) to the size of the gel. If the paper overlaps the edge of the gel, the current will short-circuit the transfer and bypass the gel, preventing efficient transfer (see Note 9).
3. Soak the PVDF membrane for 15 s in methanol then immerse it with the absorbent paper in the transfer buffer for at least 5 min. Alternatively, float the nitrocellulose membrane on the surface of deionised water, and allow it to wet from beneath

by capillary action. Then, submerge the membrane in the water for at least 5 min to displace trapped air bubbles. Soak six pieces of absorbent paper in transfer buffer for at least 30 s (see Notes 10–12).

4. Wearing gloves set up the transfer apparatus as follows: lay the bottom plate of the apparatus (the anode) flat on the bench, graphite side up. Place on the electrode three sheets of absorbent paper that have been soaked in transfer buffer. Stack the sheets one on top of the other so that they are aligned exactly. Using a glass pipette as a roller, squeeze out any air bubbles. Place the membrane on the stack of absorbent paper. Make sure the filter is exactly aligned and that no air bubbles are trapped between it and the absorbent paper. Transfer the SDS–polyacrylamide gel produced in Subheading 3.2, briefly to a tray of deionised water, and place exactly on the top of the membrane. Place the final three sheets of absorbent paper on the gel, again making sure that they are exactly aligned and that no air bubbles are trapped.
5. Carefully place the upper electrode (the cathode) on the top of the stack, graphite side down. Connect the electrical leads (positive the anode to the bottom graphite electrode) and commence transfer. Running time is 45 min to 1.5 h with a current of 0.8 mA/cm^2 of gel. Avoid extending the running time, as this can cause drying out.
6. After transfer, disconnect the power source. Carefully disassemble the apparatus. Mark the membrane to allow orientation (usually by snipping off lower left-hand corner, the number one lane) (see Notes 13–16).
7. The polyacrylamide gel can now be stained with Coomassie brilliant blue (see Subheading 3.3) to verify transfer. The prestained markers can serve as internal markers for transfer and molecular weight.
8. Let the blot air-dry (1 h at 37°C or 2 h at room temperature) to improve the protein binding and then process the membrane for staining with Ponceau S (to assess the quality of your transfer) or blocking as appropriate.
9. When using a discontinuous buffer system: Use Tris-CAPS buffer with 15% v/v methanol in the filter paper on the anode and Tris-CAPS plus 0.1% w/v SDS in the filter paper on the cathode. Wet the PVDF membrane in methanol, and then equilibrate in bottom/anode buffer for at least 30 min. To this buffer add one sheet of extra-thick blotting paper. Equilibrate the acrylamide gel in top/cathode buffer. Soak a second piece of extra-thick blotting paper in this buffer. Prepare for transfer by making the gel sandwich as follows: to the bottom (platinum) anode add extra-thick filter paper in bottom/anode

buffer, equilibrated PVDF membrane, equilibrated gel, and the second piece of extra thick filter paper in top/cathode buffer. Secure the top stainless steel cathode cover, and run at a constant current of 1.5 mA per square centimetre of gel (e.g., 120 mA for a small 8 cm × 10 cm gel) overnight.

3.5. Wet Electrophoretic Transfer of the Separated Polypeptides to a Membrane Support

For submerged or wet transfer no single set of transfer conditions offers complete and even transfer of proteins with good retention on the membrane. However, the common transfer techniques are usually considered adequate.

1. Wearing gloves, cut four sheets of absorbent paper (Whatman 3MM or equivalent) and one sheet of membrane (PVDF or nitrocellulose) to the size of the gel.
2. Soak the PVDF membrane for 15 s in methanol then immerse it with the absorbent paper in the transfer buffer for at least 5 min. Alternatively, float the nitrocellulose membrane on the surface of ionised water, and allow it to wet from beneath by capillary action then submerge the membrane in the water for 2 min. Move the membrane to soak in transfer buffer for 5 min. Wet the absorbent paper by soaking in transfer buffer (see Note 10).
3. Immerse the fibre pads in transfer buffer to ensure they are thoroughly soaked. Be careful to exclude air bubbles.
4. Assemble the transfer sandwich. One half of the gel holder cassette is usually black (this will be on the negative electrode, the cathode side). Build the sandwich in the following order, with the black plate on the bottom, therefore ensuring that the negatively charged proteins will migrate from the gel towards the anode, positive electrode, and thus onto the transfer membrane. Keep all the components wet and make sure the sandwich is tightly assembled. Place a fibre pad on the bottom of the cassette holder. Place two sheets of absorbent paper that have been soaked in transfer buffer. Stack the sheets one on top of the other so like that they are exactly aligned. Using a glass pipette as a roller, squeeze out any air bubbles. Place the SDS–polyacrylamide gel produced in Subheading 3.2, on top of the absorbent paper then put the membrane on top of the gel. Make sure the filter is exactly aligned and that no air bubbles are trapped between layers. Place two sheets of absorbent paper on the top of the stack. Then put a fibre pad on top of the absorbent paper.
5. Lock the sandwich in the cassette holder and place it in the tank blotting apparatus so that the black side of the cassette holder with the gel is facing the cathode (–). Add enough transfer buffer to the blotting apparatus to cover the cassette holder.

6. Allow the transfer to proceed for between 1 and 16 h. Thicker gels and higher-molecular-weight proteins require longer transfers. The following conditions have been optimised to permit transfer and retention of a very wide range of molecular weight proteins. For proteins over 100,000 kDa on a 15 cm × 15 cm × 0.1 cm gel, transfer at 30 V for approximately 1 h, then at 85 V for 14–16 h. For proteins under 100,000 kDa on a 15 cm × 15 cm × 0.1 cm gel, transfer at 30–60 V for 1–16 h. During transfer, the temperature will rise substantially during the run, and it is essential that a cooling coil is inserted into the apparatus prior to running the transfer. Alternatively, run the transfer in a cold room. If the system is allowed to warm up during the transfer, gas bubbles are generated within the sandwich.
7. After transfer, disconnect the power source. Remove the cassette holder from the blotting apparatus. Open the cassette holder and remove the fibre pad and filter paper with forceps. Mark the membrane to allow orientation (usually by snipping off lower left-hand corner, the number one lane).
8. The polyacrylamide gel can now be stained with Coomassie brilliant blue (see Subheading 3.3) to verify transfer. The prestained markers can serve as internal markers for transfer and molecular weight.
9. Let the blot air-dry (1 h at 37°C or 2 h at room temperature) to improve the protein binding and then process the membrane for staining with Ponceau S (to assess the quality of your transfer) or blocking as appropriate (see Notes 13–16).

3.6. Staining the Blot with Ponceau S

Ponceau S is applied in an acidic aqueous solution. Staining is rapid but not permanent; the red stain will wash away during the subsequent processing. Because the binding is reversible, the stain is compatible with most antigen visualisation techniques. Therefore, Ponceau S can be used routinely to verify gel loading and transfer, and to locate molecular weight markers (if not using prestained marker).

1. If the membrane has been dried, float it on the surface of a tray of deionised water and allow it to wet from beneath by capillary action. Submerge the membrane in the water for at least 5 min to displace trapped air bubbles.
2. Transfer the membrane to a tray containing a solution of Ponceau S stain. Incubate for 1–5 min with gentle agitation.
3. When the bands are visible, wash the membrane in several changes of deionised water at room temperature. Process until the colour of the membrane background becomes almost white.
4. Mark the positions of proteins used as molecular-weight standards with permanent black ink (if you are not using a prestained marker) and process the membrane for blocking (see Notes 17 and 18).

3.7. Blocking Non-specific Binding Sites on the Membrane

Place the membrane in a shallow tray that is slightly larger than the membrane or in a sealed plastic bag with 50–100 ml of blocking buffer. Gently agitate on a platform shaker for 1–3 h at room temperature or overnight in the cold room (see Notes 19–23).

3.8. Addition of the Antibodies

1. Dilute the primary antibody in blocking buffer according to manufacturer's guidelines or as suggested in Subheading 2.7 (see Notes 24–30). Incubate the membrane with this primary antibody solution for 1–2 h at room temperature or overnight in a cold room, with continuous agitation on a platform shaker. This stage is best performed in a heat-sealed plastic bag or even, if antibody is scarce, by putting 1 ml diluted antibody in the centre of a small glass plate and laying the membrane onto this. Ensure that the side of the membrane that was in contact with the gel is now facing the antibody. Cover the glass plate with a sheet of Parafilm slightly bigger than the membrane to prevent drying (this process is not suitable for overnight incubation).
2. Cut open the plastic bag or remove the membrane from the glass plate, and discard the blocking solution and antibody. Place the membrane in a large tray and wash it four times (10 min each time) with 200 ml of washing buffer, on a platform shaker (see Note 31).
3. Immediately incubate the filter with the secondary immunological reagent. Dilute the secondary antibody according to manufacturer's guidelines in blocking buffer and repeat the process described above for the primary antibody.
4. Cut open the plastic bag or remove the membrane from the glass plate, and discard the blocking solution and antibody. Place the filter in a large tray and wash it once for 15 min and then four times (5 min each time) with 200 ml of washing buffer, on a platform shaker (see Note 31).
5. A final wash should be performed in PBS without Tween 20 as this may interfere with certain chromogenic assays.

3.9. Detection

1. Remove membrane from the final wash and perform the enhanced chemiluminescent detection according to the manufacturer's instructions. This protocol will generally involve mixing of at least two different substrates and short incubation with the membrane under gentle incubation. Drain the membrane of excess detection reagent and wrap it in Saran Wrap and gently smooth out air pockets. Place the membrane, protein face up, in the film cassette. Switch off the lights of a dark room and carefully place a sheet of autoradiography film on top of the membrane, close the cassette, and expose the film for 15 s. Remove the film and develop it. On the basis of its appearance, estimate how long to continue the exposure of a

second piece of film. The second exposure can vary from 1 min to 1 h (see Notes 32–36).

2. The membrane can be stripped and re-probed several times to either clarify or confirm results or when small or valuable samples are being analysed (9). Sequential re-probing of membranes with a variety of antibodies is possible following the steps below. To strip the membrane wash in distilled water for 5–10 min. Wash in 0.2 M sodium hydroxide for 5 min, then wash again in distilled water. Process to the blocking of the membrane as previously (see Note 37).

4. Notes

1. Apparatus for pouring and running SDS–PAGE are available from a range of suppliers.
2. The methanol concentration will need to be optimised as it depends on the method of transfer, the type of protein gel used, and the manufacturer's recommendations. Methanol increases protein's affinity for a membrane by removing the SDS from the protein. This increases the available hydrophobic sites on the protein for binding to the membrane support. It is for this reason methanol is often used in transfer buffers. While advantageous for binding of the protein to the membrane, methanol causes the pores in the gel to constrict. This makes it mechanically more difficult for the protein to exit the gel.
3. SDS is used in Western blotting transfer buffer to aid in the elution of proteins from the gel matrix. While it inhibits the binding of the protein to the membrane, it is often necessary to facilitate complete transfer of proteins from the gel (12).
4. The cells are suspended in lysis buffer to prevent the formation of an insoluble mass when the SDS sample buffer is added. This step should be carried out as quickly as possible, using ice-cold lysis buffer, to minimise proteolytic degradation. Most types of mammalian tissue can be rapidly teased apart with forceps or cut into small pieces with scissors or scalpel beneath the surface of the lysis buffer. Mammalian tissues can be equally well dispersed mechanically and then dissolved directly in SDS-gel-loading buffer. Mammalian cells in tissue culture may be lysed gently with detergents as described in the Subheading 3.1 or alternatively lysed directly in SDS sample buffer, if the target antigen is resistant to this type of extraction.
5. After thawing, samples that have been stored at −20°C should be centrifuged at 12,000 × *g* for 5 min at 0°C in a microfuge. This removes aggregates of cytoskeletal elements.

6. Prepare numerous gels. Wrap unused gels in a moistened paper towel and plastic wrap. Store at 4°C for up to 1 week. Stacking gel length should be 1 cm from well bottom to top of separating gel. Band resolution can be improved by doubling the salt concentration in stacking and separating gels, but the gel must be run at lower voltages.
7. Use fresh APS and high-quality acrylamide. Prepare clean plates prior to mixing acrylamide and work fast.
8. Microwave methods of staining and destaining gels may shorten time considerably, but vapours are harmful. Hot solvents may harm oven, and the gel can fracture. Avoid excessive heating of gels.
9. It is important to wear gloves when handling the gel, 3MM papers, and membrane. Oil secretions from the skin will prevent the transfer of proteins from the gel to the filter.
10. The PVDF membrane is hydrophobic and offers a uniformly controlled pore structure with high binding capacity for proteins. When compared to a nitrocellulose membrane, it has improved handling characteristics and staining capabilities, increased solvent resistance, and a higher signal-to-noise ratio for enhanced sensitivities.
11. Antigens may bind differently to different types of membranes. To maximise detection sensitivity, test different types of membranes with the particular antigen–antibody combination to be used.
12. Occasionally lot-to-lot variability may be seen with some membranes.
13. After the transfer step, membranes can be wrapped in plastic and stored at 4°C for up to 3 months. But once subsequent detection procedures are started, do not let the membrane dry out.
14. Low transfer pH (<8) or high amounts of SDS (>0.1% SDS) in the gel or buffer can lead to inefficient binding of the antigen.
15. If the Western blot transfer efficiency is low, try longer transfer times or higher transfer voltages.
16. Drying the membrane is claimed to improve the retention of proteins on the filter and to reduce the non-specific binding of antibodies on the membrane during subsequent processing. However, it may also result in further denaturation and consequent alteration in immunoreactivity. Drying may therefore be advantageous for some protein/antibody combinations and disadvantageous for others. This can be established empirically for the target proteins of interest.
17. Ponceau S staining could be used as well on a blocked and immunologically probed membrane, allowing a useful rescue or verification of old blots.

18. Because the pink–purple colour of Ponceau S is difficult to capture photographically, the stain does not provide a permanent record of the experiment.
19. Membranes may be left in the blocking solution overnight in a cold room if more convenient.
20. Elimination of Tween 20 from buffers can lead to increased background.
21. Greater than 0.1% v/v Tween 20 in buffers may elute proteins from PVDF membranes.
22. Some blocking proteins may reduce antigen recognition by certain antibodies. If this is suspected, try a different blocking protein (BSA, gelatin, or casein, use at 1% w/v in PBS/Tween).
23. For applications using an AP conjugate, a blocking buffer in TBS should be selected because PBS interferes with AP. The ideal blocking buffer will bind to all potential sites of nonspecific interaction, eliminating background without altering or obscuring the epitope for antibody binding.
24. Incubation times and temperatures will vary and should be optimised for each antibody. The conditions indicated are recommended starting points.
25. Some antibodies (particularly monoclonal antibodies) recognise epitopes that may be buried or denatured when the antigen is bound to surfaces such as nitrocellulose. This effect may be enhanced when blotting protein antigens from SDS-containing gels, possibly eliminating antibody recognition.
26. When using a low-titre antibody, increase signal by eliminating Tween 20 in buffers, increasing the incubation time or increasing the concentration of the primary antibody solution.
27. The background will usually increase when using conjugated secondary antibodies at dilutions lower than 1:2,500.
28. A few IgG and other immunoglobulins (e.g., IgM) will tend to stick non-specifically to membranes, resulting in high background.
29. Localised background can result from primary antibody recognition of epitopes shared by other protein species in the sample or a primary antiserum containing a mixture of antibodies with multiple specificity. The latter problem is sometimes overcome by pre-adsorbing the antisera to remove cross-reacting antibodies.
30. Improperly stored antibodies will lose activity over time, making detection results inconsistent. Primary antisera should be stored in aliquots at –20°C to avoid freeze/thaw cycles and prevent contamination. Secondary antibodies should be stored at 4°C or in aliquots at –20°C. Sodium azide (0.02%) or thimerosal (0.05%) can be added as a preservative. Azide will inhibit horseradish peroxidase (HRP) activity.

31. As a general rule, as large a volume as possible of washing buffer should be used each time.
32. The volume of the detection solution should be sufficient to cover the membranes. The final volume required is about 0.125 ml/cm^2 membrane.
33. Drain off excess detection reagent by holding the membrane vertically and touching the edge of the membrane against a tissue paper. Gently place the membrane, protein side down, on the Saran Wrap. Close the Saran Wrap to form an envelope avoiding pressure on the membrane.
34. Ensure that there is no free detection reagent in the film cassette; the film must not get wet.
35. Do not move the film whilst it is being exposed.
36. If background is high, the membrane may be washed twice for 10 min with wash buffer and re-detected following the same method with slight loss of sensitivity. If overexposure occurs because of high light emission resulting from high target antigen concentration, leave blots in the cassette for 5–10 min before re-exposing to film.
37. Check blocking agent for residual AP activity because it could lead to high background when using an AP-based detection systems.

References

1. Laemmli EK (1970) Cleavage of structural proteins during the assembly of the head of bacteriophage T4. Nature 227:680–685
2. Towbin H, Staehelin T, Gordon J (1979) Electrophoretic transfer of proteins from polyacrylamide gels to nitrocellulose sheets: Procedure and some applications. Proc Natl Acad Sci 76:4350–4354
3. Johnson DA, Gautsch JW, Sportsman JR, Elder JH (1984) Improved technique utilizing nonfat dry milk for analysis of proteins and nucleic acids transferred to nitrocellulose. Gene Anal Tech 1:3–8
4. Batteiger B, Newhall WJ, Jones RB (1982) The use of Tween 20 as a blocking agent in the immunological detection of proteins transferred to nitrocellulose membranes. J Immunol Methods 55:297–307
5. Bradford MM (1976) A rapid and sensitive method for the quantitation of microgram quantities of protein utilizing the principle of protein–dye binding. Anal Biochem 72:248–254
6. Lowry OH, Rosebrough NJ, Farr AL, Randall RJ (1951) Protein measurement with Folin phenol reagent. J Biol Chem 193: 265–275
7. Harlow E, Lane D (1999) Using antibodies: a laboratory manual. Cold Spring Harbor Laboratory Press, New York
8. Ornberg RL, Harper TF, Liu H (2005) Western blot analysis with quantum dot fluorescence technology: a sensitive and quantitative method for multiplexed proteomics. Nat Methods 2:79–81
9. Peterson GL (1983) Determination of total protein. Methods Enzymol 91:95–119
10. Sambrook J, Fritsch EF, Maniatis T (1989) Molecular cloning a laboratory manual, 2nd edn. Cold Spring Harbor Laboratory Press, New York
11. Kyhse-Andersen J (1984) Electroblotting of multiple gels: a simple apparatus without buffer tank for rapid transfer of proteins from polyacrylamide to nitrocellulose. J Biochem Biophys Methods 10:203–209
12. Gershoni JM, Palade GE (1983) Protein blotting: principles and applications. Anal Biochem 131:1–15. Review

Chapter 7

2-DE-Based Proteomics for the Analysis of Metastasis-Associated Proteins

Miriam Dwek and Diluka Peiris

Abstract

Two-dimensional electrophoresis (2-DE) is a high-resolution technique for analysis and comparison of complex protein mixtures. With the advent of recent technical developments, its application has become significant in a wide range of fields. This chapter describes a proteomic approach for the analysis of metastasis-associated proteins using pre-fractionation of glycosylated proteins via lectin (HPA) affinity chromatography prior to separation by 2-DE. Guidelines for the preparation and storage of buffers, experimental conditions and protocols of affinity chromatography, isoelectric focussing, and SDS-PAGE conditions are provided. Critical parameters associated with the different steps of 2-DE are discussed.

Key words: Metastasis, Proteomics, Lectin, 2-DE, Affinity chromatography, SDS-PAGE

1. Introduction

Two-dimensional electrophoresis (2-DE) is a powerful and widely used technique for protein separation prior to protein identification. The system allows the fractionation of complex protein mixtures, according to their charge by isoelectric focussing (IEF) in the first dimension and according to the independent parameter of molecular mass by SDS-PAGE in the second dimension (1). This technique can separate a thousand different proteins using a single gel with each detected protein "spot" on the resulting 2-D gel potentially corresponding to a single protein in the sample. Information, such as p*I*, the apparent molecular weight, and the amount of each protein, can be obtained. Protein "spot patterns" generated by 2-DE-based separation methods can be compared to identify the levels of proteins that differ under unique cellular conditions. Protein species of interest can be identified by mass spectrometry.

Miriam Dwek et al. (eds.), *Metastasis Research Protocols*, Methods in Molecular Biology, vol. 878,
DOI 10.1007/978-1-61779-854-2_7, © Springer Science+Business Media, LLC 2012

Since the introduction by O'Farrell in 1975 (2), a large number of applications of 2-DE have been recognised in a wide range of fields, including analysis of the proteome, for the detection of disease markers, drug discovery, cancer research, and protein purity checks. These applications have become more significant as a result of a number of technical developments in the field, including the utilisation of immobilised pH gradients (IPGs) (3), pre-fractionation of proteins prior to 2-DE (4), development of protocols to separate highly basic proteins (5), increased solubilisation of hydrophobic proteins (6), and introduction of differential in-gel electrophoresis (DIGE) by Unlü et al. (7). In addition, recent developments in mass spectrometry, availability of reliable, less expensive 2-DE image analysis software, and easy access of rapidly growing databases with protein sequences have also contributed to the tremendous growth in the popularity of 2-DE.

Despite some inherent limitations, in recent years, proteomics has proven to be a powerful tool in cancer research. For example, a number of studies have been carried out to compare the protein expression levels of normal tissues with cancer tissues (8–11). In addition, comparisons have also been made between cells with metastatic phenotype with non-metastatic phenotypes using the 2-DE-based approaches (12, 13). The major limitation in comparison of protein profiles of whole cell/tissue lysate or serum is the masking of less abundant proteins by highly abundant ones. An effective proteomic analysis will naturally require the separation of highly abundant proteins from the proteins of interest in order to bring the latter into the detectable range. Such pre-fractionation strategies allow improved resolution and increase the possibility of discovering proteins of diagnostic or therapeutic interest.

Selecting the pre-fractionation step depends on the nature of the sample, the experimental goal, and most importantly the characteristics of the protein of interest. Since there is well-documented evidence of changes in protein glycosylation with metastasis formation (14), the pre-fractionation method of choice could be based on separation of glycosylated proteins from the total mixture of proteins. A number of studies have demonstrated the utility of the carbohydrate-binding protein (lectin) from the Roman snail *Helix pomatia* (HPA) for the identification of metastatic cancer, including that of breast, colon, stomach, and oesophagus (15). Proteomic approaches using HPA affinity chromatography and 2-DE have been proven as useful for the analysis of proteins in a model of metastatic and non-metastatic human colorectal cancer cells (16). In this chapter, we describe a method to perform comparative analysis of glycoprotein levels in metastatic cancer tissues with non-metastatic cancer tissues using lectin affinity chromatography followed by 2D gel analysis of the glycoprotein-enriched protein fraction.

2. Materials

2.1. Sample Preparation

1. Lectin buffer: 25 mM Tris base, 150 mM NaCl, 2 mM $MgCl_2$, and 2 mM $MgCl_2$. Dissolve in deionised water and adjust pH to 7.6 with 1 M HCl.
2. Urea buffer: 7 M Urea, 2 M thiourea, Pharmalyte 2% v/v (GE Healthcare), CHAPS 4% w/v, and DTT 1% w/v. The urea buffer has to be prepared fresh. Alternatively, store in aliquots of 500 μl at –80°C and add DTT before use.

2.2. First Dimension: Isoelectric Focussing

1. Rehydration buffer: 6 M urea, 2 M thiourea, 0.5% w/v CHAPS, 0.4% w/v DTT, 0.5% v/v Pharmalyte 3-10. Store aliquots of 500 μl at –20°C.
2. Equilibration buffer: 6 M urea, 30% v/v glycerol (see Note 1), 2% w/v SDS, 0.05 Tris–HCl buffer, pH 8.8. Store in aliquots of 5 ml at –20°C and add DTT 1% w/v and iodoacetamide 4% w/v before use.
3. Agarose sealing buffer: Agarose 0.5% w/v, Tris base 25 mM, glycine 192 mM, SDS 0.1% w/v, bromophenol blue in trace amounts. Use microwave to dissolve the agarose by heating the solution until boiling (approximately 30 s).

2.3. Second Dimension: SDS-PAGE

1. Immobiline dry strips and IEF electrode strips (GE Healthcare or Bio-Rad).
2. Resolution gel buffer: 1.5 M Tris–HCl, pH 8.8; 20% w/v SDS; acrylamide/bis-acrylamide 30% and 0.8% respectively; 10% w/v ammonium persulphate (APS); and deionised water. Make APS fresh each time and add both APS and *N,N,N′,N′*-tetramethylethylenediamine (TEMED) just before casting the gel.

2.4. Protein Stain

2.4.1. Colloidal Coomassie Blue G-250 Staining (Blue Silver)

1. 10% w/v ammonium sulphate; 10% v/v phosphoric acid; 20% v/v anhydrous methanol; 0.12% w/v Colloidal Coomassie blue G-250.
2. The dye solution is prepared by sequentially adding the ingredients to water. First, add phosphoric acid and then ammonium sulphate into water (1/10 of the final volume) to obtain a final concentration of 10% of each component.
3. When the ammonium sulphate has dissolved, add Coomassie blue G-250 powder (to a final concentration of 0.12% w/v).
4. When all solids have dissolved, add water to 80% v/v.
5. While stirring, add anhydrous methanol to reach 20% v/v final concentration.
6. Store the dye solution in a brown bottle at a room temperature.

2.4.2. Silver Staining

Use a silver staining kit (for example, the Silver Stain Plus system from Bio-Rad, UK).

3. Methods

3.1. Protein Extraction

1. Homogenise the tissue samples in lectin buffer using a hand-held homogeniser and centrifuge at 15,000 × *g* for 5 min and collect the supernatant.
2. Quantify the protein; in our laboratories, we use the Bradford assay (BioRad) or the Qubit system (Qiagen, UK).

3.2. Affinity Purification of HPA-Binding Proteins

A range of lectins with different glycan binding specificities are available and are detailed in Chapter 2, Brooks, in this volume. It is important to select a lectin which has been shown to detect glycosylation changes associated with metastatic behaviour of cancer cells. We routinely use affinity columns coupled with the lectin HPA to purify the glycoproteins prior to 2-DE analysis, but other lectins are available and the principle underlying their use remains the same. Remember, however, that the appropriate competing sugar needs to be selected for the appropriate lectin.

1. Couple 10 mg of HPA lectin (available from Sigma, E-Y Labs, or other suppliers) to a 5 ml HiTrap NHS-activated Sepharose column (GE Healthcare) using the method described by the manufacturer (see Note 1).
2. Wash and equilibrate the column with 10-column volumes of lectin buffer containing 0.1% w/v CHAPS.
3. Load 2 mg of cancer proteins onto the equilibrated HPA affinity column. If assessing patient specimens, it may be appropriate to use pooled protein samples, for example from tumours known to be metastatic to the lymph nodes, and pooled samples from non-metastatic tumours.
4. Wash with 5-column volumes of lectin buffer containing 0.1% w/v CHAPS to remove unbound proteins (see Note 1). The flow rate should be approximately 0.5 ml/min and should be maintained at a constant rate throughout all the experiments. It is important not to compact the gel and, therefore, the flow rate should be maintained at 0.5 ml/min throughout the experiment.
5. Elute the bound proteins with freshly prepared lectin buffer containing 0.1% v/v CHAPS and 0.25 M *N*-acetylglucosamine (GalNAc).
6. After use, wash the column with lectin buffer and store the column at 4°C in lectin buffer with 0.001% w/v sodium azide.

7. Dialyse the eluted proteins overnight at 4°C (see Note 1) and freeze-dry prior to reconstituting in 100 μl of urea buffer. The thiourea should be added to the sample only after protein quantification has been completed as this reagent commonly reacts with the protein dyes used for protein quantitation.
8. Quantify the proteins using a Bradford assay or, for example, the Qubit system (Qiagen, UK).

3.3. Immobilised pH Gradient Strip Rehydration

1. Mix the protein samples with rehydration buffer to give a final volume of 130 μl.
2. Load the mixture onto a 7-cm Immobiline IPG strip in the desired pH range. A maximum loading of 100 μg may be used with a 7-cm Immobiline strip. This can be conveniently undertaken in a strip swelling tray (GE Healthcare).
3. After loading, peel the protective cover off from the IPG strip and insert into the groove of an IPG swelling tray, gel side down.
4. Cover the strip with silicone oil (Sigma-Aldrich, UK) and rehydrate overnight.

3.4. First Dimension: Isoelectric Focussing

We use a Multiphor II (GE Healthcare) for the IEF, but a range of systems are available for this.

1. The following day, pre-cool the IEF electrophoresis unit to 20°C using a thermostatic circulator (for example, Multitemp III, GE Healthcare).
2. Rinse the IPG strip with deionised water for 1 s and place on a sheet of water-saturated filter paper, gel side up. Wet a second filter paper with deionised water and place over the surface of the IPG strip and blot gently to remove excess rehydration solution (see Note 2).
3. Pour a little silicone oil onto the ceramic cooling block and place the aligner tray onto the block. Place the strip – gel side up – onto the dry strip aligner tray.
4. Soak two IEF electrode strips in deionised water and place one at the anode end and one at the cathode end of the IPG strip.
5. Place the electrodes over the electrode strips and attach the tray and electrode holders to the IEF electrophoresis unit. Cover the IPG strip and IEF electrode strips with silicone oil.
6. Focus the IPG strip. The running conditions for this depend on the strip length and the pH range of the IPG strip (e.g. 7 cm, pH 4–7 IPG strip 11.5 kVh) (see Note 2).
7. After IEF, store the IPG strip at –80°C; if the second dimension is to be carried out on a different day, alternatively, proceed directly to the second dimension separation.

3.5. Equilibration of the IPG Gel Strips

1. Defrost IPG strip at room temperature for 5 min.
2. Equilibrate each IPG strip in 5 ml of equilibration buffer containing 1% w/v DTT in a polystyrene tube for 15 min.
3. Pour off the equilibration buffer, add fresh equilibration buffer with 4% w/v iodoacetamide into the tube, and equilibrate for another 15 min.
4. After the second equilibration step, rinse the IPG strip with deionised water and place on a filter paper at one edge to drain off any excess equilibration buffer.

3.6. Second Dimension: SDS-PAGE

The second dimension can be separated using home-made or commercially produced SDS-PAGE (see Chapter 6 in this volume, by Blancher and McCormick, for detailed instructions related to SDS-PAGE).

1. Pour the resolving gel solution into the assembled glass plates (pre-cleaned using deionised water and ethanol). Ensure that 0.5 cm is left at the top of the gel and overlaid with water-saturated butanol. Leave the gels to set at room temperature (see Note 3).
2. Once the gels have set, pour off the water-saturated butanol and clean the top surface of the gel with deionised water.
3. Immerse the IPG strip in SDS running buffer for a few seconds and place on top of the gel with the acidic end of the strip towards the left hand side of the gel. The gel side of the strip should be orientated such that it faces towards the thin glass plate.
4. Carefully press the IPG strip with a spatula onto the surface of the SDS gel to achieve complete contact between the strip and the gel.
5. Load molecular weight markers onto a small piece of filter paper (Whatman 3MM is ideal, cut to 0.3 cm width). Place the molecular weight markers at the acidic end of the strip ensuring that there is contact between the filter paper containing the markers and the surface of the SDS gel.
6. Overlay the strip with molten 1% w/v agarose. Allow the agarose to solidify for 5–7 min.
7. Set up the gel running apparatus and run the SDS-PAGE gel until the dye front has migrated from the lower end of the gel. This takes approximately 2 h for 12% mini gels when run at 120 V (see Note 3).

3.7. Protein Visualisation

Electrophoretically separated proteins can either be visualised by pre-conjugation of a fluorescent dye (for example, using the DIGE system, GE Healthcare) or a radioactive tracer or by staining after the sample has been separated by 2-DE. We use a post-detection approach with either the modified Neuhoff's colloidal Coomassie blue G-250 (17) or silver staining depending on the level of sensitivity required.

1. For colloidal Coomassie blue G-250 staining, wash the gel in deionised water and leave the gel in staining solution overnight to ensure approximately 50 ml of dye per mini gel.
2. The following day, pour off the staining solution and wash the gels several times in deionised water.
3. For silver staining (see Note 4): Fix the gels for 20 min in fixative enhancer solution with gentle agitation. Use 200 ml solution per mini gel.
4. Add the fixative enhancer solution: 50% v/v reagent-grade methanol, 10% v/v reagent-grade acetic acid, 10% v/v fixative enhancer concentrate, and 30% v/v deionised water. Decant the fixative enhancer solution and wash the gel in water for 10 min with gentle agitation. Repeat the washing step twice.
5. The staining solution has to be prepared within 5 min of use. Place 35 ml of deionised water into a glass beaker and stir with Teflon-coated stirring bar. Add 5 ml of silver complex solution, reduction moderator solution, and image development reagent to the beaker in the order given. Add 50 ml of development accelerator solution to the beaker immediately before use. Quickly add the staining solution to the staining vessel and stain with gentle agitation. After the desired staining intensity has been reached (may take about 15 min), decant the staining solution and place the gels immediately in a 5% v/v acetic acid solution to stop the reaction. Leave the gels in the stop solution for a minimum of 15 min and rinse the gels in deionised water for 5 min.

3.8. Image Analyses and Protein Identification

After staining, the gels can be scanned and the data recorded using a densitometer (for example, using the BioRad UK, GS-800). Protein density readings are obtained as a function of pixel intensity and may be recorded using the Quantity 1 software (BioRad, UK).

The resulting gel images are analysed using image analysis software, for example using the Progenesis PG 240 SameSpots software (Nonlinear Dynamics, UK).

During computer-based gel analysis, a master gel is created by combining all the gel images in the experiment; this forms an "average gel" which contains all the proteins separated (as spots) from each gel. The protein species (spots) in different gels are then compared with each other by aligning them with the master gel. The relative amount of a protein in a given spot is calculated based on the pixel intensity and volume of the spot. To reflect the quantitative variations in the intensity of protein spots between gel images of different groups, the spot volumes are normalised and described as a percentage of the total volume of all the protein spots on a corresponding gel image. Spots are ranked by their *p*-value from the one-way ANOVA analysis and fold changes are calculated using normalised spot volumes.

Induced or differential levels of proteins may be analysed either by matrix-assisted laser desorption ionisation time-of-flight MS (MALDI-TOF) or tandem MS (MS/MS). The mass spectra generated by MS are searched against human databases to obtain the identity of the proteins.

4. Notes

To analyse the data statistically, it is important to have an adequate number of replicates in the 2-DE-based experiments. In such experiments, two types of replicates are possible, biological and analytical. Biological replicates are always better than the analytical replicates. If the total number of gels is constrained due to expense, time, or any other reason, technical replicates should be ignored completely if these occur at the expense of biological replicates. Three is the minimum number of biological replicates required in a given experiment. However, if there are analytical replicates included in the experiment, there should be minimum of two analytical replicates, thereby making a total of six replicates.

1. It is advisable to measure the efficiency after the coupling of HPA to the affinity matrix. If it is lower than 80%, experiments should be undertaken to optimise the loading parameters varying, for example, temperature, pH, and flow rate. If there are extra proteins in the eluant along with GalNAc-containing glycoproteins, this indicates non-specific protein binding to the affinity resin. This can be prevented by washing the column more thoroughly before eluting the GalNAc-containing glycoproteins. Alternatively, the column can be washed with a more stringent second wash buffer (for example, buffer with higher salt). High salt concentrations in the sample are one of the major problems with the IEF step. This can cause the IPG strip to burn during the focussing step. Desalt the sample in deionised water or an appropriate low-salt buffer prior to the focussing step. If the current is still too high, change the electrode strips after 1–2 h of IEF. Alternatively, start IEF with a very low voltage to keep the initial current low (e.g. 150 V for 30 min).
2. Detergents solubilise hydrophobic proteins and minimise aggregation. They must have zero-net charge. Use only non-ionic and zwitterionic detergents, such as CHAPS, Triton X-100, or NP-40. Select Pharmalyte (or IPG buffer) with the same pH intervals as the Immobiline dry strip. Rinse the

rehydrated gel strip to remove excess rehydration solution; or urea crystallization on the surface of the IPG gel strip may occur and disturb the IEF pattern. If the temperature of the cooling plate is too low, this could lead to urea crystallization on the IPG gel surface. Keep the temperature of IEF unit at 20°C. When using the re-swelling tray for rehydration, the sample volume has to be limited so that it does not extend physically beyond the gel strip. For better sample entry, start the IEF at a low voltage and then gradually increase before raising to 3,500 V. The focussing time required to achieve the best-quality IEF separation pattern has to be optimised for each protein mixture, the pH range, and the length of IPG strip used. If the volt hours are insufficient, this will result horizontal streaking in the gel. On the other hand, over-focusing should also be avoided to prevent distorted protein patterns. 6 M Urea and 30% v/v glycerol in equilibration buffer help to diminish electroendosmotic effects which are responsible for reduced protein transfer from the first to the second dimension (18).

3. APS solution should be prepared fresh. A 40% w/v solution can be retained for up to 3 days if it is stored in the refrigerator. Less concentrated solutions should be prepared on the same day as they are due to be used. If the surface of the SDS gel is not flat, it may cause a distorted 2-DE pattern. This can be prevented by overlaying the surface with water-saturated butanol immediately after pouring the gel. Alternatively, a 10% w/v SDS solution may be used. Distorted gel patterns can also be caused by unevenly polymerised gels resulting from incomplete or too rapid polymerisation. Polymerisation can be slowed by decreasing the amount of APS and TEMED. Alternatively, use gel buffers at 4°C and leave the gels (mini gels) at room temperature for at least 2 h before use. Large format gels take about 6 h for complete polymerisation. If there is poor transfer of the protein from the IPG strip onto the SDS gel, initially run the gels at a low voltage (for example, at 80 V for 15 min) or perform the entire SDS PAGE at very low field strength.

4. For silver staining, use clean glass or plastic vessels. Vessels can be cleaned with 50% v/v nitric acid and rinsed thoroughly with high-quality deionised water. Always use reagent-grade chemicals and the conductivity of deionised water should not exceed 18.2 MΩ. If the proteins are to be identified by mass spectrometry, it is important to prevent the introduction of contaminating proteins into the gel before and after 2-DE separation. Filter sterilise all the buffers and if possible use a lamina flow hood when casting the gels. Do not handle the gel without wearing gloves, as it is important not to introduce keratin (human keratin is the most common type of contaminating protein) into the gels. Always use thoroughly cleaned and dust-free vessels and wear gloves or use forceps for manipulating the gels.

References

1. Görg A,Boguth G, Drews O, Köpf A, Luck C, Reil G, Weiss W (2003) Two-dimensional electrophoresis with immobilized pH gradients for proteome analysis. A laboratory manual. www.weihenstephan.de/blm/deg
2. O'Farrell PH (1975) High resolution two-dimensional electrophoresis of proteins. J Biol Chem 250:4007–4021
3. Bjellqvist BEK, Righetti PG, Gianazza E, Görg A, Westermeier R, Postel W (1982) Isoelectric focusing in immobilised pH gradients: principle, methodology and some applications. J Biochem Biophys Methods 6:317–339
4. Görg A, Postel W, Günther S (1988) The current state of two-dimensional electrophoresis with immobilised pH gradients. Electrophoresis 9:531–546
5. Hoving S, Gerrits B, Voshol H, Müller D, Roberts RC, van Oostrum J (2002) Preparative two-dimensional gel electrophoresis at alkaline pH using narrow range immobilised pH gradients. Proteomics 2:127–134
6. Luche S, Santoni V, Rabilloud T (2003) Evaluation of nonionic and zwitterionic detergents as membrane protein solubilizers in two-dimensional electrophoresis. Proteomics 3:249–253
7. Unlü M, Morgan EM, Minden JS (1997) Difference gel electrophoresis: a single method for detecting changes in protein extracts. Electrophoresis 18:2071–2077
8. Dwek MV, Alaiya AA (2003) Proteome analysis enables separate clustering of normal breast, benign breast and breast cancer tissues. Br J Cancer 89:305–307
9. Li W, Li JF, Qu Y, Chen XH, Qin JM, Gu QL, Yan M, Zhu ZG, Liu BY (2008) Comparative proteomics analysis of human gastric cancer. World J Gastroenterol 14:5657–5662
10. Okusa H, Kodera Y, Oh-Ishi M, Minamida Y, Tsuchida M, Kavoussi N, Matsumoto K, Sato T, Iwamura M, Maeda T, Baba S (2008) Searching for new biomarkers of bladder cancer based on proteomic analysis. J Electrophor 52:19–24
11. Shi-Shan D, Tian-Yong X, Hong-Ying Z, Ruo-Hong X, You-Guang L, Bin W, Shang-Qing L, Hui-Jun Y (2006) Comparative proteome analysis of breast cancer and adjacent normal breast tissues in human. Genomics Proteomics Bioinformatics 4:165–172
12. Liao Q, Zhao L, Chen X, Deng Y, Ding Y (2008) Serum proteome analysis for profiling protein markers associated with carcinogenesis and lymph node metastasis in nasopharyngeal carcinoma. Clin Exp Metastasis 25:465–476
13. Vydra J, Selicharová I, Smutná K, Šanda M, Matoušková E, Buršíková E, Prchalová M, Velenská Z, Coufal D, Jiráček J (2008) Two- dimensional electrophoretic comparison of metastatic and non-metastatic human breast tumors using in vitro cultured epithelial cells derived from the cancer tissues. BMC Cancer 8:107
14. Dwek MV, Brooks SA (2004) Harnessing changes in cellular glycosylation in new cancer treatment strategies. Curr Cancer Drug Targets 4:425–442
15. Brooks S (2000) The involvement of *Helix pomatia* lectin (HPA) binding *N*- acetylgalactosamine glycans in cancer progression. Histol Histopathol 15:143–158
16. Saint-Guirons J, Zeqiraj E, Schumacher U, Greenwell P, Dwek M (2007) Proteome analysis of metastatic colorectal cancer cells recognised by the lectin *Helix pomatia* agglutinin (HPA). Proteomics 7:4082–4089
17. Candiano G, Bruschi M, Musante L, Cantucci L, Ghiggeri GM, Garnemolla B, Orecchia P, Zardi L, Rigetti PG (2004) Blue silver: a very sensitive colloidal Coomassie G-250 staining for proteome analysis. Electrophoresis 25:1327–1333
18. Görg A, Postel W, Weser J, Günther S, Strahler SR, Hanash SM, Somerlot L (1987) Elimination of point streaking on silver stained two-dimensional gels by addition of iodoacetamide to the equilibration buffer. Electrophoresis 8:122–124

Chapter 8

Assessment of Gelatinases (MMP-2 and MMP-9) by Gelatin Zymography

Marta Toth, Anjum Sohail, and Rafael Fridman

Abstract

Gelatin zymography is a simple yet powerful method to detect proteolytic enzymes capable of degrading gelatin from various biological sources. It is particularly useful for the assessment of two key members of the matrix metalloproteinase family, MMP-2 (gelatinase A) and MMP-9 (gelatinase B), due to their potent gelatin-degrading activity. This polyacrylamide gel electrophoresis-based method can provide a reliable assessment of the type of gelatinase, relative amount, and activation status (latent, compared with active enzyme forms) in cultured cells, tissues, and biological fluids. The method can be used to investigate factors that regulate gelatinase expression and modulate zymogen activation in experimental systems. The system provides information on the pattern of gelatinase expression and activation in human cancer tissues and how this relates to cancer progression. Interpretation of the data obtained in gelatin zymography requires a thorough understanding of the principles and pitfalls of the technique; this is particularly important when evaluating enzyme levels and the presence of active gelatinase species. If properly used, gelatin zymography is an excellent tool for the study of gelatinases in biological systems.

Key words: Gelatinase, Matrix metalloproteinase, Zymography, Gelatin, Protease, Gel electrophoresis, Enzyme activity

1. Introduction

The gelatinases, MMP-2 (gelatinase A) and MMP-9 (gelatinase B), are two members of the matrix metalloproteinase (MMP) family of zinc-dependent endopeptidases (1) that have consistently been shown to be associated with cancer progression. Many studies have demonstrated that the expression and activity of gelatinases are elevated in malignant human tumors and, in most cases, these parameters have been shown to correlate with tumor angiogenesis and cancer progression (2–4). MMP-2 and MMP-9 hydrolyze a variety of extracellular matrix (ECM) and non-ECM proteins

Miriam Dwek et al. (eds.), *Metastasis Research Protocols*, Methods in Molecular Biology, vol. 878,
DOI 10.1007/978-1-61779-854-2_8, © Springer Science+Business Media, LLC 2012

(5–7); therefore, their role in cancer is rather complex with both tumor-promoting and -suppressing activities in a substrate- and context-dependent manner (8–11).

Since their discovery more than 20 years ago, the gelatinases continue to be the focus of intense research in the area of cancer and several other pathological conditions. Much of the cancer biology research focuses on factors which modulate the malignant phenotype both in vitro and in vivo. This often includes assays to monitor the expression, activation, and activity of gelatinases and their corresponding effect on the malignant phenotype. Gelatine zymography is a simple, reliable, and sensitive method that is commonly used to monitor gelatinase expression and activation in a variety of biological samples (cells, tissues, and fluids).

1.1. Biochemical Basis of Gelatin Zymography and Overview of the Method

The inherent ability of gelatinases to efficiently hydrolyze denatured collagen I (also referred to as gelatin) is the basis for the detection of these MMPs by gelatin zymography (5, 12, 13). This property of the enzymes is the main reason why these MMPs are described as gelatinases. The gelatinases are multidomain proteases, and they comprise a pro-domain, catalytic domain, gelatin-binding domain, linker, and hemopexin-like domain (1, 14, 15, 16). Although other members of the MMP family share many of these domains, gelatinases are unique as they possess a gelatin-binding domain, a stretch of approximately 175 amino acid residues in the catalytic domain that consists of three tandem copies of fibronectin type II-like modules (16). This domain exhibits high-affinity binding for gelatin and is, in part, responsible for the proteolytic activity against gelatin. The gelatin-binding domain facilitates effective purification of gelatinases by means of gelatin affinity chromatography.

Gelatin zymography enables identification of gelatinolytic activity in biological samples (for example in culture media, cell extracts, tissues extracts, and biological fluids) using sodium dodecyl sulfate (SDS)-polyacrylamide gels impregnated (co-polymerized) with gelatin (17, 18). To maintain the enzymatic activity, the samples are electrophoresed under nonreducing conditions. Removal of the SDS from the gel by Triton X-100 and incubation in a calcium-containing buffer cause partial renaturation of the enzymes, which are subsequently able to hydrolyze the embedded gelatin in situ. After staining the gel with Coomassie Blue, areas cleared of gelatin are detected as transparent bands in a blue background. The transparent areas correspond to the relative molecular mass of the active gelatinases, and also to the mass of latent inactive gelatinases, as, paradoxically, gelatin zymography reveals gelatinolytic activity generated by the latent inactive zymogen form of the enzymes (pro-gelatinase). Under physiological conditions, and in solution, pro-gelatinases do not exhibit enzymatic activity because the amino terminal pro-domain blocks binding of the catalytic zinc (5, 19). During gelatin zymography, the presence of SDS results in an unfolding of the tertiary structure of the enzyme and an irreversible dissociation of the pro-domain from the active site. Once the active site is exposed, the enzyme can initiate the degradation of gelatin after partial renaturation. The presence of SDS also disrupts non-covalent interactions between active gelatinases and their endogenous inhibitors, the tissue inhibitors of metalloproteinases (TIMPs) (20). While TIMPs are potent inhibitors of active gelatinases (20, 21), electrophoretic dissociation of gelatinase activity enzyme–inhibitor complexes under the conditions of gelatin zymography allows the detection of gelatinase activity even in biological samples enriched with TIMPs. Therefore, gelatin zymography does not provide information on the net gelatinase activity (including TIMP-free activity) in a given sample as it does not enable analysis in the presence of functional endogenous inhibitors. Therefore, additional activity assays are required to obtain quantitative data on the overall net gelatinase activity.

1.2. Uses and Limitations of Gelatin Zymography

Gelatin zymography provides reliable information on the type, relative amounts, and activity of gelatinase(s) under various physiological and pathological conditions in different cell types/tissues. However, caution must be exercised when estimating the levels of gelatinase expression between samples as zymography is essentially a semiquantitative technique and, due to the inherent characteristics of the method, has many variables that require consideration (see Note 11). Another important application of gelatin zymography is the assessment of the activation status of gelatinases in a sample. Gelatinase activation usually involves the release of the pro-domain (approximate size: 10 kDa) via proteolytic and non-proteolytic clearage mechanisms. This process causes a reduction in the molecular mass of the active enzyme species relative to the zymogen, which can be readily detected by gelatin zymography by virtue of the difference in relative molecular mass. Therefore, gelatin zymography can be used to study mechanisms and factors that regulate the activation of pro-gelatinases in various experimental systems; for example, gelatin zymography is a powerful technique enabling the study of the activation of pro-MMP-2 by membrane-type MMPs (22, 23). Gelatin zymography should, however, be used with extreme caution when searching for active MMP-9 species in biological samples as the results may give rise to the erroneous identification of the active species (see Note 10). In summary, a detailed knowledge of the limitations of this technique is an essential prerequisite for rigorous interpretation of the data from complex biological samples. Here, we provide a detailed protocol on how to set up gelatin zymography and how to use it in samples relevant for the study of tumor cell invasion and metastasis. We also provide guidance for the interpretation of the results obtained with this system.

2. Materials

1. Glass plates with 1.0-mm-thick spacers or disposable plastic gel casting cassettes, 1.0 mm thick, 10×10 cm.
2. Acrylamide/bis-acrylamide stock: Prepare 200 ml of 30% w/v acrylamide (electrophoresis grade) and 0.8% w/v bis-acrylamide solution in distilled water (dH_2O) (see Note 1). Use a surgical mask and weigh acrylamide powder in a fume cupboard. Store the acrylamide solution in a dark bottle at 4°C, where it is stable for at least 6 months. Un-polymerized acrylamide is a neurotoxin. Always handle with gloves! For more information on SDS-PAGE, please refer to Chapter 6, by Blancher and McCormick, in this volume.
3. Separating gel buffer stock: Prepare 200 ml of 1.88 M Tris/HCl buffer, pH 8.8. Autoclave and store at room temperature. The buffer is stable for at least 6 months.
4. Stacking gel buffer stock: Prepare 200 ml of 1.25 M Tris/HCl buffer, pH 6.8. Autoclave and store at room temperature. The buffer is stable for at least 6 months.
5. 1% w/v gelatin (Sigma, St. Louis, MO, Catalogue number G-8150): Prepare 5 ml of a 1% w/v solution of gelatin in dH_2O. Heat the solution to 60°C in a water bath for at least 20 min, and vortex well. Make sure that the gelatin is completely dissolved. Cool down the gelatin solution to room temperature before use. Prepare a fresh solution each time.
6. 20% w/v SDS: Prepare 200 ml of 20% w/v SDS in dH_2O. Use a surgical mask when weighing SDS. Store at room temperature. It is stable for 1 year.
7. 10% w/v ammonium persulfate (APS): Prepare 1–5 ml of 10% w/v APS in dH_2O, calculating the volume required based on the number of gels to be prepared. Store at 4°C for no longer than 2 weeks.
8. N,*N*,*N*′,*N*′-tetramethylethylenediamine (TEMED): Store in a dark bottle at 4°C.
9. Running buffer stock (10×): Prepare 1 l of 0.25 M Tris base and 1.92 M glycine, pH 8.3. The pH should be correct without adjusting. Store at room temperature, stable for several months under these conditions.
10. SDS gel running buffer: Dilute the running buffer stock (10×) with dH_2O to make 1 l and supplement with 5 ml of 20% w/v SDS to a final concentration of 0.1% w/v SDS. Store at room temperature, stable for several months under these conditions.

11. Sample buffer (4×): Prepare 10 ml of 0.25 M Tris/HCl, pH 6.8, 40% v/v glycerol, 8% w/v SDS, and 0.01% w/v bromophenol blue. Store at −20°C in 0.5-ml aliquots. Before use, warm up to room temperature to dissolve the SDS.
12. Renaturing solution stock (10×): Prepare 200 ml of 25% v/v Triton X-100 in dH_2O. Store at room temperature, stable for several months under these conditions.
13. Developing buffer stock (10×): Prepare 1 l of 0.5 M Tris–HCl, pH 7.8, 2 M NaCl, 0.05 M $CaCl_2$, and 0.2% Brij 35. Store at 4°C for 6 months.
14. Staining solution: Prepare 1 l of 0.5% w/v Coomassie Blue R-250, 5% v/v methanol, and 10% v/v acetic acid in dH_2O. Filter and store at room temperature. This solution is reusable.
15. Destaining solution: Prepare 1 l of 5% v/v methanol and 10% v/v acetic acid in dH_2O. Store at room temperature for months.
16. Phosphate-buffered saline (PBS): Prepare 1 l 0.01 M phosphate buffer, pH 7.1, 0.137 M NaCl, and 2.7×10^{-3} M KCl. Autoclave and store at room temperature, stable for several months under these conditions.
17. Lysis buffer: Prepare 100 ml of 0.025 M Tris–HCl, pH 7.5, 0.1 M NaCl, and 1% v/v Nonidet P-40 (NP-40). Store at 4°C for 6 months. Immediately before use, add the following protease inhibitors: 10 μg/ml aprotinin, 2 μg/ml leupeptin, and 4×10^{-3} M benzamidine or Protease Inhibitor Cocktail-EDTA Free (Roche).
18. Tris-buffered saline (TBS): Prepare 1 l of 0.05 M Tris–HCl, pH 7.5, and 0.15 M NaCl. Store at 4°C, where it is stable for several months.
19. TBS-CM: Prepare 200 ml of TBS supplemented with 1×10^{-3} M $CaCl_2$ and 1×10^{-3} M $MgCl_2$. Store at 4°C and use within a week.
20. TBS-B: Prepare 50 ml of TBS containing 5×10^{-3} M $CaCl_2$ and 0.02% Brij-35. Store at 4°C for 1 month or until a visible precipitate appears.
21. TBS-CM-Triton X-114: Prepare 50 ml of 1.5% v/v Triton X-114 in TBS-CM. Store at 4°C, where it is stable for several months. Immediately before use, add protease inhibitors: 10 μg/ml aprotinin, 2 μg/ml leupeptin, and 4×10^{-3} M benzamidine or Protease Inhibitor Cocktail-EDTA Free (Roche).
22. Gelatin–agarose beads (Sigma Catalogue number G-5384): Wash the beads (the amount needed for the experiment) twice with TBS-B before use to remove preservative solution. Prepare a 50% v/v suspension of gelatin–agarose beads in TBS-B.

3. Methods

3.1. Gel Preparation

The following protocol is for the preparation of eight gelatin zymography gels (10% polyacrylamide co-polymerized with 0.1% w/v gelatin) (see Notes 2 and 3).

1. Separating gel: Mix 18.0 ml of dH_2O, 5 ml 1% w/v gelatin, 16.6 ml acrylamide/bis-acrylamide stock, 10 ml separating gel buffer stock, 0.25 ml 20% w/v SDS, and 150 μl 10% w/v APS in a 125-ml Erlenmeyer flask or a 50-ml disposable centrifuge tube at room temperature.
2. Stacking gel: Mix 11 ml of dH_2O, 2 ml acrylamide/bis-acrylamide stock, 2 ml stacking gel buffer stock, 0.1 ml 20% w/v SDS, and 75 μl 10% w/v APS in the same apparatus as above.
3. Add 30 μl of TEMED to the separating gel solution to initiate the polymerization process. Swirl the solution rapidly without causing bubble formation or aeration.
4. Immediately, pipette 6.2 ml of the separating gel solution into each cassette, avoiding the formation of bubbles.
5. Carefully, overlay the separating gel solution with dH_2O filling up to the top of the cassette using a syringe or a pipette. Do not disturb the surface of the separating gel solution.
6. Allow the gel to polymerize for at least 1 h at room temperature. Polymerization is complete when a discrete line of separation can be observed between the gel and the overlaid water.
7. Decant the water from the top of the separating gel and allow the gel to remain upside down on a bench top for a few minutes to drain away all the water.
8. Add 10 μl of TEMED to the stacking gel solution, swirl rapidly, and immediately pipette the solution onto the polymerized separating gel until it reaches the top of the front plate.
9. Rapidly insert the appropriate comb (this usually has prongs to allow the formation of ten wells) into the liquid stacking gel taking care to ensure that no bubbles remain trapped under the comb. Allow the stacking gel to polymerize at room temperature (about 30–60 min).
10. Store the gels (comb on top) in a sealed plastic bag or a container containing 1× running buffer to keep them moist. The gels can be stored at 4°C for up to 2–3 weeks (see Note 3).

3.2. Sample Preparation

3.2.1. Serum-Free Conditioned Media

The gelatinases are secreted enzymes (5); therefore, in cultured cells, a significant proportion of the gelatinase "pool" is in the incubation media. Since the serum in which cells are typically grown contains gelatinases, it is necessary to prepare serum-free conditioned media for gelatin zymography (Fig. 1).

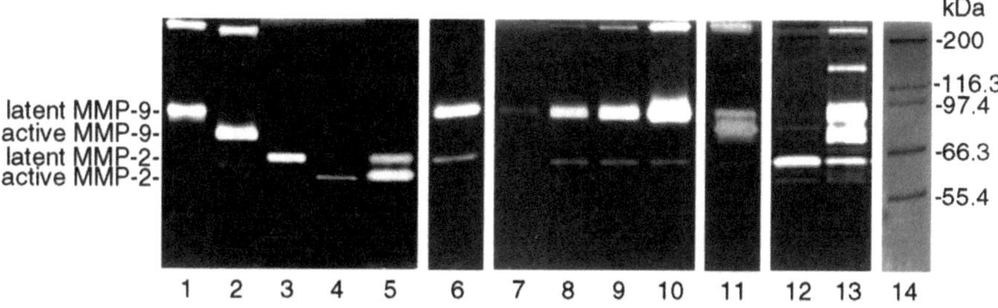

Fig. 1. Gelatin zymogram of biological samples from different sources. *Lanes 1–5*, purified recombinant human MMP-9 (*lane 1*, latent form; *lane 2*, active form) and MMP-2 (*lane 3*, latent form; *lane 4*, active form; *lane 5*, mixture of active and latent forms). Note the presence of MMP-9 dimeric forms (~200 kDa). *Lane 6*, serum-free conditioned media of HT1080 human fibrosarcoma cells. *Lanes 7–10*, serum-free conditioned media of nonmalignant human breast epithelial MCF10A cells untreated (*lane 7*) or treated with increasing concentrations (10, 25, and 50 ng/ml) of tumor necrosis factor-α (TNF-α) (*lanes 8–10*). Note the induction of pro-MMP-9 expression and secretion in response to TNF-α compared to the untreated cells. *Lane 11*, cell lysate of TNF-α-treated MCF10A cells. Note the presence of the intracellular, lower molecular weight (~83–85 kDa), precursor (partially glycosylated) form of latent MMP-9 (32). *Lanes 12* and *13*, tissue extracts (30 μg of protein) from the benign (*lane 12*) and carcinoma (*lane 13*) section of a breast biopsy. Note the induction of MMP-9 and a ~130-kDa band of unknown origin in the carcinoma. Also, note the appearance of a ~80–85-kDa form in the carcinoma. Note that it is difficult to determine whether this is active MMP-9 or the intracellular precursor form. *Lane 14*, unreduced molecular weight standard (Novex).

1. Grow the cells to be tested to approximately 80% confluence in complete growth media.
2. Wash the cell monolayer twice with sterile PBS or serum-free media to remove the serum completely.
3. Incubate the cells in serum-free media at 37°C in a CO_2 incubator for at least 12–16 h (see Note 4).
4. Collect the media and centrifuge (400 × *g* for 5 min at 4°C) to remove floating cells and debris. Retain the supernatant.
5. Mix 75 μl of the clarified conditioned media with 25 μl of 4× sample buffer, and vortex. Allow the sample to stand at room temperature for 10–15 min. Do not heat the sample! Load 30 μl per lane on the gelatin zymography gel.
6. If the intensity of the gelatinolytic bands is very low, the conditioned media can be concentrated using commercially available concentrators (10 kDa cutoff) or subjected to gelatin–agarose purification as described below.

3.2.2. Purification of Gelatinases Using Gelatin–Agarose Beads

1. In a microcentrifuge tube, mix 1 ml of clarified conditioned media with 30 μl of the gelatin–agarose bead suspension and rotate at 4°C for at least 1 h.
2. Centrifuge the tubes in a microcentrifuge for approximately 1–2 min and carefully aspirate the supernatant.
3. Wash the beads (containing the bound gelatinases), at least twice, with 1 ml of cold TBS-B.

4. After the last wash, carefully aspirate the supernatant. Add 20 μl of 1× sample buffer to the beads to elute the bound enzymes. Do not heat the samples! Centrifuge the tubes again and load the supernatants onto the gel.

3.2.3. Cell Lysate

In addition to secretion into media, gelatinases are also found intracellularly and are cell surface-associated. Intracellular enzymes represent the gelatinases trafficking through the secretory pathway, where they may be modified posttranslationally. In contrast, the extracellular gelatinases present are enzymes associated with the cell surface/matrix. Analysis of cell lysate, therefore, can provide valuable information on the rate of synthesis, extent of glycosylation (in the case of MMP-9), cell surface association, and activation status of the surface-associated gelatinases. Cell lysates are prepared as follows.

1. Wash the cells twice with cold PBS.
2. Add cold lysis buffer. Use 2 ml of lysis buffer per 150-mm^2 dish to obtain a final total protein concentration of approximately 2–3 mg/ml.
3. Scrape the cells into the lysis buffer with a rubber policeman, collect the lysate, and incubate on ice for at least 15 min.
4. Vortex and centrifuge (at 16,000×*g*) the lysate for 20 min at 4°C in a microcentrifuge. Collect the supernatant and measure protein concentration.
5. Mix 75 μl of supernatant with 25 μl of 4× sample buffer. Do not heat the sample. Load up to 30 μl per lane on the gelatin zymography gel. In the minigel system described here, do not load more than 40 μg of total protein per lane, as excessive amounts of protein will disrupt the band separation and adversely affect resolution (see Note 9).

If the gelatinase activity is low, phase partition with Triton X-114, as described below, can be used to concentrate the cell-associated gelatinases.

3.2.4. Phase Extraction with Triton X-114

Triton X-114 is a biphasic detergent that is soluble in aqueous buffers at 4°C. However, above 20°C samples containing Triton X-114 separate into two distinct phases (aqueous and detergent). In general, hydrophilic proteins partition into the aqueous phase while hydrophobic proteins partition into the detergent phase. However, anomalous distributions of proteins have been reported in the Triton X-144 system. Being hydrophilic in nature, gelatinases are mostly found in the aqueous phase. However, gelatinases (most evident with MMP-2) have also been found in the detergent phase (24, 25), suggesting a strong association with the plasma membranes. Phase partition is carried out as follows.

1. Wash the cells twice with cold PBS. During all the following steps, keep samples on ice unless otherwise stated.
2. Add 2 ml of cold TBS-CM-Triton X-114 solution containing protease inhibitors per 150-mm^2 dish. This should yield approximately 2–3 mg of total protein per ml.
3. Scrape the cells into the solution and transfer to a tube.
4. Incubate the extract on ice for at least 15 min and then centrifuge (16,000 × *g* for 20 min) at 4°C.
5. Collect the supernatant, transfer to a new tube, and incubate for 2 min in a 37°C water bath.
6. Centrifuge (16,000 × *g*) the sample at room temperature for approximately 5 min to obtain the lower detergent and upper aqueous phases.
7. Carefully collect and transfer the aqueous phase into a new tube without disturbing the detergent phase. Cool down the tubes briefly. Keep the tube with the detergent phase on ice.
8. Add 30 μl of the gelatin–agarose beads to 500 μl of the aqueous phase and proceed as described in Subheading 3.2.2 above.
9. Add 2 volumes of TBS-B and 1 volume of 4× sample buffer to the detergent phase (4× dilution) (see Note 5). Do not heat the sample. Load 30 μl per lane for gelatin zymography.

3.2.5. Preparation of Tissue Extracts

Fresh tissue biopsies derived from tumor samples are an important source for examining gelatinase expression during tumor progression (Fig. 1).

1. Cut the tissue of interest (~50 mg) into small pieces. Remove any visible fat.
2. Add approximately 500 μl of cold lysis buffer with protease inhibitors.
3. Homogenize the tissue on ice with a pestle (Kontes, Vineland, NJ, Catalogue number 749520-0000) in a microcentrifuge tube (see Note 6).
4. Centrifuge (16,000 × *g*) the homogenate for 10 min at 4°C. Collect the supernatant and measure the protein concentration.
5. Adjust the protein concentration to 1 μg/μl of 1× sample buffer. Load equal amounts of protein per lane for the gelatin zymography.

3.2.6. Gelatinase Standards

It is important to include gelatinase standards in each zymogram to accurately determine the type and activation status of the enzyme(s) expressed in a given sample. Conditioned medium from HT1080, human fibrosarcoma cells (American Type Culture Collection, CCL-121), is optimal since it contains both MMP-2 (72 kDa) and MMP-9 (~92 kDa) (Fig. 1). Conditioned media obtained from HT1080 cells treated for 16 h with 1×10^{-4} M 12-*O*-tetradecanoylphorbol-13-acetate or 10 μg/ml of concanavalin A also contain active MMP-2 (approximately 62 kDa) and can be used as a reference for active MMP-2 (26). If available, purified natural or recombinant gelatinase enzymes can also be used (Fig. 1) (see Note 7).

1. Dilute the purified gelatinases to a final concentration of 1 ng/μl of 1× sample buffer. Do not heat the sample.
2. Load 1–5 ng of the enzyme per lane on the zymography gel.

3.3. Running and Developing the Gel

1. Gently pull the comb out of the stacking gel and peel off the tape from the bottom of the cassette. Place the cassette into the gel apparatus and fill the buffer chambers with 1× SDS gel-running buffer.
2. Load the samples and run the gel at constant voltage (125 V, starting current should be approximately 30–40 mA/gel) until the bromophenol blue tracking dye reaches the bottom of the gel (approximately 90 min). These running conditions will prevent overheating of the gel (see Note 2).
3. Carefully remove the gel from the cassette and place it in a plastic tray containing 100 ml 1× diluted renaturing solution. Incubate the gel for 30 min at room temperature with gentle agitation.
4. Decant the solution and rinse the gel at least once with 300 ml of dH_2O.
5. Incubate the gel at room temperature for an additional 30 min in 100 ml of 1× developing buffer with gentle agitation.
6. Decant the developing buffer and replace it with 100 ml of fresh 1× developing buffer. Incubate the gel at 37°C for approximately 16 h in a closed tray (see Note 8).
7. Decant the developing buffer and stain the gel in staining solution for at least 1 h or until the gel is uniformly dark blue. The staining solution can be collected and reused. However, it will require a longer staining time.
8. Destain the gel with destaining solution until areas of gelatinolytic activity appear as clear sharp bands over the blue background (see Note 9).

4. Notes

1. For reliable results during electrophoresis, it is crucial that the acrylamide/bis-acrylamide stock solution is prepared from electrophoresis-grade reagents. Ready-to-use 30% acrylamide/bis-acrylamide 37.5:1 (2.6%C) solution can be purchased from Bio-Rad Laboratories (Hercules, CA; Catalogue number

161-0158), thereby obviating the need to handle acrylamide powder. A range of ready-made gelatin zymography gels (Novex® 10% Zymogram) are available from Invitrogen, allowing reproducible results.

2. We commonly use 10% polyacrylamide gels for separating gelatinases. However, the percentage of acrylamide and the thickness of the separating gel can be varied depending on the aim of the separation. For instance, to better visualize the dimeric form of MMP-9 (~200 kDa) (Fig. 1) and/or to obtain a better resolution of bands with similar molecular weight (latent and active forms), a lower percentage (7–8%) polyacrylamide solution can be used. It should be noted, however, that by taking this approach the gelatinolytic bands observed will be less sharp. Alternatively, the gel can be run for an additional 30 min after the tracking dye has reached the bottom of the 10% PAGE.
3. It is very convenient to prepare the gels in advance since the polymerized gels can be stored at 4°C for 2–3 weeks without any effects on resolution. Avoid bacterial contamination of buffers and solutions as bacterial proteases may result in the appearance of nonspecific gelatinolytic bands. This can be minimized by filter sterilization of buffers and stock solutions and storage at 4°C, as indicated.
4. Preparation of conditioned media: Long incubation times (48–72 h) of cells in serum-free media can affect viability. Incubate the cells in a minimum amount of serum-free media (~50–60% of the amount of growth media) to obtain more concentrated samples.
5. Triton X-114 extraction and detergent phase: Due to the high sensitivity of gelatin zymography, the presence of gelatinases in the detergent phase may result from contamination with enzymes from the aqueous phase. To minimize such cross-contamination, repeat again the extraction procedure by adding ice-cold TBS-CM to the detergent phase. Vortex and repeat the phase partition as described in section 3.2.4, steps 5–7. Discard the upper phase and continue as described in section 3.2.4, step 9.
6. Preparation of tissue extracts: The volume of lysis buffer should be varied depending on the tissue being used. Be aware that inherent differences in tissue structure, between and within specimens, as well as protein extractability may vary (18). Therefore, the results should be carefully interpreted! After centrifugation, the tissue homogenate may contain floating lipids. Repeat the centrifugation step to obtain a clear homogenate.
7. Gelatinase standards: Purified gelatinases can be obtained from commercial sources. To activate purified gelatinases, incubate

the enzymes in 1×10^{-3} M *p*-aminophenylmercuric acetate (APMA) (toxic!). This should be prepared from a fresh stock of 0.01 M APMA in 0.05 M NaOH and diluted tenfold in TBS-B. Incubate the purified latent enzymes (2 ng/μl) with APMA in a 37°C water bath. Latent MMP-2 is readily activated to the 62 kDa species after approximately 30 min while latent MMP-9 is partially activated to the 82 kDa form after 2 h. The presence of TIMPs in the enzyme preparation will inhibit or slow down the activation process, causing generation of intermediate inactive forms (approximately 64 kDa for MMP-2 and 85 kDa for MMP-9) (20, 27, 28). On the other hand, long incubation times with APMA of TIMP-free gelatinases will cause the appearance of low-molecular-weight active forms: an approximate 45 kDa form for MMP-2 (29) and an approximate 67 kDa form for MMP-9 (30).

8. The incubation time for the gel in the developing buffer is critical. Since the presence of the gelatinolytic bands is the result of enzymatic activity, varying the incubation time will affect the intensity of the gelatinolytic bands. For most conditions, an overnight incubation will provide optimal resolution and reproducible results. Therefore, for better resolution, it is preferable to increase or decrease the amount of sample loaded into the gel rather than changing the incubation time.
9. Gelatinolytic bands: Under optimal conditions, gelatinolytic bands should be sharp and well defined (Fig. 1). This is generally true with samples derived from conditioned media or with purified enzymes. Cell lysates and tissue extracts, in contrast, may produce distorted bands due to the high protein content. This can be minimized by reducing the amount of protein loaded per lane or concentrating the sample using phase partition and agarose-bead purification. Since other proteases, including serine proteases, can also exhibit gelatinolytic activity (31), it is important to ascertain the type (metallo vs. nonmetallo) of enzyme detected by addition of a chelating agent. Incubating the gel in developing buffer containing 0.02 M EDTA will inhibit metalloproteases and cause the disappearance of the gelatinolytic bands they produce. A reliable assessment of the nature of the enzyme present in the samples can be made by comparing the molecular weight of the gelatinolytic bands with known gelatinase standards (for example by using the conditioned media from HT1080 cells) and the use of EDTA. Note that other major MMPs showing gelatinolytic activity are stromelysin 1 (MMP-3) and interstitial collagenase (MMP-1) (19). However, the intensity of the bands produced by these MMPs is significantly lower than that elicited by gelatinases. In addition, their molecular mass is different.

10. Latent vs. active species: When using gelatin zymography to study gelatinase activity, it is important to be aware of the pitfalls of the methodology. Gelatinases are usually activated in a sequential process involving the generation of an inactive intermediate species. These are then processed to generate the full mature active form. The small differences in molecular mass (approximately 2–4 kDa) between intermediate and active species require evaluation of optimal sample amounts, running conditions, and incubation times; otherwise, it will not be possible to discriminate between the different forms. Therefore, if good band separation and molecular mass determination are not achieved due to poor running/incubation conditions and/or overloaded samples, then discrimination between the intermediate (inactive) and fully active species will be challenging. Under these conditions, the intermediate species may easily be confused with the active forms. In addition, due to the differences in glycosylation between the precursor (intracellular) and the mature (secreted) form of latent MMP-9 (32), the intracellular precursor form of pro-MMP-9 (approximately 85 kDa) can be easily mistaken for the active species. This is most likely to occur when using samples derived from cell lysates or tissue extracts, all of which may contain a pool of intracellular pro-MMP-9 (Fig. 1). Inclusion, in the same gel, of active enzymes as standards and/or reliable non-reduced molecular weight markers will aid in the discrimination between latent (intermediate, precursor) and active species. When in doubt, additional methods including immunoblot analysis, using antibodies capable of discrimination between latent and active species, and/or enzymatic assays need to be conducted to further verify if the samples contain truly active MMP-9.
11. In relation to the use of gelatin zymography for quantitative assessments: Although gelatin zymography has been claimed to be a quantitative technique, major problems arise during the process of method standardization due to the many variables involved in this technique. These include (1) variations in the amount of co-polymerized gelatin; (2) nature, source, preparation, and amount of the sample; (3) incubation time and temperature; (4) washing conditions; (5) staining and destaining conditions; and (6) source and condition (latent, active) of the standards. These diverse variables make it very difficult to establish reproducible assay conditions when analyzing multiple samples and parameters. Even when careful and reproducible assay conditions are established, the results obtained are at best semiquantitative. Therefore, it is our recommendation that the user refrains from using gelatin zymography as the only tool for measuring the levels of gelatinase activity in biological samples. If quantification is the goal, ELISA assays

are preferred. Nevertheless, when used properly, gelatin zymography provides a wealth of information regarding gelatinase activity in experimental and biological systems and is a simple, reliable, and reproducible method for analysis of this class of enzymes.

Acknowledgements

The authors would like to thank all the ex-members of the Fridman lab for suggestions and tips throughout the years. This work has been supported by an NIH/NCI grant (R01 CA-61986) to R.F.

References

1. Massova I, Kotra LP, Fridman R, Mobashery S (1998) Matrix metalloproteinases: structures, evolution, and diversification. FASEB J 12:1075–1095
2. Handsley MM, Edwards DR (2005) Metalloproteinases and their inhibitors in tumor angiogenesis. Int J Cancer 115:849–860
3. Turpeenniemi-Hujanen T (2005) Gelatinases (MMP-2 and -9) and their natural inhibitors as prognostic indicators in solid cancers. Biochimie 87:287–297
4. Deryugina EI, Quigley JP (2006) Matrix metalloproteinases and tumor metastasis. Cancer Metastasis Rev 25:9–34
5. Murphy G, Crabbe T (1995) Gelatinases A and B. Methods Enzymol 248:470–484
6. McQuibban GA, Gong JH, Tam EM, McCulloch CA, Clark-Lewis I, Overall CM (2000) Inflammation dampened by gelatinase A cleavage of monocyte chemoattractant protein-3. Science 289:1202–1206
7. McCawley LJ, Matrisian LM (2001) Matrix metalloproteinases: they're not just for matrix anymore! Curr Opin Cell Biol 13:534–540
8. Hamano Y, Zeisberg M, Sugimoto H, Lively JC, Maeshima Y, Yang C, Hynes RO, Werb Z, Sudhakar A, Kalluri R (2003) Physiological levels of tumstatin, a fragment of collagen IV alpha3 chain, are generated by MMP-9 proteolysis and suppress angiogenesis via alphaV beta3 integrin. Cancer Cell 3:589–601
9. Chen X, Su Y, Fingleton B, Acuff H, Matrisian LM, Zent R, Pozzi A (2005) An orthotopic model of lung cancer to analyze primary and metastatic NSCLC growth in integrin alpha1-null mice. Clin Exp Metastasis 22:185–193
10. Bonfil RD, Sabbota A, Nabha S, Bernardo MM, Dong Z, Meng H, Yamamoto H, Chinni SR, Lim IT, Chang M, Filetti LC, Mobashery S, Cher ML, Fridman R (2006) Inhibition of human prostate cancer growth, osteolysis and angiogenesis in a bone metastasis model by a novel mechanism-based selective gelatinase inhibitor. Int J Cancer 118:2721–2726
11. Deryugina EI, Quigley JP (2009) Pleiotropic roles of matrix metalloproteinases in tumor angiogenesis: contrasting, overlapping and compensatory functions. Biochim Biophys Acta 1803:103–120
12. Collier IE, Wilhelm SM, Eisen AZ, Marmer BL, Grant GA, Seltzer JL, Kronberger A, He CS, Bauer EA, Goldberg GI (1988) H-ras oncogene-transformed human bronchial epithelial cells (TBE-1) secrete a single metalloprotease capable of degrading basement membrane collagen. J Biol Chem 263: 6579–6587
13. Wilhelm SM, Collier IE, Marmer BL, Eisen AZ, Grant GA, Goldberg GI (1989) SV40-transformed human lung fibroblasts secrete a 92-kDa type IV collagenase which is identical to that secreted by normal human macrophages. J Biol Chem 264:17213–17221
14. Briknarova K, Gehrmann M, Banyai L, Tordai H, Patthy L, Llinas M (2001) Gelatin-binding region of human matrix metalloproteinase-2: solution structure, dynamics, and function of the COL-23 two-domain construct. J Biol Chem 276:27613–27621

15. Morgunova E, Tuuttila A, Bergmann U, Tryggvason K (2002) Structural insight into the complex formation of latent matrix metalloproteinase 2 with tissue inhibitor of metalloproteinase 2. Proc Natl Acad Sci USA 99: 7414–7419
16. Bode W (2003) Structural basis of matrix metalloproteinase function. Biochem Soc Symp 70:1–14
17. Heussen C, Dowdle EB (1980) Electrophoretic analysis of plasminogen activators in polyacrylamide gels containing sodium dodecyl sulfate and copolymerized substrates. Anal Biochem 102:196–202
18. Woessner JF Jr (1995) Quantification of matrix metalloproteinases in tissue samples. Methods Enzymol 248:510–528
19. Nagase H (1997) Activation mechanisms of matrix metalloproteinases. Biol Chem 378: 151–160
20. Murphy G, Willenbrock F (1995) Tissue inhibitors of matrix metalloendopeptidases. Methods Enzymol 248:496–510
21. Olson MW, Gervasi DC, Mobashery S, Fridman R (1997) Kinetic analysis of the binding of human matrix metalloproteinase-2 and -9 to tissue inhibitor of metalloproteinase (TIMP)-1 and TIMP-2. J Biol Chem 272:29975–29983
22. Toth M, Bernardo MM, Gervasi DC, Soloway PD, Wang Z, Bigg HF, Overall CM, DeClerck YA, Tschesche H, Cher ML, Brown S, Mobashery S, Fridman R (2000) Tissue inhibitor of metalloproteinase (TIMP)-2 acts synergistically with synthetic matrix metalloproteinase (MMP) inhibitors but not with TIMP-4 to enhance the (Membrane type 1)-MMP-dependent activation of pro-MMP-2. J Biol Chem 275:41415–41423
23. Zhao H, Bernardo MM, Osenkowski P, Sohail A, Pei D, Nagase H, Kashiwagi M, Soloway PD, DeClerck YA, Fridman R (2004) Differential inhibition of membrane type 3 (MT3)-matrix metalloproteinase (MMP) and MT1-MMP by tissue inhibitor of metalloproteinase (TIMP)-2 and TIMP-3 regulates pro-MMP-2 activation. J Biol Chem 279:8592–8601
24. Azzam HS, Arand G, Lippman ME, Thompson EW (1993) Association of MMP-2 activation potential with metastatic progression in human breast cancer cell lines independent of MMP-2 production. J Natl Cancer Inst 85: 1758–1764
25. Lewalle JM, Munaut C, Pichot B, Cataldo D, Baramova E, Foidart JM (1995) Plasma membrane-dependent activation of gelatinase A in human vascular endothelial cells. J Cell Physiol 165:475–483
26. Gervasi DC, Raz A, Dehem M, Yang M, Kurkinen M, Fridman R (1996) Carbohydrate-mediated regulation of matrix metalloproteinase-2 activation in normal human fibroblasts and fibrosarcoma cells. Biochem Biophys Res Commun 228:530–538
27. Toth M, Chvyrkova I, Bernardo MM, Hernandez-Barrantes S, Fridman R (2003) Pro-MMP-9 activation by the MT1-MMP/MMP-2 axis and MMP-3: role of TIMP-2 and plasma membranes. Biochem Biophys Res Commun 308:386–395
28. Toth M, Osenkowski P, Hesek D, Brown S, Meroueh S, Sakr W, Mobashery S, Fridman R (2005) Cleavage at the stem region releases an active ectodomain of the membrane type 1 matrix metalloproteinase. Biochem J 387:497–506
29. Fridman R, Fuerst TR, Bird RE, Hoyhtya M, Oelkuct M, Kraus S, Komarek D, Liotta LA, Berman ML, Stetler-Stevenson WG (1992) Domain structure of human 72-kDa gelatinase/type IV collagenase. Characterization of proteolytic activity and identification of the tissue inhibitor of metalloproteinase-2 (TIMP-2) binding regions. J Biol Chem 267:15398–15405
30. O'Connell JP, Willenbrock F, Docherty AJ, Eaton D, Murphy G (1994) Analysis of the role of the COOH-terminal domain in the activation, proteolytic activity, and tissue inhibitor of metalloproteinase interactions of gelatinase B. J Biol Chem 269:14967–14973
31. Pineiro-Sanchez ML, Goldstein LA, Dodt J, Howard L, Yeh Y, Tran H, Argraves WS, Chen WT (1997) Identification of the 170-kDa melanoma membrane-bound gelatinase (seprase) as a serine integral membrane protease. J Biol Chem 272:7595–7601
32. Toth M, Gervasi DC, Fridman R (1997) Phorbol ester-induced cell surface association of matrix metalloproteinase-9 in human MCF10A breast epithelial cells. Cancer Res 57:3159–3167

Chapter 9

Determination of Cell-Specific Receptor Binding Using a Combination of Immunohistochemistry and In Vitro Autoradiography: Relevance to Therapeutic Receptor Targeting in Cancer

Michael R. Dashwood and Marilena Loizidou

Abstract

Mapping of receptor binding to specific structures, or cells within tissue samples, provides valuable information regarding biological and pathological mechanisms. Such information may potentially be translated into targeted therapies, especially in the field of cancer treatment. In this chapter, a receptor localization technique is described which utilises frozen sections of human tissue and combines immunohistochemistry (IHC) and micro-autoradiography. IHC utilises antibodies tagged to an enzymatic complex to identify specific cell types (such as epithelial cells or fibroblasts) within the tissue under investigation; this step is immediately followed by the second technique which is based on the use of radiolabelled compounds (radioligands) that selectively bind to preselected membrane receptors. This approach allows visualisation of cells of interest by immunohistochemical staining of tissue sections (colour product) in combination with the use of radiolabelled compounds that are detected following exposure to radiation-sensitive film or emulsion to produce a map of receptor distribution or localisation of cell-specific receptor binding. The system described has been used to compare receptor binding to cells in normal human colorectal tissue with that in colorectal cancer specimens.

Key words: Cell type-specific immunohistochemistry, Receptor micro-autoradiography, Receptor binding

1. Introduction

1.1. Autoradiography: Background

The term autoradiography (ARG) denotes the visualisation of the distribution pattern of radioactivity in either biological or non-biological material. The principles of ARG are the same as those for conventional photography, where the image is formed using an emulsion usually containing a lattice of silver halide embedded in

Miriam Dwek et al. (eds.), *Metastasis Research Protocols*, Methods in Molecular Biology, vol. 878,
DOI 10.1007/978-1-61779-854-2_9, © Springer Science+Business Media, LLC 2012

gelatine. For regional "mapping", X-ray or similar films (macro-ARG) are used and a nuclear emulsion is required in order to visualise receptor binding at the microscopic level (micro-autoradiography, micro-ARG). The method by which the radioligand binding is captured and detected is affected by the activity of the radioactive source used. Radiolabelling is generally achieved using low-activity beta emitters (for example [^{3}H] or [^{14}C]) or high-activity gamma emitters (for example [^{125}I]), compounds such as ^{3}H-naloxone or ^{125}I-endothelin, respectively. The low-activity-labelled compounds require a longer exposure period to film or emulsion but theoretically achieve a higher degree of resolution than iodinated compounds, which generate autoradiographs following short exposure times but often with limited resolution.

The theoretical background underlying the principles of ARG will not be covered in this chapter; these are adequately explained in various methodological books on the technique (1–4). Originally, in vivo ARG was used to identify receptor binding in experimental animals, where a radioligand for a given receptor was injected intravenously, the animal sacrificed, after the appropriate time, and the tissue of interest removed, sectioned, and exposed to film or emulsion. There are various disadvantages associated with in vivo ARG, in particular the large amount of radioligand required as its concentration is diluted greatly in the circulation. Consequently, to reduce running costs, small animals such as rats and mice are mostly used for such studies. Also, when radiolabelled material is administered systemically and subsequently localised in tissue sections, one cannot confidently distinguish receptor binding from (radiolabelled) drug distribution/diffusion. Indeed, intravenous administration of radiolabelled drugs developed by the pharmaceutical industry is routinely used as a means of investigating drug distribution, drug metabolism, and placental diffusion.

A major advance in the ability to identify and localise radioligand receptor binding was introduced over 30 years ago by Young and Kuhar (5) who showed that fresh, unfixed, frozen sections retained the ability to bind radiolabelled materials, thus providing a method of studying receptor characteristics in vitro. This technique offered many advantages over in vivo ARG, in particular the ability to control binding conditions. For example, incubation buffers could be tailored to suit the compounds under investigation (for example, addition of NaCl increases ^{3}H-naloxone binding to rat brain sections); much lower concentrations of radioligand were required (pico- or nanomolar amounts); and binding to specific receptor types could be established since the degree of non-specific binding could be revealed by "saturating" binding to the receptor under investigation in paired sections and incubating in the presence of excess unlabelled ligand. For details of pharmacological receptor characterization, see ref. (2). Furthermore, using in vitro ARG, it has become possible to study compounds that do not readily pass the blood–brain barrier, enabling researchers to generate receptor maps for many novel compounds or neurotransmitters in the brain.

Receptor localization in Young and Kuhar's original technique was achieved (after incubation) by exposing the slide-mounted tissue sections to coverslips pre-coated with a nuclear emulsion (for example, Ilford K2 or K5 emulsions) and subsequently processing using standard photographic solutions (for example, Kodak D19 high contrast developer followed by Ilford Hypam fixative). Once ARGs have been produced, the underlying tissue histology is confirmed using traditional stains, such as haematoxylin and eosin or toluidine blue, so that receptor binding to identifiable structures is achieved at the microscopic level. A major advance in the field of quantitative receptor ARG occurred with the introduction of tritium-sensitive film that not only produces "receptor maps" (originally in brain sections) but, more importantly, allows densitometric evaluation to be performed (6). For ligands labelled with high energy, gamma emitters, such as ^{125}I-endothelin, regular X-ray film, or Hyperfilm™ MP, can be used. However, this film technique lacks the resolution achieved using nuclear emulsion and limits the level at which receptor binding can be detected (see Note 1).

1.2. Combining Autoradiography and Immunohistochemistry

Macro- and micro-autoradiographic techniques each have their specific advantages. Since ARGs generated on film provide "receptor maps" and enable receptor binding to be quantified, this method offers a useful means of comparing experimentally induced or disease-induced alterations in both experimental animal models and human tissue. Where ARGs are produced using nuclear emulsion, binding can be localised at the microscopic level. When using either of these methods, once the ARG image has been obtained, the histology of the underlying tissue sections is examined using traditional stains, such as haematoxylin and eosin. While this may be perfectly adequate in many cases, conventional stains will not allow specific cell types to be identified with confidence. This is particularly important when studying pathological tissue, where a variety of infiltrating cells may appear, ranging from those associated with inflammatory conditions to those involved in extracellular matrix formation, cell proliferation, and apoptosis. In an attempt to overcome this problem, a technique combining the immunohistochemical identification of cell type with ARG localisation of radioligand binding was developed and originally used in coronary tissue (7).

As with ARG, the theoretical background underlying the principles of immunohistochemistry (IHC) will not be covered here as the technique is well established and information on the theory and application is easily sourced (see also chapter 2 on immunohistochemistry by Brooks and Hall in this volume).

1.3. Applications to Cancer Studies

We have used the combined ICH/ARG technique to identify cell types within colorectal cancer which expressed endothelin-1 (ET-1) binding sites. ET-1 was originally described as an endothelium-derived peptide with potent constrictor, proinflammatory, and mitogenic activity. Two receptors for this peptide have been identified and cloned: ET_A (ET_AR) and ET_B (ET_BR) and many receptor-selective antagonists have been developed. Various groups have reported changes in receptor levels within solid tumours, such as ovarian and prostate cancers: notably, an over-expression of ET_AR and a down-regulation of ET_BR. Since ET_AR is the main receptor that propagates proliferation and survival signals, this has a direct impact on cancer growth and progression (8). Furthermore, specific ET_AR antagonists have been approved for human use (for example, Atrasentan, Zibotentan). Therefore, information on the distribution of compounds binding to these receptors within cancer tissues provides useful information regarding the potential use of ET_AR antagonists in treatment regimens for colorectal cancers.

In this chapter, the application of conventional IHC to determine endothelial and epithelial cells, nerves, and fibroblasts within frozen tissue sections, followed by micro-ARG to localise receptor binding to these cells specifically as applied to cancer tissue is described. The focus is on the logistics of combining IHC and micro-ARG for successful receptor/cell co-localisation. The model system comprises normal colorectal tissues and colorectal cancer specimens and the investigation of the cell type-specific binding sites of ET-1 (9).

1.4. Methodological Advantages

Although immunostaining (histochemistry/fluorescence) "double-labelling" techniques are commonly used to localise cell-specific receptors and may be evaluated by, for example, confocal microscopy, the difference between receptors and their binding sites is often overlooked. While recombinant antibodies may provide the possibility of identifying specific receptor types, they provide no information regarding the membrane binding characteristics of agonists or antagonists. Physiologically, endogenous factors (agonists) act on membrane-binding sites/receptors to trigger specific events. Pharmacologically, synthetic compounds (antagonists) are developed that bind to such receptors and block these effects. Although IHC identifies and localises receptors at the microscopic level, it cannot be used to study receptor binding and assess drug/receptor interaction. As outlined above, this is possible using ARG, where radiolabelled compounds are used, not only to localise receptor binding using nuclear emulsion but also to quantify the receptor binding by image analysis or monitoring the levels of radioactivity incorporated. In this way, regional variations, disease-induced changes, receptor characterisation, and to some extent receptor function may be studied (Chapter 10 in Leslie and Altar (2)).

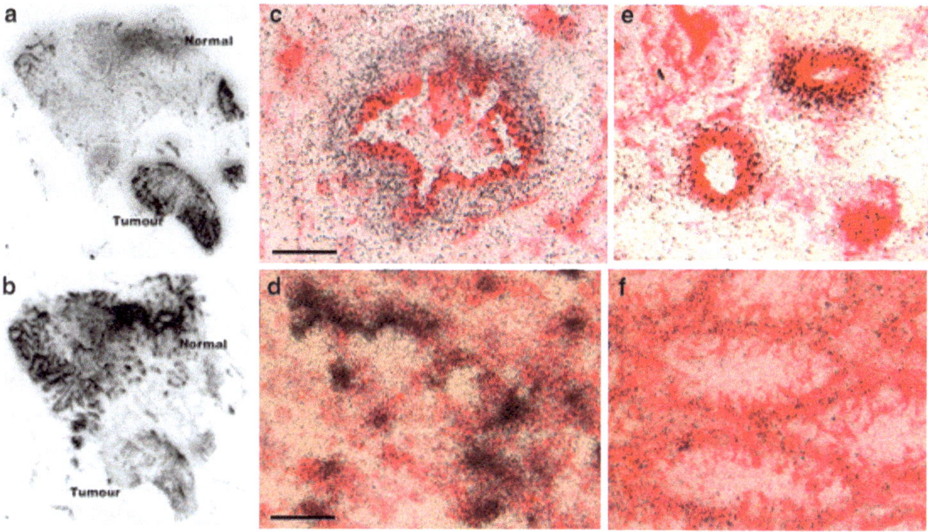

Fig. 1. Whole tissue: Autoradiographs showing binding (*dark grains*) to ET_AR (**a**) and ET_BR (**b**) in serial sections of tissue specimen containing both cancer (tumour) and normal colon—the autoradiographs were generated on film. Blood vessels: Combined IHC/ARG using an ET_AR-selective radioligand/CD31 in colorectal cancer (**c**) and normal colon (**e**). CD31 staining (*red*) identifies blood vessels. Binding to ET_AR (*dark grains*) co-localises with CD31. Fibroblasts: Combined IHC/ARG using ET_AR-selective radioligand/AS02 in colorectal cancer (**d**) or ET_BR-selective radioligand/AS02 in normal colon (**f**). AS02 staining (*red*) identifies fibroblasts. Binding to ET_AR or ET_BR (*dark grains*) co-localises with AS02. *Scale bar* = 200 μm in (**c**), (**e**); 50 μm in (**d**), (**f**).

A limitation with previously published ARG studies has been the inability to identify, with confidence, the specific cells within tissue sections. In such cases, researchers rely on commonly used histological stains. In order to address this situation, we developed the combined IHC/ARG method to localise binding to endothelin receptor subtypes located on specific cells involved in human colorectal cancers.

Using this approach, we have demonstrated the presence of ET_AR and ET_BR in colorectal cancer (9) and confirmed previous findings (by densitometric assessment of film ARGs) that a statistically significant increase in ET_AR and decrease in ET_BR binding are seen in colorectal cancer tissue as compared to normal colon (10, 11). The consistent increase in ET_AR binding in cancer tissue was partly co-localised to the epithelial cells lining the cancer glands, but was also present in stromal structures. ET_BR binding was decreased in cancer tissue, and localised to epithelia and other structures (Fig. 1). Since the most striking ET_AR binding was observed in the stroma, the specific cell types capable of binding the two radioactive ligands was assessed, focussing specifically on endothelial cells and fibroblasts (using antibodies for CD31 and AS02). ET_ARs are known to be expressed predominantly by vascular smooth muscle cells, with ET_BR expressed mainly by endothelial cells. The combined IHC/ARG system showed that ET_AR binding (black grains) was to the outer region (consistent with vascular smooth muscle) of blood vessels (red staining). This was more pronounced in the colorectal cancer sections compared to normal colon (Fig. 1). The opposite was true for ET_BR which was the predominant receptor subtype in normal tissue and was observed to be decreased in colorectal cancer. The results obtained using the combined IHC/ARG system as applied to colorectal cancer tissue contrast with those published by others (12) who described increased mRNA and protein levels for both ET_AR and ET_BR in tumour compared with normal colon. It is important to consider that alterations in receptor mRNA or protein do not necessarily reflect the degree of receptor binding, a measure of ligand/receptor or agonist/antagonist interaction that may be assessed using in vitro ARG. Interestingly, in the same study, increased ET_AR/ET_BR levels found in colon tissue extracts using molecular techniques were supported by the results generated from tissue sections by in vitro ARG (12). This discrepancy may be explained by the different protocols used where, rather than localising cell-specific binding using combined IHC/ARG, receptor binding was "suggested" based on the use of traditional histological staining of adjacent tissue sections. Furthermore, slide-mounted sections that had been incubated for ARG had been fixed in 4% formaldehyde/PBS, conditions known to have an adverse effect on receptor binding (2, 5). These observations underscore the need to identify cell-specific binding in tissue sections by combining IHC and ARG and also highlight the need to adhere to the conditions described in the early original papers and technique books describing various aspects of in vitro ARG (1–5).

2. Materials

2.1. Frozen Tissue Sectioning

1. Cut the tissue with a cryostat, and mount using Tissue-Tek (OCT) compound (VWR, Cat No 25608-903).
2. Frozen sections should be thaw-mounted onto polylysine-coated microscope slides (BDH, Cat No M0823).

2.2. Cell Type-Specific Immunohistochemistry

1. The following primary antibodies have been tested successfully using this method: CD31 (endothelial cells): Dako (Clone JC 70A); Thy-1: Dianova, Hamburg, Germany (clone AS02, fibroblasts), alternative clones available from Millipore (e.g. clone F15-42-1); Ber-EP4 (epithelial cells): Dako (Cat No M0804); and NF200 (nerves): Sigma (Cat No N-0142).
2. Vectastain® Universal ABC-AP kit (Vector Laboratories, Cat No AK-5200).

3. "Vector Red"/Alkaline Phosphatase Substrate Kit 1 (Vector Laboratories, Cat No SK-5100).
4. 0.01 M Phosphate-buffered saline (PBS), pH 7.4.

2.3. Micro-Autoradiography

1. Radioligands, ^{125}I-PD151242 and ^{125}I-BQ3020, to identify ET_A and ET_B receptors, respectively (custom labelling available from various companies, including Perkin Elmer; specific activity of both ~2,200 Ci/mmol).
2. K2 nuclear emulsion (Ilford, Cheshire, UK, Cat No 135 5099).
3. D19 high-contrast developer (Kodak).
4. Hypam fixative and Ilfostop solution (both Ilford).

3. Methods

3.1. Frozen Tissue Sectioning

1. Transport frozen tissues from liquid nitrogen bank or −70°C freezer storage to cryostat (see Note 2).
2. Remove tissues from cryotubes and secure on cryostat chucks using OCT compound.
3. Leave samples to equilibrate to approximately −25°C (temperature inside cryostat) for about 20 min.
4. Cut frozen sections at 10 μm onto polylysine-coated microscope slides (see Note 3).
5. Leave to air-dry (20 min) and store at −70°C (see Note 4).

3.2. Cell Type-Specific Immunohistochemistry

This is a "rapid staining procedure" described in the instructions for immunohistochemical staining accompanying Vector Laboratories' Vectastain® Universal ABC-AP kit.

All steps are carried out at room temperature.

1. Remove the slide-mounted tissue sections from −70°C storage and allow to equilibrate to room temperature for 20 min. Do not fix the tissue.
2. Pre-incubate the mounted sections in normal horse serum (diluted 1:100 in 0.01 M PBS) for 20 min.
3. Incubate with a cell type-specific primary antibody: e.g. CD31 (1:100 in PBS) to identify endothelial cells; AS02 (1:100 in PBS) for fibroblasts; although not described in this study, Ber-EP4 (1: 200 in PBS) can be used to identify epithelial cells; and NF200 (1:200 in PBS) can be used to identify nerves (see Note 5).
4. Wash 3× with PBS.
5. Incubate with biotinylated Universal Secondary Antibody (1:50 dilution in 1:100 normal horse serum in 0.01 M PBS) for 30 min.

6. Wash 3× with PBS.
7. Incubate for 30 min with Vectastain ABC-AP reagent 1:100 dilution solution A in 0.01 M PBS + 1:100 dilution solution B mix to be make up at the same time as step 5 in order to allow solution to stand for 15 to 30 min prior to use.
8. Wash 3× with PBS.
9. Incubate in alkaline phosphatase substrate solution (1:50 dilution of solution 1, with 1:50 dilution of solution 2 and 1:50 dilution solution 3, all in 5 mL 200 mM Tris–HCl, pH 8.2) for about 20–30 min. Observe the slides for the appearance of red staining—we use "Vector Red". Although other coloured substrates are available (for example, Vector Blue), the red stain was used as it contrasts well with the dark autoradiographic grains; Fig. 1.
10. Rinse in 0.01 M PBS for approximately 5 min.
11. Blow-dry slide-mounted sections at room temperature for approximately 15 min.

Immediately following the immunohistochemistry step, continue to the micro-ARG.

3.3. Micro-Autoradiography

1. Receptor-specific binding is determined by incubating the slide-mounted sections for 2 h at room temperature in a 50 mM Tris–HCl buffer, pH 7.4, 0.1% w/v bovine serum albumin, 5 mM $MgCl_2$, and 100 KIU aprotinin containing 150 pmol/L [^{125}I]-PD-151242 to identify ET_AR-binding sites and [^{125}I]-BQ3020 to identify ET_BR-binding sites. The specific activity of both compounds was 2,200 Ci/mmol/L (GE Healthcare, UK). The degree of non-specific binding for each radioligand has been established by incubating paired sections in the presence of 1 μmol/L unlabelled ET-1 (Bachem Fine Chemicals, Basel, Switzerland) (see Note 6).
2. After incubation, the sections are washed two times for 10 min in Tris buffer at 4°C and then dipped in distilled water at 4°C to remove buffer salts.
3. The slide-mounted sections are dried in a stream of warm air for approximately 15 min followed by approximately 15 min in cool air.
4. Leave to dry overnight at room temperature.
5. Fix the slides by placing them in an evacuated desiccator containing paraformaldehyde powder that vaporises at 80°C for 2 h (see Note 7).
6. The fixed sections should be allowed to cool under darkroom conditions.
7. Next, warm K2 nuclear emulsion diluted 1:1 with 2% v/v glycerol in distilled water to 42°C by placing in a container immersed

in a water bath; once the liquid has reached temperature, dip the slides in (see Note 8).

8. The emulsion-coated slides should be allowed to dry in the dark overnight and may be stored for up to 14 days at 4°C in lightproof boxes containing desiccant.
9. Finally, the emulsion is processed according to the manufacturer's instructions (5 min in Kodak D19 developer, dipped briefly in Ilfostop solution, 10 min in Ilford Hypam fixative diluted 1:3 in tap water).
10. Tissue sections are then stained with haematoxylin, dehydrated, and processed for histological examination (refer to chapter 1 on immunochemistry, by Brooks in this volume).

3.4. Densitometry of Macro-Autoradiographs

There are various scanning systems and imaging software available which can be used or adapted for quantitative assessment of autoradiographs (for robust methodology, see ref. 6). A basic in-house method involves digitising the images (Hewlett Packard scanner) and analysing by reading (and comparing) grey densities of the histograms in two different systems: Adobe PhotoShop and Corel Draw. In this densitometric analysis, each cm^2 can be divided into 255 points and each point given a comparative value from 0 to 250 (0 = black, 250 = white).

4. Notes

1. A higher degree of resolution is achieved with emulsion for two main reasons: (1) unlike film, the autoradiographic image remains in register with the underlying tissue and (2) the silver bromide crystal of emulsion is of a uniform shape and 0.2 μm diameter, whereas in film the crystals are irregular in shape and approximately 1.6 μm in diameter.
2. Transport specimens on dry ice or in liquid nitrogen to avoid unnecessary thawing followed by refreezing which destroys tissue architecture.
3. Polylysine coating enhances adherence of tissues on slides: this is important since the double-staining procedure is harsh and tissue lifting from the slide and sample loss may occur.
4. Store slides in pairs, placed back to back with tissue facing outwards; wrap tightly in aluminium foil, place in sealable polythene bag, label, and store at −70°C. This reduces tissue dehydration and/or water absorption.
5. Antibody titrations—for optimum staining, antibody concentrations were titrated in preliminary experiments.

Negative controls were incubated with PBS in the absence of the primary antibody.

6. Radioistopes must be handled and stored according to local regulations. Ensure that you are fully trained in the use of these chemicals; before ordering reagents, appropriate approval and risk assessments must be undertaken. The radioligand concentrations used in these experiments were determined using a classical pharmacological approach, where the preliminary saturation analysis was carried out to obtain the radioligand affinity (K_D) for the receptor-binding site under investigation.
7. Fixation in paraformaldehyde vapour preserves the tissue morphology and prevents diffusion of the radioligand from its receptor when dipping in molten nuclear emulsion.
8. In some cases, macro-autoradiographs may be generated on film for further densitometric analysis. Here, slide-mounted tissue sections are treated as described in the Notes 1–5 above. The slides were then apposed to Hyperfilm™ MP (GE Healthcare, UK) in X-ray cassettes for up to 14 days at 4°C in lightproof conditions and processed following the manufacturer's instructions. Sections may subsequently be stained with haematoxylin and eosin.
9. In combining these two techniques, a few methodological changes are made. In conventional IHC, tissue is generally fixed before proceeding with incubation in the primary antibody and subsequent processing. In our combined technique, fixation is performed in paraformaldehyde vapour after completion of IHC/ARG as fixation will reduce or abolish receptor binding to tissue sections if performed before ARG incubations. Also, in conventional ARG, an initial pre-incubation step is employed before incubating in radioligand. The purpose of this is to reduce endogenous transmitter levels that would interfere with radioligand binding to the receptor under investigation. Since tissue sections have been washed many times during the IHC process before the ARG, this step is no longer necessary.

References

1. Boast CA, Snowhill EW, Altar CA (1986) Quantitative receptor autoradiography, vol 19, Neurology and neurobiology. Alan R Liss, New York
2. Leslie FM, Altar CA (eds) (1988) Receptor biochemistry and methodology, vol 13, Receptor localization, ligand autoradiography. Alan R Liss, New York
3. Stewart MG (ed) (1992) Quantitative methods in neuroanatomy. Wiley, West Sussex
4. Wharton J, Polak JM (eds) (1993) Receptor autoradiography principles and practice. Oxford University Press, New York
5. Young WS III, Kuhar MJ (1979) A new method for receptor autoradiography: [3H] Opiod receptors in rat brain. Brain Res 179:255–270
6. Unnerstall JR, Niehoff DL, Kuhar MJ, Palacios JM (1982) Quantitative receptor autoradiography using [^{3}H] Ultrofilm: application to

multiple benzodiazepine receptors. J Neurosci Methods 6:59–73

7. Dashwood MR, Timm M, Muddle JR, Ong ACM, Tippins JR, Parker R, McManus D, Murday AJ, Madden BP, Kaski JC (1998) Regional variations in endothelin-1 and its receptor subtypes in human coronary vasculature: pathophysiological implications in coronary disease. Endothelium 6:61–70
8. Grant K, Loizidou M, Taylor I (2003) Endothelin-1: a multifunctional molecule in cancer. Br J Cancer 88:163–166
9. Hoosein MM, Dashwood MR, Dawas K, Ali HM, Grant K, Savage F, Taylor I, Loizidou M (2007) Altered endothelin receptor subtypes in colorectal cancer. Eur J Gastroenterol Hepatol 19:775–782
10. Inagaki H, Bishop AE, Eimoto T, Polak JM (1992) Autoradiographic localization of endothelin-1 binding sites in human colonic cancer tissue. J Pathol 168:263–267
11. Ali H, Dashwood M, Dawas K, Loizidou M, Savage F, Taylor I (2000) Endothelin receptor expression in colorectal cancer. J Cardiovasc Pharmacol 36:S69–S71
12. Egidy G, Juillerat-Jeanneret L, Jeannin JF, Korth P, Bosman FT, Pinet F (2000) Modulation of human colon tumor-stromal interactions by the endothelin system. Am J Pathol 157: 1863–1874

Chapter 10

Fluorescence *In Situ* Hybridization for Cancer-Related Studies

Lyndal Kearney and Janet Shipley

Abstract

Cytogenetic analysis of tumour material has been greatly enhanced over the past 30 years by the application of a range of techniques based around fluorescence *in situ* hybridization (FISH). Fluorescence detection for *in situ* hybridization has the advantage of including the use of a multitude of fluorochromes to allow simultaneous specific detection of multiple probes by virtue of their differential labelling and emission spectra. FISH can be used to detect structural (translocation/inversion) and numerical (deletion/gain) genetic aberrations. This chapter will deal with FISH methods to detect and localize one or more complementary nucleic acid sequences (probes) within a range of different cellular targets including metaphase chromosomes, nuclei from cell suspension, and formalin-fixed paraffin-embedded FFPE tissue sections. Methods for the efficient localization of probes to FFPE tissue cores in tissue microarrays (TMAs) are also described.

Key words: Fluorescence *in situ* hybridization, Formalin-fixed paraffin-embedded tissue, Tissue microarrays

1. Introduction

In situ hybridization describes the annealing of a labelled nucleic acid to complementary nucleic acid sequences in a fixed target (e.g., chromosomes, free nuclei, nuclei in tissue sections, and DNA) followed by visualization of the location of the probe. Since its development 40 years ago (1, 2) it has transformed into a highly effective and rapid technique of use in a range of studies such as characterizing chromosomal aberrations, gene mapping, and marker ordering, as well as expression studies. *In situ* hybridization originally used radioactively labelled probes, but the strict regulations on handling radioactivity, long exposure times, and

Miriam Dwek et al. (eds.), *Metastasis Research Protocols*, Methods in Molecular Biology, vol. 878,
DOI 10.1007/978-1-61779-854-2_10, © Springer Science+Business Media, LLC 2012

some practical difficulties with the use of radioactive labels limited the application of the technique. In the 1980s several methods using non-radioactive labelling were developed (3–7). The ease and effectiveness of fluorescence methods [fluorescence in situ hybridization (FISH)] in particular have now almost rendered the radioisotopic-based techniques obsolete. FISH has been developed to incorporate "chromosome painting" and analysis of the whole genome for aberrations using approaches of comparative genomic hybridization (CGH), multi-fluor FISH (M-FISH), and spectral karyotyping (SKY).

In FISH, the probes are labelled either directly with fluorochromes such as fluorescein isothiocyanate (FITC) or tetramethyl rhodamine that fluoresce at different wavelengths when excited by UV light, or indirectly with biotin or digoxygenin. After hybridization, excess probe is washed off at specific stringency, this is then followed by detection of indirectly labelled probes and the location of the probes using an epifluorescence microscope. The advantage of fluorescence detection for in situ hybridization includes the ability to use multiple fluorochromes thereby allowing simultaneous specific detection of multiple probes. This technique allows probes to be used with metaphase chromosomes, interphase nuclei, and released chromatin. FISH also enables numerical (deletion/gain) (8) and structural (translocation/inversion) aberrations to be detected.

This chapter deals with DNA probes hybridized to DNA targets primarily for the localization of genes and markers for refining cytogenetic analysis. We will describe the methods of one or more colour FISH on metaphase chromosomes, interphase nuclei, and nuclei in paraffin sections including tissue microarrays (TMAs) (9). M-FISH (10) and SKY (11) probes and analysis systems are currently commercially available from Vysis and Applied Spectral Imaging, Ltd. As these companies provide detailed protocols their methodologies will not be described here.

1.1. Probes

A wide variety of probe types are used to hybridize to target molecules of different preparations in a variety of FISH applications. Probes contain sequences homologous to either specific repetitive or unique regions of the genome. Probes that detect tandemly repeated sequences for centromere and telomere visualization (alpha satellites, beta satellites, and telomere probes) are useful for deducing aneuploidy and aberrations in tumour material either in metaphase chromosomes or, in some cases, interphase nuclei. Chromosome-specific paints are probes which hybridize to the whole or part of the chromosome and are generally derived from chromosomes of cells sorted on a fluorescence activated cell sorter (FACS) machine or chromosomes microdissected from fixed tissue preparations. These are commercially available, and the many unique regions in the probe bind specifically to targets along the length of the chromosome, while the non-specific repetitive regions are suppressed through the addition of unlabelled non-specific repetitive DNA, such as Cot-1 DNA. This approach is known as chromosomal in situ supression or competitive in situ suppression (CISS) (12, 13). Suppression of non-specific repeats is also required for unique sequence probes which contain such elements, for any genomic probe greater than a few kilobases (kb) in size. Unique sequence probes include cloned cDNA or more usually genomic DNA which vary according to the capacity of the cloning vector to accept different lengths of sequence. The following are the approximate insert sizes of commonly used probes for FISH:

- Plasmids 500 bp to 2 kbp.
- Phages 5–10 kbp.
- Cosmids 30–40 kbp.
- Fosmids-up to 50 kbp.
- PACs-P1-artificial chromosomes 130–150 kbp (14).
- BACs-bacterial artificial chromosomes 100 kbp to 1 Mbp (15).

The degree of homology between probe and target, and potentially less homologous targets, is a consideration in determining the stringency of hybridization and washing. Standard conditions are fairly stringent, but, for example, for cross-species hybridization the stringency may be lowered. Another consideration for unique sequence probes is the size of the insert as this impacts on the amount of probe required. As the insert size gets smaller, the detection of the signal gets more difficult. Mapping probes smaller than 2–3 kbp is difficult and needs statistical analysis.

Stronger signals may be obtained through indirect detection and extra rounds of signal amplification following labelling with small molecules such as biotin and digoxigenin.

1.2. FISH on Metaphase Chromosomes

FISH on metaphase chromosomes can be used to localize a new probe to a specific chromosome band, to reveal characteristic chromosomal abnormalities, for example translocations, to check the correct chromosomal localization of a probe, or to determine the linear order of probes on a chromosome. In relation to cancer and metastasis, metaphase FISH can be used to characterize specific chromosomal aberrations in tumour cell cultures and cell lines. To help researchers who are not familiar with karyotyping, the probe can be co-hybridized to a chromosome with a differentially labelled marker or centromere-specific probe. For linear ordering there should be at least 1–3-Mb distance between the two probes (16, 17): this allows clear visualization and, as chromatin folding may cause problems with relatively close probes, several metaphases should be analyzed or a high-resolution approach taken.

1.3. FISH on Interphase Nuclei

Interphase FISH can be applied to cases where metaphase material is not available, for example fresh or frozen solid tissue tumours and archival material embedded in paraffin wax. Since chromatin of interphase nuclei is less condense than that of metaphase chromosomes the mapping resolution is higher. Interphase FISH analysis of many nuclei can be used to determine probe order in the 100–1,000-kb range. The signal patterns change according to chromatin folding, the degree of condensation, and whether the cells have passed through the DNA-synthesis stage of the cell cycle when they were fixed in methanol:acetic acid. If the cells are captured before cell division, each chromosome will have only one chromatid, thus one would observe hybridization signals as single dots. After replication of DNA during the S-phase, the probe will hybridize to two sister chromatids per chromosome, giving signal doublets.

1.4. FISH on Formalin-Fixed Paraffin-Embedded Material Including Tissue Microarrays

Formalin-fixed paraffin wax embedded (FFPE) tissue is frequently the most convenient, readily available material from solid tumours. TMAs consisting of multiple cores from different FFPE tumour samples on one slide are increasingly used to facilitate high throughput analysis of hundreds of samples simultaneously (9). Control tissues can also be incorporated into the array. Consecutive sections from TMAs can be screened using in situ hybridization techniques for particular genomic changes previously identified using approaches such as array CGH analyses. Analysis of multiple genetic and expression changes makes it possible to build up complex information of molecular markers in the same set of tumour specimens and rapidly identify and validate clinico-pathological correlations.

1.5. Multi-fluor FISH and Spectral Karyotyping

Multi-fluor FISH (also known as multiplex FISH) and SKY are technological developments that allow the independent recognition of multiple chromosomes simultaneously. Previously only a few fluorochromes were available for FISH, thus the number of probes that could be tested per experiment by chromosome painting was limited. In 1990 Nederlof et.al. (18) described a method for detecting more than three target DNA sequences using only three fluorochromes, and this technique was termed "combinatorial labelling" (19, 20) or "ratio labelling" (21). A probe can be labelled with one or more fluorochrome-conjugated nucleotides and detected with one or more optical filters. The number of useful Boolean combinations of N fluorochromes is 2^{N-1}. Following this idea Speicher et al. (10) developed epifluorescence filter sets and computer software to detect and discriminate 27 combinatorially labelled DNA probes hybridized simultaneously in a method known as M-FISH. Whole chromosome paints for the twenty-two autosomes and two sex chromosomes are labelled combinatorially and hybridized to metaphase spreads. In this way, simple and complex translocations, interstitial deletions and insertions, chromosomal aneuploidy, and double minute chromosomes can be identified rapidly.

Spectral karyotyping is a similar approach where each human chromosome is observed in a different colour (11). SKY uses a spectral imaging approach that combines Fourier spectroscopy, charge-coupled device (CCD) imaging, and optical microscopy to measure simultaneously at all points in the sample emission spectra in the visible and near-infrared spectral range. The main difference between SKY and M-FISH is that in SKY the image is captured with a triple bandpass filter set that allows all dyes to be excited and measured together without image shift.

2. Materials

2.1. Growing Clones and Extracting Probe DNA

1. PAC, BAC or fosmid clones for the region of interest can be identified using the UCSC Human Genome Browser (http://genome.ucsc.edu) or Ensembl (http://www.ensembl.org/index.html) databases and obtained from the Sanger Institute Clone ordering system (http://www.sanger.ac.uk) or the BACPAC Web site at the Children's Hospital Oakland Research Institute (http://bacpac.chori.org/order_clones.php).
2. Luria Broth (LB) agar plates: LB agar is stored as a solid, melt the agar, and cool to 55°C, add appropriate antibiotic and pour into Petri dishes. Dry in a tissue culture hood for 15 min before use.
3. LB broth: store at 4°C until use.
4. Appropriate antibiotic.
5. Solution I: 50 mM glucose; 10 mM EDTA; 25 mM Tris–HCl, pH 8.0; 100 μg/ml of RNAse A; filter sterilize and cool to 4°C.
6. Solution II: 0.2 N NaOH, 1% SDS; freshly prepared.
7. Solution III: 60 ml of 5 M potassium acetate; 11.5 ml of glacial acetic acid; 28.5 ml water; filter sterilize and store at room temperature (RT). The resulting solution is 3 M with respect to potassium and 5 M with respect to acetate.
8. Phenol: Tris-buffered (Invitrogen Ltd., UK, Catalogue no. 15513-039). Phenol is highly corrosive. Wear gloves when using. Dispose of using approved waste-disposal protocol. After contact with skin, wash immediately with plenty of soapy water. Make phenol-chloroform immediately before use.
9. Chloroform–isoamyl alcohol mix: 24 volumes of chloroform to one volume of isoamyl alcohol (Fisher Scientific, Catalogue no. A/6960/08). Make immediately before use. Chloroform

(VWR International Ltd., Catalogue no. 100776B) is toxic if swallowed, is a respiratory system irritant, and is a contact hazard; wear gloves when handling and use in a chemical fume hood. Dispose of waste using approved disposal protocol.

10. 100% Isopropanol.
11. 70% Ethanol.
12. RNAse A: 10 mg/ml stock concentration.

2.2. Probe Labelling

2.2.1. Nick Translation

1. 1 μg probe DNA.
2. RNase A (Sigma-Aldrich, UK, Catalogue no. R-4642).
3. For direct labelling: 25 nmol Cy3 dUTP, Cy5 dUTP (GE Healthcare, UK), Spectrum Green™ dUTP (Vysis).
4. For indirect labelling: 1 mM biotin-16-dUTP or digoxigenin-11-dUTP (Roche Applied Science, UK).
5. 10× nick translation buffer: 0.5 M Tris–HCl, pH 8.0; 50 mM $MgCl_2$; 0.5 mg/ml nuclease-free bovine serum albumin (BSA).
6. 10× dNTP mix: 0.5 mM each dATP; dCTP; dGTP; and 0.1 dTTP (Roche Applied Science). 100 mM dithiothrietol (DTT).
7. DNase I stock solution: 10,000 U/ml (RNase free Grade 1 pure; Roche Applied Science, Catalogue No 4716728001).
8. DNase I dilution buffer: 50% v/v glycerol; 0.15 M NaCl; 20 mM sodium acetate pH 5.0.
9. 10 U/μl DNA polymerase I (New England BioLabs).
10. 0.5 M EDTA, pH 8.0.

2.2.2. Random Primer Labelling

1. BioPrime DNA Labelling System (Invitrogen Ltd., Catalogue no. 18094-011). The kit comprises: 2.5× random primers solution; 10× dNTP biotinylate mixture; control DNA; Klenow fragment DNA polymerase; stop buffer and distilled water. The kit may be stored at –20°C for up to a year.
2. 10× Digoxigenin dNTP random priming labelling buffer: 1 mM dGTP; 1 mM dATP; 1 mM dCTP; and 0.65 mM dTTP in distilled water. Aliquot into 50 μl amounts and store at –20°C for up to a year.
3. Digoxigenin-11-dUTP (25 nmol; Roche Applied Science, Catalogue no. 11093088910). Store at –20°C.
4. 1 μg/μl Human Cot-1 DNA (Invitrogen Ltd., Catalogue no. 15279-011) plus 10 μl salmon sperm DNA (Sigma-Aldrich, D7656): Mix 500 μl Cot-1 DNA (1 μg/μl) plus 10 μl salmon sperm DNA (10 μg/μl). Reduce the volume from 500 to 50 μl using a Microcon-YM30 filter (Millipore), or by precipitation. Aliquot and store at –20°C for up to 1 year.

2.2.3. Labelled Probe Purification

1. Microspin G50 columns (GE Healthcare, Catalogue no. 27-5330-02). Store at RT.
2. Sephadex column washing buffer: 10 mM Tris–HCl, pH 7.0; 1 mM EDTA.

2.3. Preparation of Metaphase Slides

1. Sodium heparin: stock concentration of 1 kU/ml, stored at 4°C.
2. Phytohaemagglutinin (PHA-M form): available dissolved in 5 ml distilled sterile water, stored in 1 ml aliquots at −20°C or at 4°C for 30 days (Gibco-Life Technologies, Catalogue no. 10576-023).
3. Culture medium: RPMI 1640, with 200 mM L-glutamine.
4. 25 cm^2 Cell culture flasks.
5. Foetal bovine serum (FBS).
6. Penicillin and streptomycin solution: prepared in a stock concentration of 10,000 IU/ml.
7. Colcemid solution: 10 μg/ml stored at 4°C.
8. Hypotonic solution: 0.075 M potassium chloride, warm to 37°C before use.
9. Fixative solution: one volume glacial acetic acid to three volumes methanol, prepare fresh and cool to 4°C before use.
10. Clean 76 mm × 26 mm glass microscope slides stored in 500 ml ethanol. Dry immediately with lint-free tissues before use.
11. Waterbath at 80°C.
12. Fine-tipped plastic pipettes.
13. Silica gel desiccant.
14. Versene (cell line culture harvest): 1:5,000 (1×) stored at 4°C (Gibco Life Science Technologies).
15. Trypsin/EDTA (cell line culture harvest): 0.5 g of trypsin and 0.2 g EDTA per litre of in modified Puck's Saline A (Gibco-Life Science Technologies).

2.4. Preparation of Interphase Nuclei from Frozen Tissue (Touch Preps)

1. Clean 76 mm × 26 mm glass microscope slides, store in 500 ml ethanol, and dry immediately with lint-free tissues before use.
2. Fixative solution: one volume glacial acetic acid to three volumes methanol, prepare fresh and cool to 4°C before use.
3. Silica gel desiccant.
4. L-15 (Leibovitz media) with glutamax-l (Sigma-Aldrich, Catalogue no. 31415-011) stored at 4°C.
5. Dulbecco's PBS: phosphate buffered saline (500 ml for ten slides) stored at 4°C.
6. Type H collagenase (Sigma-Aldrich, Catalogue no. C-8051): dissolve 1 g of powder in 10 ml L-15 and store at −20°C in 50 μl aliquots.

7. 1 M $MgCl_2$, pH 7.3; autoclaved.
8. Formaldehyde solution: store at 15–30°C (contains 10–14% methanol added as a stabilizer, Fisher Scientific, Catalogue no. F/1500/PB08). Toxic, use in fume cupboard!
9. 70, 90 and 100% v/v ethanol solution series.

2.5. Paraffin Sections

1. Freshly cut FFPE tissue sections and/or tissue.
2. Microarray sections, cut at 4–5 μm and mounted on positively charged slides to help prevent tissue detachment.
3. SPoT-Light tissue pre-treatment kit (Invitrogen Ltd., Catalogue no. 00-8401) stored at 4°C.
4. Digest All-3 pepsin solution (Invitrogen Ltd., Catalogue no. 00-3009) stored at 4°C.

2.6. Hybridization

1. 22 mm×22 mm number 1.5 coverslips.
2. Rubber cement.
3. Moist chamber: a plastic box with a sheet of paper towel moistened with water.
4. Ethanol.
5. Hybridization buffer: 10% w/v dextran sulphate; 2× SSC (see item 9 below); 50% v/v formamide; 1% v/v Tween 20, pH to 7.0. Store at −20°C as ready to use aliquots.
6. Human Cot-1 DNA (Invitrogen Ltd., Catalogue no. 15279-011)1 mg/ml.
7. Salmon sperm DNA (Sigma-Aldrich, Catalogue no. D7656) 10 mg/ml.
8. Pre-hybridization buffer: Mix 500 μl Cot-1 DNA (1 mg/ml) plus 10 μl salmon sperm (10 mg/ml). Reduce volume from 500 to 50 μl in a Microcon-YM30 (Millipore) or by precipitation. Aliquot. Store at −20°C for up to 1 year.
9. 20× SSC: 3 M NaCl, 0.3 M sodium citrate. For 500 ml: 87.6 g NaCl, 44.1 g Na citrate. Add some distilled water. Adjust to pH 7.4 with concentrated HCl before adding distilled water to 500 ml. Autoclave. Can be stored for several months at 4°C.
10. Deionized formamide (NBS Biologicals Ltd., UK, Catalogue no. 0606-500ML). Formamide is a carcinogen. Use in a fume hood. Wear gloves when handling. Dispose of using approved waste-disposal protocol.
11. Hybridization buffer solution: 60% v/v deionized formamide; 12% w/v dextran sulphate; 20× SSC 2.4×; 0.7 mM EDTA (stock 500 mM), pH 8.0; 400 μg/ml salmon sperm DNA (stock: 10 mg/ml). It is important the pH of the hybridization mix is adjusted to between pH 7.0–7.5. At less than pH 7.0 the sensitivity is significantly reduced, while at pH higher than

7.5 background staining is significantly increased. Mix at 4°C overnight and heat to 42°C with stirring for 20 min. Aliquot and store at −20°C for up to 6 months.

12. Deionized formamide should be used as it contains fewer impurities.

2.7. Post-hybridization Washes

1. Place 50% w/v formamide, 2× SSC pH 7.0 in a Coplin jar. Warm the solution in the Coplin jar to 42°C in a waterbath prior to use.
2. 2× SSC pH 7.0, autoclaved. Warm in a Coplin jar to 42°C in a waterbath prior to use.
3. PBS solution with 0.001% v/v of Igepal: 500 μl of Igepal detergent in 500 ml of PBS solution. Warm to dissolve Igepal, mix well, store at 4°C.
4. 70, 90, and 100% v/v ethanol.
5. DAPI solution: Vectashield mounting medium with DAPI, 4′,6-diamidino-2-phenylindole (1.5 μg/ml, Vector Laboratories Ltd., UK).
6. 22 mm × 50 mm number 0 coverslips.

2.8. Fluorescent Detection of Probes Labelled with Digoxigenin or Biotin

1. Blocking buffer: SSCTM. To 10 ml SSCT add 0.5 g nonfat milk powder. Solubilize at 42°C. Filter sterilize through a 0.4-μm filter. Use fresh.
2. Anti-digoxigenin–fluorescein 200 μg (Roche Applied Science, Catalogue no. 11207741910). Reconstitute with distilled water, aliquot into 0.5-ml volumes in microcentrifuge tubes. Can be stored with light excluded at −20°C for 1 year. Stable for a few months at 4°C. Before use, centrifuge at 10,000 × *g* for 5 min to remove any protein complex precipitates which may appear as speckled background staining and bright spots on the tissue when screening.
3. For each 22 mm × 50 mm detection area, take 0.75 μl carefully from the surface of the anti-digoxigenin-fluorescein and dilute in 150 μl SSCTM, to a dilution of 1:200. Make this dilution on ice no more than 5 min before use.
4. Streptavidin-Cy3 conjugate 1 mg/ml (Sigma-Aldrich, Catalogue no. S6402). For each 22 mm × 50 mm detection area, mix 0.75 μl streptavidin-Cy3 into 150 μl SSCTM. Make this dilution on ice no more than 5 min before use.
5. Antifade mounting medium and nuclear stain: Vectashield mounting medium with 1.5 μg/ml DAPI stain, 4′,6′-diamidino-2-phenylindole (Vector Laboratories Ltd., UK, H-3404). Protect from light. Store at 4°C.
6. Microscope coverslips:# 0 thickness 22 mm × 50 mm.

2.9. Image Capture and Analysis

Epifluorescence microscope equipped with appropriate filters in combination with a cooled CCD camera and suitable computer hardware and software.

3. Methods

3.1. Growing Clones and Extracting Probe DNA (See Note 1)

1. Streak bacteria on LB agar plates supplemented with appropriate antibiotic and incubate inverted at 37°C overnight.
2. Inoculate a single isolated bacterial colony into 10 ml of LB broth in a 50-ml Falcon tube, supplemented with appropriate antibiotic, and incubate with shaking at 225–300 rpm at 37°C for 16 h.
3. Spin culture at 1,500 × *g* for 10 min.
4. Discard supernant and drain pellet by inverting tube onto a paper towel (take care that the pellet does not slide down the side of the tube).
5. Re-suspend pellet by vortexing in 400 μl of ice-cold solution I. At this stage the cell suspension can be transferred to a 1.9-ml microcentrifuge tube.
6. Add 800 μl of solution II and invert the tube very gently to mix the contents. Leave at RT for 5 min. The appearance of the suspension should change from very turbid to almost translucent.
7. Add 600 μl of solution III slowly while gently shaking the tube. A thick white precipitate of protein and *Escherichia coli* DNA will form. Place tube on ice for at least 5 min.
8. Centrifuge at 1,500 × *g* for 10 min at 4°C.
9. Remove tubes from centrifuge and put on ice. Transfer supernatant into a fresh microcentrifuge tube. Try to avoid picking up any white precipitate material.
10. Add an equal volume of phenol/chloroform: isoamyl alcohol. Mix and spin at maximum speed in a microcentrifuge for 5 min.
11. Transfer the supernatent into a fresh microcentrifuge tube and add an equal volume of chloroform: isoamyl alcohol, mix well and spin at maximum speed in a microcentrifuge for 5 min.
12. Transfer the supernant into a fresh microcentrifuge tube.
13. Add isopropanol, 0.7 of the volume, and place tube on ice for at least 5 min or leave at –20°C overnight.
14. Spin at maximum speed for 10 min at 4°C.
15. Discard the supernatant and add 0.5 ml of 70% v/v ethanol, invert the tube several times to wash the DNA pellet. Spin at a maximum speed for 5 min at 4°C. The DNA pellet should face toward the outside of the centrifuge.

16. Discard as much of the supernatant as possible. Occasionally, pellets will become dislodged from tube so it is better to aspirate off the supernatant carefully rather than pour it off.
17. Air-dry the DNA pellet at RT.
18. Re-suspend the DNA pellet in 100 μl of water with occasional tapping. DNA can be stored at −20°C if not used immediately.
19. Incubate the tube at 65°C for 15 min. After it cools down, measure the DNA concentration (see Note 2).

3.2. Probe Labelling

3.2.1. Nick Translation (See Note 3)

1. Firstly, treat the DNA with RNase as follows: For each 1 μg DNA add 200 ng RNase A. Incubate for 30 min at 37°C in a waterbath. Place the tube on ice.
2. Add the following (in order) to a 1.5-ml Eppendorf tube on ice:
   - 1 μg Probe DNA (RNA free).
   - 1 μl (1 nmol) Cy3, Cy5 dUTP or 1 μl (1 mM) biotin-16-dUTP, digoxigenin-11-dUTP, or 2.5 μl Spectrum Green or Spectrum Orange™ dUTP.
   - 5 μl 10× nick translation buffer.
   - 5 μl dNTP mix.
   - 5 μl 100 mM DTT.
   - Sterile distilled water to make up to a final volume of 50 μl.
   - 2–4 μl DNase I (dilute stock 1/20 in water just before use: discard after use).
   - 1 μl 10 U/μl DNA Polymerase I.
3. Mix well.
4. Incubate the tubes at 16°C for 2 h.
5. Stop the reaction by placing the tubes on ice.
6. Run a 5 μl aliquot on a 2% TBE agarose gel at 100 V for 1 h, with base-pair marker (e.g., *Phi*Xi74 *Hae*III). The optimum size of the labelled fragments is 100–500 bp (average 300). If the DNA is larger than this, re-incubate with DNase for a further 30 min. If the DNA is smaller than 100 bp, it needs to be re-labelled, using a lower concentration of DNase 1. When the DNA is the correct size, stop the reaction by adding 1.5 μl 0.5 M EDTA and place on ice.
7. Store at −20°C until required.

3.2.2. Random Primer Labelling (See Note 4)

1. Labelling with biotin is carried out with the reagents included in the BioPrime DNA system. All DNA probes and labelling reagents should be kept on ice and centrifuged briefly before use.
2. Label each DNA probe as follows: to a 0.2-ml reaction tube add 300 ng amplified probe, adjusted to 24 μl with distilled water. On ice, add 20 μl 2.5× random primers solution. Mix well.

Denature the DNA mixture by heating in hot block at 95°C for 5–10 min. Complete DNA denaturation is essential for efficient labelling. Cool immediately on ice to retain DNA as single strands.

3. Briefly centrifuge the tubes and then, on ice, add 5 μl 10× dNTP biotinylate mixture to label the denatured DNA with biotin, or add 1.75 μl Digoxigenin-11-dUTP pH 7.5 and 5 μl 10× digoxigenin dNTP buffer for digoxigenin labelling. Adjust volume to 50 μl with distilled water.
4. Mix thoroughly but carefully to avoid shearing the DNA.
5. Add 5 U Klenow fragment DNA polymerase. Mix gently but thoroughly then centrifuge at 10,000 × *g* for 15–30 s.
6. Incubate at 37°C for 3 h.
7. Add 5 μl stop buffer and mix carefully.
8. The fragment size and incorporation of the labelled product can now be checked using a 0.7% w/v agarose gel (see Note 3). Product quantification should be carried out after the purification step.

3.2.3. Purification of Labelled Probe

Purify the labelled probe using the commercially available Microspin G50 columns following manufacturer's instructions:

1. Carefully pipette the labelled DNA to the centre of the prepared column. It is important to ensure that all the liquid passes through the Sephadex matrix so that all unincorporated nucleotides are removed.
2. Following centrifugation at 400 × *g* for 5 min collect the purified probe in a pre-labelled tube.
3. For probes labelled by random priming, the concentration can be checked using a 0.7% w/v agarose gel. Probe concentration and quality can also be determined using a nanodrop spectrophotometer. Probes labelled using random priming would be expected to yield 5–10 μg DNA representing a 10- to 20-fold increase in DNA. If there is no amplification, the labelling reaction has not worked.

3.3. Preparation of Target DNA

3.3.1. Metaphase Chromosomes from Lymphocytes

Appropriate precautions should be taken before starting work with biological material. All the steps prior to fixing the cells should be carried out in a tissue culture hood.

1. Add 10 ml of FBS, 1 ml penicillin and streptomycin solution, 1 ml PHA into 100 ml of RPMI medium and mix.
2. Take 5 ml of normal or test blood from a volunteer into a tube containing 50 μl of sodium heparin, roll tube gently to mix. Five millilitre of blood would give ten flasks of culture from which approximately 200 slides can be made.

3. Aliquot 10 ml of media into each cell culture flask and add 0.5 ml blood. Mix.
4. Gas all flasks briefly (10–20 s) with 5% CO_2.
5. Incubate cultures at 37°C for 72 h.
6. Between 30–90 min prior to harvesting (see Note 5), add 100 μl of colcemid solution and return to 37°C (colcemid solution arrests cell division).
7. Warm 0.075 M KCl to 37°C. Make fresh fixative and cool to 4°C.
8. After incubation with colcemid, transfer cultures to capped 10 ml centrifuge tubes and centrifuge at 500×*g* for 5 min.
9. Discard supernatant using a plastic pipette and re-suspend the pellet in the remaining supernatant by flicking gently.
10. Add 10 ml pre-warmed hypotonic 0.075 M KCl (hypotonic solution causes the cells to swell by osmosis making them burst easily when dropped onto slides), mix by inverting and incubate at 37°C for 8 min.
11. Centrifuge at 500×*g* for 5 min and discard supernatant, leaving the interphase layer which contains some of the white cells.
12. Flick the tube gently to re-suspend the pellet. Continue flicking the tube while adding cold fixative, dropwise, to pellet. After carefully adding the first 5–10 drops, fill the tube up to 10 ml with fixative. Mix by inverting the tube gently. Make sure the pellet is re-suspended and does not have any cell clumps since after fixation single cells cannot be retrieved from the clump.
13. Spin at 500×*g* for 10 min.
14. Discard supernant with pipette (into appropriate fixative waste, not into chloros).
15. Repeat steps 12–15 up to a further six times.
16. Re-suspend the pellet in a few drops of fixative (0.5–1 ml). After this, the fixed cells can be kept at 4°C for a maximum of a week or at -20°C for a month. Before making slides after storage, the cells should be washed in fresh fixative at least one more time.
17. Take a slide out of ethanol and wipe dry with lint-free tissue (e.g. Kimwipes).
18. Hold the slide over an 80°C waterbath for 2 s.
19. Take away from the waterbath and hold the slide at an approximate 40° angle.
20. Drop the fixed cell suspension on to both halves of the moistened slide surface from a distance of about 30 cm (i.e. two separate drops onto each slide). See Note 6 for alternative conditions for slide making. Usually the quality of metaphase

preparations change from one flask to next, thus it is advised to make separate batches of slides from each flask.

21. Allow the slides to dry at RT.
22. Observe one slide made from each tube under a phase contrast microscope (×20 objective) to assess the approximate number of metaphases and amount of cytoplasm on each half of the slide. 5–10 metaphases per field with no cytoplasm around them is an ideal slide. If there are too many metaphases on the slide, more fix can be added to the tube until the required density is achieved (see Note 7 for reducing cytoplasm).
23. Repeat steps 17–21 until the required number of slides have been prepared.
24. Slides should be stored at −20°C in a box containing desiccant (e.g., silica gel).

3.3.2. Metaphase Chromosomes from Cell Line Cultures

1. Add 25 μl colcemid solution into a cell culture flask (25 cm^2 flasks with 5 ml media) and incubate at 37°C for up to 2 h.
2. Warm 0.075 M KCl to 37°C. Make fresh fixative and cool to 4°C.
3. Steps 3–6 are for adherent cell cultures. For suspension cell cultures proceed directly to step 8.
4. Transfer the medium from the flask into a 10 ml centrifuge tube avoiding the cell layer.
5. Put 2–3 ml of versene solution into the flask, replace the cap, and swirl to wash off the serum from all the surfaces of the flask (serum stops the action of trypsin/EDTA). Discard the versene.
6. Put 2–3 ml trypsin/EDTA solution into flask, swirl, and incubate at 37°C for 3 min.
7. Tap flask 4–5 times (observe under phase contrast microscope to see that all cells are detached from the flask). Pour the contents of the centrifuge tube back into the flask (to stop the action of trypsin).
8. Transfer the suspension into the centrifuge tube.
9. Spin at 500 × *g* for 5 min.
10. From here onward the fixation and slide making is the same as above (see section for metaphase chromosomes from lymphocytes: steps 9–24 inclusive).

3.3.3. Preparation of Interphase Nuclei from Suspension Cultures

This method is usually used in cases where tumour material cannot be cultured to obtain metaphase chromosomes e.g. pleural effusion. The same procedure is followed as in the method for fixing metaphase chromosomes.

1. Centrifuge the sample at 500 × *g* for 5 min.

2. Follow steps 9–24 in the section for metaphase chromosomes from lymphocytes: steps 9–24 inclusive.

3.3.4. Touch Preparations from Frozen and Fresh Tissue

1. Take slides out of the 100% ethanol and wipe dry with lint-free tissue.
2. For frozen tissue: keep the tissue on dry ice until use. With alcohol-cleaned forceps remove the tissue from the vial and cut off a small piece with a sterile scalpel. With forceps take the tissue and apply gently to areas of hybridization on clean slides. Try to rotate the tissue so that large clumps of thawed tissue are not deposited onto the slide. If the tissue sample is stuck in the vial, immerse the vial in liquid nitrogen and break the vial (after freezing) with a pestle and mortar.
3. Touch the tumour sample lightly onto the slides at RT.
4. Immediately place the slides in cold methanol for 20 min. Be careful not to let slides dry out. Let slides air-dry for 1 min. Fix tissue for 20 min in a Coplin jar filled with fixative cooled to –20°C.
5. Transfer the slides into fresh fixative at RT for 20 min.
6. Allow to air-dry. When observed under phase contrast microscope, try to find dull grey cells which are good for hybridization.
7. Store slides at –20°C in a box containing silica gel until processing.
8. Processing: mix 50 μl of collagenase solution in 50 ml of L-15 medium and warm to 37°C in a Coplin jar.
9. Treat the slides by immersing in collagenase for 5 min at 37°C.
10. Rinse in Dulbecco's PBS with 50 mM $MgCl_2$, pH 7.3; for 10 min at RT, twice.
11. Fix the slides in 1% formaldehyde in PBS with 50 mM $MgCl_2$, pH 7.3; for 10 min at RT. Formaldehyde is toxic therefore should be used in a fume cupboard.
12. Rinse the slides in PBS with 50 mM $MgCl_2$ for 2 min at RT three times.
13. Dehydrate by processing through 70 ethanol, 90 ethanol, and 100% v/v ethanol for 2 min each. Air-dry.
14. The slides are now ready for denaturation. They can be stored at –20°C with desiccant until use.

3.3.5. Preparation of Target Formalin-Fixed Paraffin-Embedded Tissue Slides

1. The pre-treatment of FFPE tissue, including TMA slides, can be carried out either using SPOT-Light tissue pretreatment and Digest-All kits (as described below taking about 3 h) or by a standard pre-treatment and protease digestion method (see Note 8).

2. Tissue pre-treatment is important to maximize the accessibility of the target to the probe (9). Pre-treatment will help break DNA cross links formed during tissue fixation. However, variations in the processes of fixation and embedding mean that some tissue samples will not give good FISH results regardless of how the pre-treatments and proteolytic enzyme digestions are adjusted.
3. It is important that the tissue section does not dry, unless specified, during the following procedures.
4. Pre-heat 50 ml xylene to 55°C in a lidded Coplin jar, and a further 200 ml of xylene in a bottle to 55°C. Use a waterbath within a fume hood. It is important that the temperature reaches 55°C.
5. Slides should be placed on a hotplate set at 100°C for 30 min, to melt the paraffin wax, immediately prior to deparaffinization in xylene. The complete removal of paraffin wax is important as residual wax can lead to weak probe signals.
6. Assemble the equipment for the pre-treatment of the slides. Put approximately 1 L of water in a 2 L glass beaker. Add a stirrer bar and a raised plastic platform that supports one or two Coplin jars. Fill the Coplin jar with distilled water and place on the platform above the stirrer bar. Cover the beaker with foil and place on the hot-plate/stirrer. Set hot-plate to maximum heat. It is important that the temperature of the water in the Coplin jar reaches at least 95°C.
7. Meanwhile de-paraffinize the slides by washing 2–3 times for 5 min each in xylene pre-heated to 55°C. With this, and all procedures, it is essential that slides placed against the edge of the Coplin jar have the tissue section facing inwards.
8. Wash twice for 3 min in 100% ethanol at RT.
9. Place each slide individually on a hot-block at 98°C, pipette 250 μl ethanol on to the tissue area, covering immediately with a 22 mm × 50 mm #1.5 glass coverslip. Heat for 30 s, or until the ethanol has evaporated, when the tissue will appear white. This procedure helps to chemically age the slide but it is important not to over dry the slide as the coverslip will stick to the tissue.
10. Remove the coverslip and cool the slide briefly before rehydrating the tissue by soaking in a Coplin jar of distilled water at RT for 2 min. Wash twice more in distilled water for 2 min at RT. The slides can be kept moist in the emptied Coplin jar for a short while. The tissue should not be allowed to dry out.
11. Microwave 50 ml of SPOT-Light pre-treatment buffer to boiling in a 200 ml bottle with lid resting loosely on top. Heating will take approximately 1–1.5 min. Carefully empty the hot

water from the Coplin jar into the beaker of boiling water and replace with the heated pre-treatment buffer, cap and allow to heat for 5 min. It is important not to boil the pretreatment buffer for too long before adding the slides as the buffer deteriorates at this temperature.

12. Incubate the slides in pre-treatment buffer at ≥98°C for 15 min. Treat a maximum of three slides at one time to avoid lowering the buffer temperature when they are added. Pre-treatment of slides needs to be optimized. Inadequate tissue permeabilization can give rise to paraffin auto-fluorescence and weak probe signals as the probes are unable to access the target DNA sequences.
13. Following incubation, wash twice for 5 min in a clean Coplin jar with distilled water. Gently dab the slide edges with a tissue to remove excess water and place the slide on a rack over a waterbath set at 28°C to warm for 5 min. At the same time warm 300 μl of Digest All-3 Pepsin solution per 22 mm × 22 mm hybridization area, in capped microcentrifuge tubes floated in the 28°C waterbath for 5 min.
14. Add 300 μl warmed pepsin solution to each slide, incubating on a level rack over the waterbath at 28°C for 4–10 min depending on tissue type (see Note 9).
15. Drain the pepsin solution from the slide and soak in distilled water for two 3 min washes. Dab off the excess water from the slide edge with a tissue. The slide is now ready for immediate hybridization.

3.4. Hybridization

3.4.1. Non-paraffin Slides

1. All slides made from fresh material, except touch imprints, should be baked at 65–68°C for 2 h before denaturation. Carry out baking before proceeding with denaturation.
2. Pre-heat denaturation solution to 73 ± 1°C in a Coplin jar in a waterbath. Make sure the temperature of the denaturation solution is above 72°C before slides are immersed.
3. Immerse the slides into denaturation solution for 2–3 min (see Note 10).
4. Transfer slides into 70% v/v ethanol for 3 min, then into 90% v/v ethanol for 3 min and 100% v/v ethanol for 3 min. Change the 100% ethanol and leave slides in for an extra 3 min.
5. Put slides onto a hotplate at 40°C until probe is ready.
6. Add the appropriate amount of probe per hybridization in to a 1.5 ml microcentrifuge tube as listed below. Prepare highly repetitive probes in a different tube than single-copy sequence probes.
   - 100–200 ng for plasmids, fosmids, PACs, and BACs (except highly repetitive sequence probes e.g. centromeric, telomeric probes).

- 80–100 ng for cosmids.
- 20–50 ng for highly repetitive sequence probes (e.g. centromeric, telomeric probes).

7. Add 5 μg (5 μl of 1 mg/ml stock) of Cot-1 DNA into each tube except highly repetitive sequence probes. Mix.
8. Add two volumes of 100% ethanol. Mix.
9. Dry probes in a rotary vacuum dryer (spin vac).
10. The total volume of probe to apply per half of the slide (per hybridization) is 10 μl. If two probes are to be used in one hybridization, then re-suspend the probes accordingly, for example, a single hybridization of centromeric probe and single-copy sequence requires that you re-suspend a single-copy sequence probe in 8 μl of hybridization buffer per hybridization and centromeric probe in 2 μl of hybridization buffer per hybridization.
11. Denature the probe DNA at 75–80°C for 5–6 min.
12. Chill the DNA on ice and centrifuge briefly to ensure the liquid is at the bottom of the tube.
13. Pre-anneal repetitive sequences by incubation at 37°C for 20 min.
14. Mix appropriate amounts of the two probes and apply onto one half of the previously denatured slide. Cover with a 22 mm × 22 mm no. 1.5 coverslip (take care not to trap any air bubbles underneath, if so ease them out using a Gilson pipette tip) and seal with rubber cement.
15. Place the slides in a moist chamber and incubate at 37°C for 24 h. The slides can be incubated for longer but there is a risk of increased background signal.
16. Proceed to post hybridization washes.

3.4.2. Paraffin FISH

1. Prepare the probes for two-colour FISH by combining 300 ng digoxigenin labelled and biotin labelled probe with 0.5 μl 10 mg/ml Cot-1 DNA; 10 mg/ml salmon sperm DNA and 9.6 μl hybridization buffer; per 22 mm × 22 mm hybridization area. The size of the hybridization area will depend upon the size of the section or array, and the amount of probe added should be adjusted accordingly. It is very important that all components are thoroughly re-suspended and mixed before use. Store on ice until the slide is ready for hybridization. If the probe is incompletely dissolved in the hybridization buffer this can give rise to bright background spots visible during the screening step.
2. Centrifuge the probe for 5 s at 10,000 × *g* and carefully apply to the centre of the 22 mm × 22 mm coverslip. Invert the slide

gently onto it. Carefully remove any air bubbles by gently depressing the coverslip. Take care not to force the hybridization buffer out of the side of the coverslip. Air bubbles will result in weak or no signal. Repeat for each hybridization area. Seal with rubber cement or place in a metal hybridization chamber humidified with 300 μl 6× SSPE or 2× SSC.

3. Co-denature the probe and target DNA on a hot block set at 98°C for 7–10 min (see Note 11). If the slides were sealed with rubber cement re-apply as this will have dried. Complete DNA denaturation is essential for annealing and efficient labelling. If using commercially labelled probes, denature for the temperature and time recommended by the manufacturer.
4. Place the rubber cement sealed slides in a humidified container on gauze or paper towel moistened with 6× SSPE or 2× SSC. Protect from light and incubate at 37°C overnight to hybridize the probe to target DNA. Hybridization can be undertaken for a longer time but this may increase the background staining.
5. Programmable denaturation and hybridization instruments are available.

3.5. Post-hybridization Washes

Posthybridization washes are performed at specific stringencies to wash off excess unbound probe. Insufficiently stringent washes will allow non-specific probe binding and background fluorescence. However, with targets exhibiting poor homology care should be taken not to wash off the probe. The washes described below are for solutions containing formamide. Alternative high-temperature, stringent SSC washes can be used (see Note 12).

1. Prepare equipment and washes in advance. Warm 50-ml Coplin jars containing freshly prepared 50% w/v formamide, 2× SSC v/v, pH 7.0; and 2× SSC v/v, pH 7.0; to 42°C in a waterbath. Check the temperature of the solution with a thermometer before use. Warm a humidified box to 37°C in an incubator. Prepare SSCT at RT and store SSCTM and antibodies on ice. Washes and incubations should be performed in subdued light to minimize background staining. Insufficiently stringent washes may give background staining or non-specific binding of probe.
2. Remove the slide from the hybridization chamber after the overnight incubation step and carefully peel off the rubber cement. The coverslip will often lift off at the same time, but if not, gently dip the slide in 2× SSC at 42°C thereby allowing the coverslip to fall off without scratching the tissue. With all washes, it is important that there is enough liquid in the Coplin jar to completely cover the slides.
3. Wash in 50% w/v formamide, 2× SSC v/v, at 42°C for 5 min, agitating the slides gently during the wash step or using a shaking waterbath. Repeat this wash twice more.

4. Monitor the wash solution temperature as it should not fall below 42°C.
5. Wash twice in 2× SSC at 42°C for 5 min, agitating as above.
6. Wash once in SSCT at RT for 3 min. Drain and blot the excess liquid from the slide edge with an absorbent tissue and proceed to the detection step.
7. For directly labelled probes apply 10 μl of DAPI solution and cover with #0 coverslip. Take care not to trap any air bubbles. Carefully ease them out using a Gilson pipette tip.
8. For indirectly labelled probes proceed to Subheading 3.6.

3.6. Indirect Fluorescent Detection of Probes Labelled with Digoxigenin or Biotin

Indirect fluorescent probe detection amplifies the probe signal. The method described here uses anti-digoxigenin-FITC and streptavidin-Cy3 for dual colour probe detection. Conjugated fluorescent dyes are light sensitive and should be used and stored in subdued light.

1. Pipette 150 μl SSCTM blocking buffer onto a 22 mm × 50 mm #1.5 glass coverslip. Invert the slide onto the coverslip with blocking buffer and immediately, carefully, re-invert. Gently remove any air bubbles. Repeat for each slide. This blocking step reduces non-specific protein binding which causes high background fluorescence.
2. Place slide in a humidified box protected from light and incubate at 37°C for 20 min. Ensure box is kept level so that the coverslip remains in place and the solution is evenly distributed across the tissue.
3. Dip the slide a few times in SSCT to remove the coverslip without damaging the tissue and then transfer the slide to a clean Coplin jar containing SSCT. Incubate for 3 min at RT.
4. Drain the slide but do not allow to dry out. Pipette 150 μl anti-digoxigenin-FITC in SSCTM on to a 22 mm × 50 mm #1.5 glass coverslip. Invert the tissue section on to it. Carefully re-invert the slide. Incubate for 20 min in a humid box at 37°C, again ensuring the box is level.
5. Remove the coverslip as above. Wash three times for 2 min each. Wash in Coplin jars containing SSCT at RT.
6. Repeat steps 4–5 with the second antibody, for example, streptavidin-Cy3. Drain the slide. Wash twice in PBS solution for 5 min at RT. Drain. Dehydrate through an ethanol as before.
7. Air-dry in subdued light e.g. in a drawer or under a foil cover. Mount in Vectashield anti-fade medium and stain with DAPI by carefully pipetting a 10-μl drop per hybridization area on to each half of a 22 mm × 50 mm #0 glass coverslip. Invert and

lower the tissue onto the mounting medium and then carefully re-invert the slide.

8. Allow at least 1 h for DAPI stain to penetrate the tissue before screening. Slides can be stored at 4°C, protected from light. The screen should usually be performed within 7 days but slides have been successfully screened after storage for several weeks.

3.7. Image Acquisition and Analysis

1. The slides are viewed using a fluorescence microscope with filters appropriate to the fluorochromes used—excitation at 543 nm induces Cy3 red emission, FITC at 530-nm green emission, with excitation at 364 nm inducing DAPI fluorescence (blue emission). Optimal filters are important to obtain maximum signal intensity. The filters can degrade with age and use and they should be checked to ensure continued maximum light transmission. The microscope should be linked to a computer system with an appropriate software package for capturing digital images. The procedure for image capture will vary accordingly. FISH images are usually captured using a ×63 or ×100 oil immersion lens (see Note 13).
2. TMAs can be screened more easily using an automated capture system. Automated systems usually capture using a low power objective (×40 as opposed to ×100) and increase the image size digitally which can give lower resolution. The length of TMA scanning time required, and therefore prolonged exposure to UV light, can cause fluorochrome bleaching and degrading of the FISH signals; even with the use of DAPI anti-fade mounting medium.
3. Screening and analysis time depends on the quality of signal obtained and the size of the sample. It is realistic to expect reliable results in excess of 80% of samples. However, if hybridization is poor, or if further probes are to be hybridized to the same slide to maximize the use of the tissue, treatments may be employed for destaining and reduction of autofluorescence.
4. A minimum of 100 distinct tumour nuclei should be scored and compared with analysis of a similar number from appropriate normal control tissue. This is important to establish cut-off values for different probes and signal patterns so as to guard against scoring events occurring by chance, or ploidy variation. Analysis problems may also include cellular heterogeneity in tissues, truncated cells, and differences in hybridization efficiencies between samples as well as tumour and control tissue. Areas where borders of individual nuclei are not clearly defined or where high cellular density causes nuclei to overlap should not be analyzed. Care should be taken not to overlook signal that may be in a different focal plane and to exclude

background fluorescence in more than one channel which may appear as signal. The results of scoring and analysis should be checked blind by a second experienced operator.

4. Notes

1. This is essentially an alkaline lysis mini-prep method that can be scaled up if necessary. It works well for plasmids, cosmids, fosmids, PACs, and BACs.
2. As the amount of probe is important, it is essential for accurate measurement of the DNA concentration. For the accurate quantitation of low concentrations of DNA a fixed wavelength fluorometer can be used. Alternatively, DNA concentration can be measured using a nanodrop spectrophotometer, or estimated on a gel against known amounts of DNA ladder.
3. One of the critical factors for success of FISH is the fragment size of the labelled probe. Nick translation labelling allows the end fragment size to be controlled by the amount of DNase I in the reaction mix. The optimum size range is 10–500 bp with an average size of 300 bp.
4. Random priming is the preferred labelling method for FFPE material as it can label even very short DNA and generate sensitive, homogeneously labelled DNA probes. The amount of newly synthesized DNA can exceed the amount of template, so effectively amplifying the amount of probe.
5. Although colcemid time can be varied between 30 and 90 min to obtain more cells in metaphase, longer incubation times will give more condensed chromosomes. Therefore 30 min of colcemid incubation should be ideal for adequate numbers of metaphase spreads.
6. The quality of the metaphase spreads is critical and depends on the humidity and temperature of the laboratory, how the material was processed and the way the slides are made. Extremely dry conditions will prevent the chromosomes spreading and very humid conditions give rise to what is known as "chromosome soup".
7. The following conditions are recommended:
   Slides can be humidified by placing in ethanol followed by iced water, frosting them in freezer, or holding them over a water-bath at 80°C before dropping the cell suspension. After the cells have been dropped, the slides can be dried by putting them onto a hotplate at 54°C, inside a 37°C incubator in a moist chamber, or leaving them at RT. We find that holding

the slide over the waterbath and then drying them at RT provides good quality metaphases.

The presence of excess cytoplasm covering the metaphase chromosomes or nuclei reduces the effect of denaturation and signal hybridization, and may increase the background signal. The fixative can be modified by increasing the ratio of glacial acetic acid up to 1.5 and reducing the ratio of methanol to 2.5. This should reduce the amount of cytoplasm on the slide. Pre-treatment with proteinase K and pepsin may also help to reduce the effect of cytoplasm.

8. Sodium thiocyanate and pepsin can be used as an alternative tissue pre-treatment step.
   (a) In this case, warm 1 M sodium thiocyanate to 45°C, warm 9% w/v NaCl pH 1.5–37°C immediately before use prepare 0.5% w/v pepsin solution using warmed 9% w/v NaCl pH 1.5.
   (b) Following de-paraffinization, incubate slides in a Coplin jar containing 1 M sodium thiocyanate solution, warmed to 45°C, for 15–30 min according to tissue type and section thickness. Agitate the slide gently during the incubation step. Discard solution after use. Rinse in distilled water.
   (c) Incubate slide at 37°C in freshly diluted 0.5% w/v pepsin solution for 15 min (pepsin is added to warmed 9% w/v NaCl pH 1.5 solution immediately before use).
   (d) Wash once in Hanks' Buffered Salt solution or PBS at RT.
   (e) Rinse in distilled water. Dehydrate through an ethanol series (70, 90 and 100% v/v) and proceed to hybridization.
9. The protease treatment times vary according to the tissue type e.g. 6 min for breast tissue, 5 min for prostate tissue, and 4.5 min for testis tissue. Pepsin digestion helps to remove cytoplasm and permeabilize tissue, allowing improved target signal detection. Tissue unmasking is one of the most critical steps for sensitive detection but in each step of the tissue preparation there are many variables that could affect sensitivity and increase the background staining. Pepsin is used as the protease treatment as it is easily inactivated and allows the reaction to be carefully controlled. It better preserves cellular morphology and nuclear structure but over-digestion may lead poor morphology, increased background, loss of signal, or loosening of the tissue from the slide.
10. The denaturation of the target DNA on the slide should be optimized. Under- or over-denaturation will lead to aberrant signal visualization and chromosome morphology, especially on metaphase chromosomes. "Swollen" chromosomes which

do not take up DAPI counterstain or nuclei with a wrinkled appearance are an indication of over-denaturation. Bright DAPI with very good chromosome morphology but poor signal indicates under-denaturation.

11. Slides and probes can be denaturated separately:
    (a) In a waterbath preheat a Coplin jar containing 50 ml 70% v/v formamide, 2× SSC, pH 7.0; DNA denaturation solution until the temperature of the liquid is 72°C.
    (b) Prepare fresh 70% v/v ethanol and cool to 4°C for use in slide dehydration.
    (c) Incubate the slide in heated denaturation solution for approximately 2 min—older slides may need denaturation of up to 4 min.
    (d) Immediately transfer to a Coplin jar containing 70% v/v ethanol at 4°C for 3 min to prevent the DNA re-annealing. Complete the dehydration through 90% v/v, and 100% v/v ethanol for 3 min each at RT.
    (e) Air-dry the slide and keep warm, e.g. on a hot-plate, at 37°C until the probe is added.
    (f) Denature the probe by heating the tubes in a preheated waterbath, or hot block, at 75–80°C for 5 min.
    (g) Chill on ice for 30 s and microcentrifuge briefly. Pre-anneal repetitive sequences at 37°C for 30 min to 3 h before applying to denatured slide.
    (h) Centrifuge the probe for 5 s at 10,000×*g* and carefully apply the probe/hybridization mix to the centre of a 22 mm×22 mm coverslip. Invert the slide gently on to the coverslip. Repeat for each hybridization area. Remove air bubbles by gently depressing the coverslip for example, with blunt forceps or a micropipette barrel. Take care not to force the hybridization buffer out of the side of the coverslip.
    (i) Seal the coverslips with rubber cement and incubate overnight at 37°C in a humid box.

12. Alternative high-temperature washes can be used for directly labelled probes, including those that have been commercially labelled.
    (a) Carefully remove coverslip, incubate slide for 1–2 min in an open jar containing 0.3% v/v NP-40, 2× SC at 72°C.
    (b) Wash for 1 min in a Coplin jar containing 0.3% v/v NP-4/2× SSC at RT.
    (c) Drain. Agitate briefly in detergent wash (PBS+1% v/v Igepal CA-630 Tween 20) at ambient temperature.

(d) Rinse twice in distilled water, agitating at ambient temperature.

(e) Drain. Dehydrate through an ethanol series of 70, 90, and 100% v/v for 3 min each.

(f) Air-dry. Mount with 10 μl Vectashield with DAPI per hybridization area on a 22 mm x 50 mm #0 glass coverslip. Invert tissue onto coverslip. Allow at least 1 h incubation.

13. Capturing cells individually is usually better than capturing them as a group. Some signals can be brighter than others, and when captured as a group, the brighter signals tend to shorten the exposure time making the weaker signals harder to see. For FISH on paraffin sections at least 100 nuclei should be captured for each experiment. Statistical analysis is performed on the data obtained.

References

1. Pardue ML, Gall JG (1969) Molecular hybridization of radioactive DNA to the DNA of cytological preparations. Proc Natl Acad Sci USA 64:600–604
2. John H, Birnstiel M, Jones K (1969) RNA–DNA hybrids at the cytological level. Nature 223:582–587
3. Bauman JGJ, Wiegant J, Borst P, van Duijn P (1980) A new method for fluorescence microscopical localization of specific DNA sequences by *in situ* hybridization of fluorochrome labelled RNA. Exp Cell Res 128:485–490
4. Langer PR, Waldrop AA, Ward DC (1981) Enzymatic synthesis of biotin-labelled polynucleotides: novel nucleic acid affinity probes. Proc Natl Acad USA 78:6633–6637
5. van Proijen-Knegt AC, van Hoek JF, Bauman JG, van Duijn P, Wool IG, van der Ploeg M (1982) *In situ* hybridization of DNA sequences in human metaphase chromosomes visualized by an indirect fluorescent immunocytochemical procedure. Exp Cell Res 141:397–407
6. Landegent JE, Jansen in de Wal N, Baan RA, Hoeijmakers JH, van der Ploeg M (1984) 2-Acetylaminofluorene-modified probes for the indirect hybridocytochemical detection of specific nucleic acid sequences. Exp Cell Res 153:61–72
7. Tchen P, Fuchs RPP, Sage E, Leng M (1984) Chemically modified nucleic acids as immunodetectable probes in hybridization experiments. Proc Natl Acad Sci USA 81:3466–3470
8. Summersgill B, Goker H, Weber-Hall S, Huddart R, Horwich A, Shipley J (1997) Molecular cytogenetic analysis of adult testicular germ cell tumors and identification of regions of consensus copy number change. Br J Cancer 77:305–13
9. Summersgill B, Clark J, Shipley J (2008) Fluorescence and chromogenic *in situ* hybridization to detect genetic aberrations in formalin-fixed paraffin embedded material, including tissue microarrays. Nat Protoc 3:220–234
10. Speicher MR, Stephen GB, Ward DC (1996) Karyotyping human chromosomes by combinatorial multi-fluor FISH. Nat Genet 12:368–375
11. Schrock E, du Manoir S, Veldman T, Schoell B, Wienberg J, Ferguson-Smith MA, Ning Y, Ledbetter DH, Bar-Am I, Soenksen D, Garini Y, Ried T (1996) Multicolor spectral karyotyping of human chromosomes. Science 273: 494–497
12. Landegent JE, Jansen in de Wal N, Dirks RW, Baao F, van der Ploeg M (1987) Use of whole cosmid cloned genomic sequences for chromosomal localization by non-radioactive *in situ* hybridization. Hum Genet 77:366–370
13. Lichter P, Tang CJ, Call K, Hermanson G, Evans GA, Housman D, Ward DC (1990) High-resolution mapping of human chromosome 11 by *in situ* hybridization with cosmid clones. Science 247:64–69
14. Ioannou PA, Amemiya CT, Garnes J, Kroisel PM, Shizuya H, Chen C, Batzer MA, Jong PJ (1994) A new bacteriophage P1-derived vector for the propagation of large human DNA fragments. Nat Genet 6:84–89

15. Shizuya H, Birren B, Kim U, Mancino V, Slepak T, Tachiiri Y, Simon M (1992) Cloning and stable maintenance of 300-kilobase-pair fragments of human DNA in *Escherichia coli* using and F-factor-based vector. Proc Natl Acad Sci USA 89:8794–8797
16. Trask BJ, Massa H, Kenwrick S, Gitschier J (1991) Mapping of human chromosome Xq28 by two-color fluorescence *in situ* hybridization of DNA sequences to interphase nuclei. Am J Hum Genet 48:1–15
17. Senger G, Ragoussis J, Trowsdale J, Sheer D (1993) Fine mapping of the MHC class II region within 6p21 and evaluation of probe ordering using interphase fluorescence *in situ* hybridization. Cytogenet Cell Genet 64:49–53
18. Nederlof PM, van der Flier S, Wiegant J, Raap AK, Tanke HJ, Ploem JS, van der Ploeg M (1990) Multiple fluorescence *in situ* hybridization. Cytometry 11:126–131
19. Nederlof PM, Robinson D, Abuknesha R, Hopman AH, Tanke HJ, Raap AK (1989) Three-colour fluorescence *in situ* hybridization for the simultaneous detection of multiple nucleid acid sequences. Cytometry 10: 20–27
20. Ried T, Baldini A, Rand TC, Ward DC (1992) Simultaneous visualization of seven different DNA probes by *in situ* hybridization using combinatorial fluorescence and digital imaging microscopy. Proc Natl Acad Sci USA 89:1388–1392
21. Dauwerse JG, Wiegant J, Raap AK, Breuning MH, van Ommen GJ (1992) Multiple colors by fluorescence *in situ* hybridization using ratio labelled DNA probes create a molecular karyotype. Hum Mol Genet 1:593–598

Chapter 11

Using a Quartz Crystal Microbalance Biosensor for the Study of Metastasis Markers on Intact Cells

Julien Saint-Guirons and Björn Ingemarsson

Abstract

The use of biosensors has become a standard method to characterize biomolecular interactions. Data obtained from biosensor studies are widely used to evaluate drug candidates, particularly in relation to their binding properties towards a selected target. The importance of measuring such interactions in a biologically relevant environment has become the new challenge for the biosensor technologies. In this chapter we describe a method for studying interactions between proteins and targets at the surface of intact cells using a quartz crystal microbalance (QCM) biosensor.

Key words: QCM biosensor, Lectins, Intact cells, Cell immobilization, Interaction rate constants, Kinetics

1. Introduction

1.1. Biosensors

The analysis of biomolecular interactions to determine the kinetics and affinity of such interactions has become essential for the development of new pharmaceutical substances. A biosensor is defined as a device that enables the measurement of interactions between one binding partner (ligand) immobilized on the surface of a transducer and a second binding partner in solution (analyte). The essential aspect of the biosensor, the transducer, senses a change in the mass associated with the interaction of interest and converts this physical modification into a measureable electric signal. Various transducers may be used and are of two main types, optical and acoustic. The quartz crystal microbalance (QCM) system is the most widespread acoustic transducer. Biomolecular interactions, especially protein–protein, protein–carbohydrate, and protein–DNA binding events are involved in key aspects of development

Miriam Dwek et al. (eds.), *Metastasis Research Protocols*, Methods in Molecular Biology, vol. 878,
DOI 10.1007/978-1-61779-854-2_11, © Springer Science+Business Media, LLC 2012

and metabolism and have been implicated in cancer metastasis. Continuous flow biosensors provide a valuable insight into the mechanism of a specific interaction by permitting the determination of the kinetic parameters; including the association rate constant (k_a), the dissociation rate constant (k_d), and the affinity of the interaction. Several other techniques, including isothermal titration calorimetry (ITC), radioimmunoassay (RIA), or enzyme linked immunosorbent assay (ELISA), may also be used to determine the affinity of a given interaction. However, these approaches determine the affinity by equilibrium analysis using reporter labels, whereas biosensors provide label-free and kinetic real-time analysis of biomolecular interactions and provide more information and enable further understanding of the molecular interactions.

There are several biosensors available on the market for studies of biomolecular interactions. These are based on either optical detection principles, for example the Biacore (GE Healthcare), ProteOn (Biorad), Octet (Fortebio) systems, or acoustic devices based on the quartz crystal microbalance technology (QCM) for instance Q sense E4 (Qsense) or Attana 200 (Attana AB). In this chapter we will focus on the use of a QCM biosensor for studies of molecular interactions at the surface of intact cells.

QCM technology is based on the piezoelectric property of quartz crystals. The piezoelectric effect was first discovered in quartz crystals by Jacques and Pierre Curie in 1880 (1). By applying an alternating voltage to an AT-cut crystal briefly, with a frequency close to the resonance frequency of the crystal, the crystal will start to oscillate at its specific resonance frequency. The key feature of the QCM technology was discovered by Sauerbrey (2) who established a proportional relationship between a change in frequency with mass added to the crystal surface.

QCM biosensors act as very sensitive balances where the mass added to, or removed from, the sensor surface (corresponding respectively to the association and dissociation of the binding partners) is monitored in real time, providing information about the kinetic rate constants, as well as the affinity that characterizes a specific interaction. Analysis using a biosensor system may provide insights into the mechanism of the interaction between binding partners.

1.2. Cell Analysis Using a QCM Biosensor

The QCM biosensor has mainly been used to study bimolecular interactions, where one partner (ligand) is immobilized on the crystal surface using covalent amine coupling, physisorption, or capturing on streptavidin-coated crystals (if the ligand is biotinylated). The second partner (the analyte) is then injected over the sensor surface and the interaction is monitored in real time by measuring the change in resonance frequency of the quartz crystal. The main applications include: off-rate screening, affinity measurement, kinetics, active concentration measurement, binding assays as well as epitope mapping.

Biosensors based on QCM technology have proven to be powerful and efficient tools for studies of molecular interactions, mainly antigen–antibody interactions (3), and also lectin–carbohydrate recognition (4). Almost all biosensor studies are performed using isolated and purified molecules. However, it has become essential for the development of new therapeutics to provide a tool for screening and characterization of drug candidates in a more biologically relevant environment, for example, at the surface of intact cells. Previous work using QCM biosensors and mammalian cells have been focused on monitoring morphological changes in response to drugs in static systems (5). More recently, a QCM biosensor was developed for studies of real-time biomolecular interactions at the cell surface. The applications include lectin–carbohydrate and antibody–receptor interactions. This approach enables kinetic evaluation of the interactions between lectins and glycoproteins, or antibodies and over-expressed receptors, in a biological context. This flow-through system provides access to real-time measurement of affinity constants in a biologically relevant context.

In this chapter we will provide a guide for generating real-time, qualitative, and quantitative analysis of biomolecular interactions at the surface of cancer cells.

2. Materials

2.1. Equipment

1. Attana Cell 200 QCM biosensor (Attana AB).
2. Upright fluorescence microscope.
3. Cell culture compatible polystyrene-coated quartz crystal sensor chips (Attana AB, product number: 3621-3033).
4. Cell culture equipment and plasticware.
5. Cell counting device (Hemocytometer).

2.2. Reagents

1. Trypsin/EDTA 0.25% v/v (store at −20°C). Heat the solution to 37°C before use.
2. Phosphate buffered saline (PBS). Dissolve in distilled water and check that the pH is 7.2. Autoclave for 20 min at 121°C.
3. DAPI-dihydrochloride nucleic acid stain (stock solution: 14.3 μM in PBS), stored at 4°C in an opaque vial.
4. Cell culture medium appropriate for the cell type used.
5. Foetal calf serum (FCS).
6. Formaldehyde solution, 3.7% v/v in PBS.
7. 70% v/v ethanol or industrial methylated spirits (IMS).

3. Methods

3.1. Preparation of the Cell-Sensor Surface

Different sensor surfaces may be used for cell immobilization these include gold or polystyrene. Typically, cell culture compatible polystyrene-coated crystals are used. Briefly, the whole quartz crystal sensor is incubated with an appropriate cell suspension for about 24 h, in order to allow the cells to attach, grow, and proliferate on the surface. The concentration of cells used may vary but it is normally adequate if about 60–80% of the sensor surface is covered. After about 24 h the cells are chemically fixed on the sensor crystals washed and subsequently placed in a flow of running buffer in the QCM biosensor. After a period of stabilization of up to several hours (typically overnight), interactions of biomolecules such as antibodies or lectin with the cells can then be evaluated.

In order to generate high-quality data, the preparation of a suitable cell-covered sensor surface is essential. For this system adherent cells that grow as monolayers are typically used. The following steps provide detailed guidelines to allow the preparation of such surfaces using the MCF7 breast cancer cell line but adjustments might be necessary depending on the cell type used and/ or the type of experiment to be performed.

1. Remove cell medium from cultivation flask (we typically use a 75-cm^2 flask).
2. Add an appropriate amount of trypsin/EDTA (for example, 2.5 ml trypsin/ 0.25% w/v EDTA) and incubate at 37°C until cells have started to lift from dish if needed.
3. Add sufficient cell medium to the detached cells to re-suspend them. This step neutralizes the trypsin activity.
4. Centrifuge the cells to pellet them (5 min at 1,500 × *g*) and re-suspend the cells in fresh medium. A cell count should now be performed.
5. Adjust the cell concentration to 10^6 cells/ml.
6. A cell culture compatible polystyrene-coated quartz crystal sensor is placed on the bottom of the well of a 24-well plate.
7. Prepare an appropriate dilution of the cell suspension to obtain between 20,000 and 70,000 cells (depending on experiment) on the sensor surface. This represents between 125,000 and 450,000 cells/cm^2. In this example, 250,000 breast cancer cells/cm^2 were incubated with the sensor crystal in a 24-well plate and then placed in a tissue culture incubator at 37°C for 24 h. The duration of this phase of the experiment can be adjusted according to the growth and proliferative properties of the cells to be studied.
8. After incubation, the cells are chemically fixed to the surface using standard fixation procedures. As an example, 10 min

incubation in 1 ml of ice cold formaldehyde solution, 3.7% v/v in PBS, may be used. After the fixation step several washes in PBS must be performed in order to remove the formaldehyde solution (typically we use five washes each 1 ml of PBS). The cell-coated crystal may then be stored in PBS at 4°C for several weeks.

9. In order to evaluate the cells on the sensor surface, the nuclei of fixed cells are stained using, for example, the fluorescent nucleic acid dye DAPI hydrochloride (2.9 μM) which is directly loaded onto the sensor surface in 50 μl volume and left to incubate for 5 min in the dark. After removal of the DAPI solution, wash three times with PBS wash. Visualize the sensor surface using an upright fluorescence microscope.
10. If the cell coverage on the sensor surface is adequate, the quartz crystal may be placed into the sensor chip housing, and introduced into the QCM biosensor. A continuous flow of an appropriate running buffer, for example PBS, supplemented with non-ionic detergent such as Tween 20 at 0.05% v/v is commenced at a flow rate of 25 μl/min.
11. Prior to undertaking measurements, the sensor needs to equilibrate for several hours in running buffer (typically an overnight equilibration is used) with a flow rate of 25–50 μl/min (see Note 1).

3.2. Interaction Measurements

A typical experiment designed to measure the interaction between a lectin/antibody and the surface of cancer cells is described.

1. After equilibration of the sensor, the crystal should oscillate at a stable frequency with a drift <0.2 Hz/min.
2. An injection of an appropriate reference sample is performed by loading the injection loop connected to the valve. The valve is switched to reference and is introduced into the line of flow allowing it to be transferred to the sensor surface. When an appropriate volume of the reference sample has been injected (usually 70% of the injection loop volume) the valve is switched to the original position (see Note 2).
3. The sample to be analyzed is injected by following step 2, above. Appropriate concentrations of analyte (lectin or antibody) should be prepared in running buffer. In this example, several concentrations of lectin are prepared (ranging from 12.5 to 100 μg/ml). During the injection step, the frequency is monitored, and increases in response to the interaction between the lectin and the cell surface, adding mass on the sensor surface. After injection of the sample, the dissociation of the lectin from the cell glycans, may be monitored for at least 200 s. The dissociation time needed is related to the stability of the interaction, i.e. longer dissociation times are required for

slowly dissociating interactions, and are needed in order to acquire enough data for reliable rate determinations.

4. The sensor surface needs to be regenerated in order to enable other interactions to be evaluated using the same sensor, therefore any remaining bound lectin (or antibody) should be desorbed. Typically, the regeneration solution consists of a 10 mM glycine solution at low pH (pH 1.0–3.0). In the example presented (Fig. 1), the high degree of avidity of binding required potent regeneration conditions. 10 mM glycine pH 1.0, supplemented with 500 mM NaCl was injected three times in order to remove all of the bound lectin from the surface of the cells. The regeneration efficiency is evaluated to ensure that the frequency recorded after the regeneration is similar to the frequency monitored before injection of the sample (see Note 3).
5. Sensor surfaces can typically be regenerated between 10 and 20 times; however, this will depend on the cell type under investigation and therefore should be determined for each cell type in turn.

3.3. Data Analysis

The binding response curves obtained for the interaction of the binding partners of interest must be processed and analyzed to extract the rate constants for the reaction.

1. The signal obtained using the reference sample needs to be subtracted from the interaction curves obtained with the sample of interest (Fig. 1a) in order to obtain referenced interaction curves (Fig. 1b).
2. This operation needs to be repeated for every interaction analyzed.
3. To determine the kinetic parameters of an interaction, three or four binding curves obtained with various concentrations of analyte should be analyzed simultaneously. In the example shown in Fig. 1b, the binding curves obtained with the lectin at 12.5, 25, 50, and 100 μg/ml were analyzed using the biosensor kinetic data analysis software ClampXP developed by Myszka and Morton (6). In this software the kinetics of the reaction are determined by fitting theoretical models, for the interaction, to the experimental data. Data-sets fitted with a 1:1 interaction model (with or without adjustment for mass transport limitations) enables evaluation of the accuracy by which the model describes the interaction of interest. Global fitting using the 1:1-model to the experimental data sets may be used to extract kinetic rate constants, such association rate constant (k_a) and dissociation rate constant (k_d). From these, the dissociation constant (K_D), characterizing the affinity of the interaction of interest, can be calculated (see Note 4).

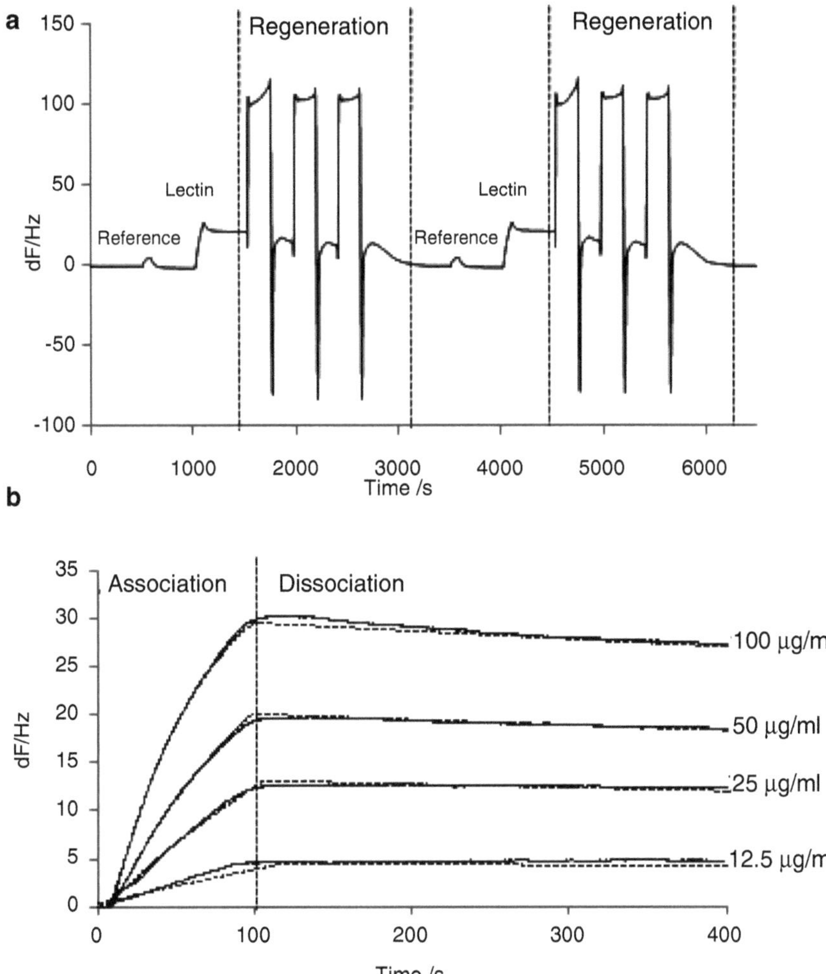

Fig. 1. (**a**) Two interaction cycles consisting of a reference injection, followed by a lectin injection and subsequent regeneration of the surface. (**b**) Kinetic evaluation of the interaction between a lectin and breast cancer cells. Referenced interaction curves at increasing concentrations (*solid dark lines*), overlaid with curves resulting from global fit using a theoretical 1:1 model (*dashed lines*). Rate constants were extracted using the Clamp XP software, ($k_a = 2.0 \times 10^4$ M^{-1}/s, $k_d = 3.0 \times 10^{-4}$ s^{-1}). The affinity of the interaction was also determined, $K_D = 15$ nm.

4. Notes

1. Preparation of crystals: The preparation step is essential in order to generate high-quality data using the QCM biosensor technology. A polystyrene-coated quartz crystal sensor must be covered with enough cells, evenly distributed over the surface, to generate a sufficient signal when the interaction of interest is monitored. However, the surface should not be saturated as the heavy load of cells may challenge the mechanical

performance of the biosensor and, hence, may decrease the quality of the data. The empirical approach adopted is to define conditions enabling cell coverage of 60–80% (as an indication) of the sensor surface. A visual evaluation using a fluorescent nuclei dye using an upright fluorescence microscope is important to ensure that cells evenly cover the sensor surface.

2. Referencing: QCM biosensors are sensitive to physico-chemical parameters such as viscosity and/or conductivity of the buffers. Hence, differences in salt composition and concentration between the running buffer and the sample injected over the surface may produce non-specific signal (typically <2–3 Hz) that may interfere with the analysis. This is especially important for weak responses, as the low density of binding sites on the sensor surface, are by definition more sensitive to interference. Appropriate reference solutions must be chosen and allowed to interact with the surface using the same injection parameters as for the sample injections. This approach makes it possible to subtract non-specific binding of the analyte interaction responses and generate high-quality data.

3. Surface regeneration: Surface regeneration is performed to remove remaining analyte bound to the cells before a subsequent analysis cycle can begin. The regeneration must desorb bound analyte, but should not affect the cells on the surface, for example it must leave the surface intact and active to enable further binding events to be reproducibly monitored. Empirical determination of appropriate regeneration conditions must be performed, starting with low concentration solution followed by higher concentration and acidic pH if needed. Examples of regeneration solutions to be used: weak 10 mM glycine/HCl pH >2.5; 10 mM HEPES/NaOH pH <9; 1 M NaCl, intermediate (10 mM glycine/HCl pH 2–2.5; 10 mM glycine/NaOH pH 9–10, strong 10 mM glycine/HCl pH <1.5 supplemented with 0.5 M NaCl.

4. Kinetic analysis of interaction data: QCM biosensors enable measurements of interactions occurring between two biomolecules, e.g. an analyte (A) and a surface bound ligand (B), to form a complex (AB), in a biological context. A single-step bimolecular interaction can be described by A + B AB and is characterized by interaction-specific association and dissociation rate constants that can be obtained by kinetic evaluation of the experimental data against a theoretical model using a global curve-fitting software (for example, Clamp XP). Rate constants relate the rate of reaction to the concentration of the binding partners. The association rate constant (several abbreviations may be found such as k_1, k_{on}, k_{fwd} or k_a) is expressed in M^{-1}/s and the dissociation rate constant (k_{-1}, k_{off}, k_{rev} or k_d) is expressed in s^{-1}. The affinity of the reaction is often referred to

as the dissociation constant, $K_D = k_d/k_a$, and is expressed in molar (M). The smaller the value for k the higher the affinity of the interaction. The use of the association constant, $K_A = 1/K_D$ expressed in M^{-1} may also be observed in some studies.

Several considerations must be taken into account when performing a kinetic study, particularly for an interaction occurring at the cell surface. Sometimes the transport/diffusion rate of the analyte from the bulk flow to the vicinity of the surface may restrict the association of the complex. In these cases, the concentration close to the interaction site will be lower than in the bulk flow due to an association rate faster than the dissociation rate, and the association rate will be determined by the diffusion rate to the surface. This is often referred to as mass transport limitation, and will result in biased results if this is not compensated for. One way to reveal such mass transport limitations is to assess the shape of the binding curve. Instead of a typical (exponential) curved response in the association phase (at least at higher concentrations), the response curves are more or less linear, with little or no curvature. If this is the case, the best solution is to increase the transport of the analyte to the surface by increasing the flow rate, or to reduce the association rate at the surface by reducing the surface density of the immobilized ligand. If this is not possible, an appropriate curve fitting model, taking the diffusion rate of the analyte into account, must be used to enable correct evaluation of the kinetic parameters of the reaction of interest. In addition, avidity of the bio molecules in the system might need to be taken into account, this is very important in the case of multivalent analytes, such as lectins. Avidity is defined as the combined strength of several bond interactions as opposed to affinity, which is determined according to the 1:1 interaction model, describing the strength of a single bond interaction. Hence, it is difficult to determine the affinities for multivalent analytes other than in relative terms.

References

1. Curie J, Curie P (1880) Piezoelectric and allied phenomena in Rochelle salt. Comput Rend Acad Sci Paris 9:294–297
2. Sauerbrey G (1959) The use of quartz oscillators for weighing thin layers and for microweighing. Z Phys 155:206–222
3. Koutsopoulos S, Unsworth LD, Nagai Y, Zhang S (2009) Controlled release of functional proteins through designer self-assembling peptide nanofiber hydrogel scaffold. Proc Natl Acad Sci USA 106:4623–4628
4. Pei Z, Anderson H, Aastrup T, Ramström O (2004) Study of real-time lectin-carbohydrate interactions on the surface of a quartz crystal microbalance. Biosens Bioelectron 21: 60–66
5. Marx KA, Zhou T, Montrone A, McIntosh D, Braunhut SJ (2007) A comparative study of the cytoskeleton binding drugs nocodazole and taxol with a mammalian cell quartz crystal microbalance biosensor: different dynamic responses and energy dissipation effects. Anal Biochem 361:77–92
6. Myszka DG, Morton TA (1998) CLAMP: a biosensor kinetic data analysis program. Trends Biochem Sci 23:149–150

Chapter 12

Cell Separations by Flow Cytometry

Derek Davies

Abstract

Flow cytometry has become a standard method for separating individual subsets of cells from a heterogeneous population. Multilaser, multicolour cell sorters are increasingly common and have become more complex in recent years increasing the number of applications available. However, a cell sorting experiment is only as good as the input sample, and the preparation of this is extremely important. This chapter describes the methods used to prepare samples for flow cytometry and how they can be adapted and optimised according to cell type.

Key words: Flow cytometry, Cell sorting, Fluorescence, Fluorescent protein, Sample preparation

1. Introduction

1.1. Cell Analysis by Flow Cytometry

Flow cytometry is a means of measuring the physical and chemical characteristics of particles in a fluid stream as they pass one by one past a sensing point. The modern flow cytometer consists of a light source, collection optics and detectors, and a computer to translate signals to data. In effect, a flow cytometer can be described as a large and powerful fluorescence microscope where the light source is of a highly specific wavelength, generally produced by a laser, and the human observer is replaced by a series of optical filters and detectors that aim to make the instrument more objective and more quantitative. As a cell passes through the laser beam, light is scattered in all directions, and also at this point any fluorochromes present on the cell are excited and emit light of a higher wavelength. Scattered and emitted light is collected by two lenses—one set in front of the light source and one set at right angles to it. By a series of beam splitters, optical filters, and detectors the wavelengths of light specific for

Miriam Dwek et al. (eds.), *Metastasis Research Protocols*, Methods in Molecular Biology, vol. 878,
DOI 10.1007/978-1-61779-854-2_12, © Springer Science+Business Media, LLC 2012

particular fluorochromes can be isolated and quantitated. Recent advances in laser technology, electronics systems, and fluorochrome technology mean that multi-laser flow cytometers that are capable of measuring 18 or more fluorescence channels are more readily available. The theory of operation of flow cytometers is well documented and there are several good general books on the subject (1–4).

1.2. Cell Sorting by Flow Cytometry

Flow sorting may be defined as the process of physically separating particles of interest from other particles in the sample. Sorting can be accomplished by two flow cytometrically based methods—the electrostatic deflection of charged droplets (so-called "stream-in-air" sorters) (2, 5) or mechanical sorting (6). Most commercially available cell sorters use the electrostatic method which is based on the principles of droplet formation, charge, and deflection analogous to those used in ink-jet printers. Any fluid stream in the atmosphere will break up into droplets but this is not a stable process. However, by applying vibration at certain frequencies it is possible to stabilise the point at which droplets break off from the stream, the droplet size and the distance between the drops. Therefore the time between the point at which the cell passes through the laser beam and is analysed until its inclusion in a droplet as it breaks from the stream—the drop delay—is known and constant (Fig. 1). By calculating this time period, the droplet containing the cell of interest can be specifically charged through the fluid stream the moment that the drop is forming. To avoid cell loss, the duration of the charging pulse can be altered to include either or both the preceding and the following drop. Charged droplets will then pass through an electrical field created by two plates—one charged positively, the other charged negatively. Droplets containing a charge will be attracted towards the plate of opposite charge and in this way will be separated from the stream.

Mechanical sorters do not use the droplet method but rather employ a motor-driven syringe to aspirate the fluid containing the cell of interest (7). From a practical point of view mechanical sorters are relatively slow (maximum sorting speed of 500 cells/s) but they do have advantages—the system is enclosed preventing both contamination and evaporation, and it is easier to set up and perform a sort so a skilled operator is not a prerequisite.

The sorting speed of stream-in-air flow cytometers varies depending on the manufacturer and the design of the machine from 5,000 to 40,000 cells/s. However, this is still relatively slow compared with bulk isolation methods such as cell filtration techniques, magnetic bead separation, or cell affinity techniques as even at top speeds no more than 10^8 cells may be sorted in an hour.

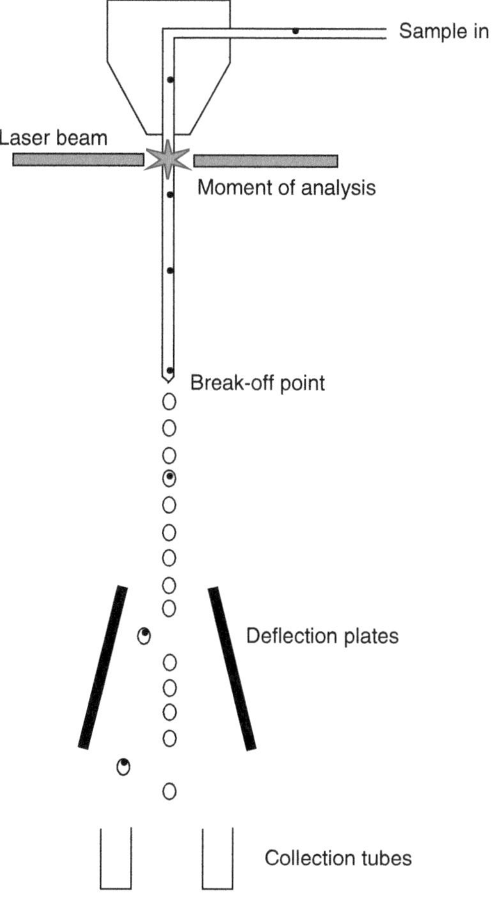

Fig. 1. Schematic diagram of a typical stream-in-air sorter. Cells are analysed at the laser intersection (Moment of analysis), enclosed in droplets at the break-off point and are charged at this point if they are to be sorted. High-voltage deflection plates attract cells of the opposite polarity.

However, in comparison with other techniques flow sorting achieves the highest cell purity and recovery. In addition such stream-in-air sorters are capable of sorting up to six subsets, which may be defined by quantitative and qualitative measurements of multiparametric cell characteristics, the number of which is limited only by the configuration of the flow sorter. Examples of this type of sorter are the MoFlo and Astrios (Beckman Coulter), FACSAria and Influx (Becton Dickinson), and Synergy (Sony iCyt). It is also possible to adjust the mode of sorting depending on whether high purity (the default mode), high recovery (if a small, precious population is needed), or high count accuracy (for single cell sorting for cloning or PCR) is required.

1.3. Applications of Flow Cytometry

Anything that can be tagged with a fluorescent marker can be examined on a flow cytometer. This can be a structural part of the cell such as protein, DNA, RNA, an antigen (surface, cytoplasmic or nuclear), or a specific cell function (apoptosis, ion levels, pH, membrane potential). As long as a specific cell population can be identified by its fluorescence characteristics it can be sorted. Examples of the applications of flow analysis and sorting are given in Table 1. Additionally, it is also possible to excite and detect a wide-range of fluorescent proteins such as cyan fluorescent protein, green fluorescent protein, yellow fluorescent protein, and red fluorescent protein variants.

The fluorochrome of choice will to a large extent depend on both the intended application and the illumination wavelengths available in the cytometer. The most common laser wavelengths and the fluorochromes that can be used with these are given in

Table 1
Examples of flow cytometric applications in mammalian cells

Application	Reference
Phenotyping (surface, cytoplasmic, or nuclear antigen)	(8–10)
Cell cycle analysis (DNA or kinetics via Bromodeoxyuridine)	(11, 12)
Functionality e.g. Calcium flux, pH, membrane potential	(13–15)
Apoptosis and cell death	(16, 17)
Enzyme activity	(18, 19)
Monitoring drug uptake	(20)
Measurement of RNA or protein content	(21)
Fluorescence In Situ Hybridisation (FISH)	(22, 23)
Measurement of telomere length	(24)
Detection of fluorescent proteins/monitoring of transfection	(25, 26)
Detection of stem cells	(27, 28)
Sterile sorting for re-culture	(2, 7)
Sorting of rare populations	(29)
Single cell sorting for cloning or PCR	(30, 31)
Sorting for protein, RNA or DNA extraction	(32)
Chromosome sorting for production of chromosome-specific paints	(33, 34)
Isolation of defined populations e.g. tumour cells from normal cells	(35)

Table 2
Common fluorochromes

Laser wavelength	Examples
UV (ca. 350 nm)	Aminomethyl coumarin DAPI (4,6-diamidino-2-phenylindole) Hoechst (33258 or 33342) Indo-1
405 nm	Pacific Blue Pacific Orange Brilliant Violet 421
488 nm	Fluorescein isothiocyanate (FITC) Alexa Fluor 488 R-Phycoerythrin (PE) PerCP (peridinin chlorophyll protein) PE-Cy5 and PE-Cy7 tandem conjugates Propidium iodide
635 nm	Allophycocyanin (APC) TO-PRO-3 Iodide Cy5 APC-Cy7 tandem conjugate

Table 2. The choice will depend on the number of cell characteristics being examined, the spectral overlap between the fluorochromes and the commercial availability of them.

The most common application of cell sorting is to separate a sub-population of cells based on their specific phenotype, whether this be, for example, tumour cells from normal cells or cells expressing a particular antigen after transfection.

The key point of flow cytometry is that measurement is performed on a cell-by-cell basis—this is the most powerful aspect of flow cytometry but, therefore, a pre-requisite for flow cytometric analysis or sorting is that the sample must be in single cell suspension. The preparation of samples for flow cytometry depends to a large extent on the source of the cells and the requirements of each cell type. In all a single cell suspension should have maximum yield and maximum viability but no one isolation technique will be applicable to all cell types. Some optimization of sample preparation is always required. Once in a single cell suspension, the cells of interest should be prepared by labelling with fluorochromes either to detect antigenic determinants, structural components, or functional status that will allow them to be specifically identified. However, it must be stressed that the success of a sort depends almost entirely on the quality of the input sample, and there are several ways that this may be optimised. So, the key to successful cell sorting is the preparation of the sample.

2. Materials

1. Trypsin/versene: 0.02% w/v EDTA (known as versene by many cell culturists; store at room temperature), 0.25% w/v trypsin (store at −20°C). Add 4 ml trypsin to 16 ml versene. This mixture can be used for up to 1 week if stored at 4°C. Warm the solution to 37°C before use.
2. Phosphate buffered saline (PBS): 8 g NaCl, 0.5 g KCl, 1.43 g Na_2HPO_4, 0.25 g KH_2PO_4. Dissolve in 1 L of distilled water. Check that the pH is 7.2. Autoclave for 20 min.
3. Propidium iodide (PI; 50 μg/ml in PBS). This is light sensitive so should be stored in an opaque container at 4°C.
4. DAPI (200 μg/ml in methanol; use at 1 μg/ml).
5. TO-PRO-3 iodide (1 μM stock in PBS; use at 50 nM).
6. Trypan Blue (0.4% w/v).
7. Cell culture medium appropriate to the cell used, both with and without Phenol Red.
8. Foetal Calf Serum (FCS).
9. 70% v/v ethanol. Take 700 ml of absolute ethanol and add 300 ml of distilled water.
10. Nylon mesh—35 μm and 70 μm (Lockertex, Warrington, Cheshire, UK; Small Parts Inc., Miami, FL, USA).

3. Methods

3.1. Preparation of Cells for Flow Cytometry (See Notes 1–6)

3.1.1. Suspension Cells E.g. Cultured or Primary Blood Cells

If the sample under investigation is a suspension cell line that grows in suspension (for example: Jurkat cells or HL-60 cells) or is naturally in suspension (for example: peripheral blood), processing is relatively straightforward although care must be taken not to damage or lose cells during the processing procedure.

1. Perform a viable cell count using PI or Trypan Blue. Live cells will exclude the dye whereas it will be taken up by cells whose membranes have been compromised.
2. Select the desired number of cells and decant into a sterile container.
3. Centrifuge at 250 × *g*. The length of centrifugation will depend on the volume of fluid, for small volumes (up to 100 ml) 10 min is sufficient, increase this for larger volumes (see Note 3).
4. Carefully pour off the supernatant taking care not to disturb the pellet.

5. Resuspend the pellet in medium at a density of approximately 10^6 cells/ml.
6. Perform antigen staining (see Subheading 3.2).

3.1.2. Adherent Cell Lines or Primary Cultures

For adherent cells, the sample preparation can be more involved. In general the usual methods used for sub-culturing of adherent cell lines are sufficient to produce a suspension of cells for flow sorting.

1. Remove culture medium by suction using a sterile pipette.
2. Add an appropriate amount of trypsin/versene (for example: 10 ml per 10-cm dish, 20 ml per 250-ml flask). Wash the fluid around and discard all but a small volume (5 ml per flask, 2 ml per dish).
3. Examine the cell monolayer microscopically at regular and frequent intervals and tap the vessel gently to aid the dispersion of cells. Incubate at 37°C if the progress is slow.
4. When the cell sheet is sufficiently dispersed, add an appropriate amount of growth medium with serum (for example: 10 ml per dish, 20 ml per flask) and carefully resuspend the cells in the medium. The addition of medium serves to neutralise the effect of the enzyme.
5. Perform a viable cell count using PI or Trypan Blue. Resuspend the pellet in medium at a density of approximately 10^6 cells/ml.
6. Centrifuge at 250 × *g* for 10 min and again resuspend the pellet in medium at a density of approximately 10^6 cells/ml.
7. Once cells are in suspension, antigen staining can be performed (see Subheading 3.2).

3.1.3. Solid Tissue

1. Place tissue in a 10-cm^2 sterile tissue culture plate and add 20 ml of enzyme (trypsin/versene) solution.
2. Leave for 15 min at 37°C checking constantly for the cells to release.
3. If cell release is slow, tease gently using sterile forceps/scalpel. Repeat this as necessary.
4. Add medium containing 10% v/v FCS to neutralise the enzymatic action.
5. Decant cells solution into a sterile 50-ml tube and centrifuge at 250 × *g* for 10 min.
6. Discard supernatant, perform a viable cell count and resuspend cells at a density of approximately 10^6 cells/ml. Repeat centrifugation step.
7. Finally resuspend at a density of approximately 10^6 cells/ml, before antigen staining (see Subheading 3.2).

3.2. Antigen Staining

3.2.1. Directly Conjugated Antibody

1. Take cells at 10^6/ml (see Note 7) and centrifuge at 250 × *g* for 10 min. Carefully pour off the supernatant.
2. Add appropriate amount of fluorochrome-labelled antibody (see Note 8); incubate for 15 min at 37°C (see Note 9).
3. Add medium to cell density of 10^6/ml. Spin, discard supernatant, and repeat.
4. Resuspend at 10^6/ml in phenol red-free medium containing a low level of serum or protein (no higher than 2% v/v; see Note 10) in a sterile container before flow cytometry.

3.2.2. Indirect Staining of Antigen

1. Take cells at 10^6/ml (see Note 7) and centrifuge at 250 × *g* for 10 min.
2. Add appropriate amount of primary antibody (see Note 8); incubate for 15 min at 37°C (see Note 9).
3. Add medium to cell density of 10^6/ml. Centrifuge at 250 × *g* for 10 min, discard supernatant, and repeat.
4. Add fluorochrome-labelled secondary antibody at a dilution of between 1:10 and 1:20 (If the primary antibody is monoclonal, this will generally be a rabbit anti-mouse antibody). Incubate for 15 min at 37°C.
5. Add medium to cell density of 10^6/ml. Spin, discard supernatant, and repeat.
6. Resuspend at 10^6/ml in phenol red-free medium containing a low level of serum or protein (no higher than 2%; see Note 10) in a sterile container before flow cytometry. Prepare tubes for cell collection (see Note 13) and pre-filter sample if necessary (see Note 14).

3.3. Preparation of the Flow Cytometer for Sterile Sorting

1. Sterilise the cytometer by passing 70% v/v ethanol through all sheath and sample lines for 60 min. Wash out by replacing ethanol in the sheath container with sterile deionised water for 30 min, then replace this with sterile sheath fluid (PBS; see Note 11).
2. Wash down all exposed surfaces—sample lines, nozzle holder, nozzle, deflection plates, tube holders—with 70% v/v ethanol.
3. Define cell population within the sample using a combination of dead cell dye exclusion, pulse width, and scatter characteristics of the particles in suspension (Figs. 2a, b and 3a–c).
4. Define the population to be sorted on the basis of fluorescence characteristics (Figs. 2c and 3d).
5. Decide whether purity, recovery, or count accuracy is the most important factor and adjust the mode of sorting accordingly (see Notes 12 and 20). Cells may be sorted into tubes, 384-, 96-, 24-, or 6-well plates or directly onto slides or any other user-defined receptacle.

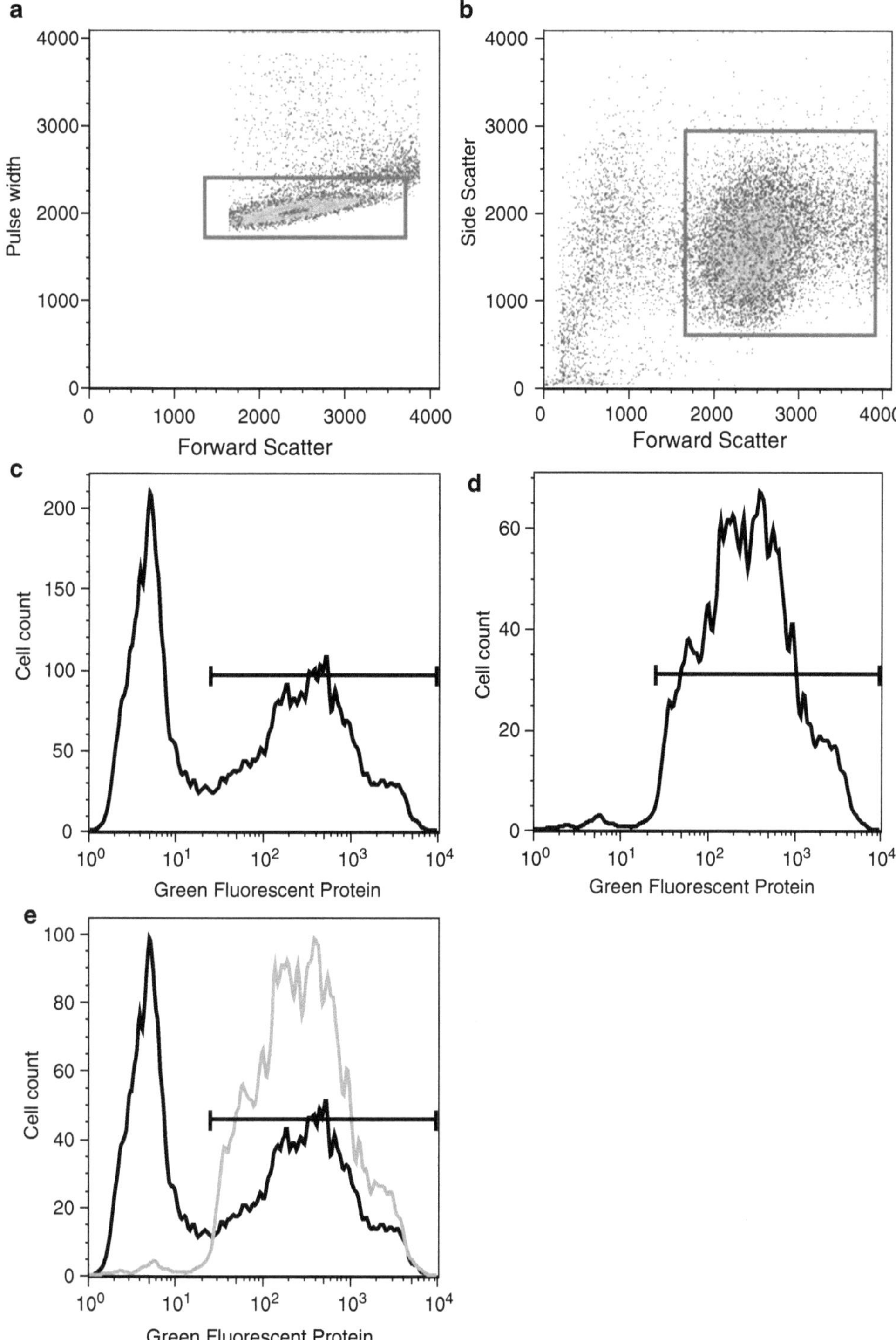

Fig. 2. A single colour sort. Here cells are expressing a Green Fluorescent Protein construct. Single cells are defined on the basis of their pulse width (**a**) and scatter characteristics (**b**). The GFP-expressing cells have then been selected (**c**) and sorted. Post-sort purity is over 99% (**d**). (**e**) shows an overlay of pre- and post-sorted populations.

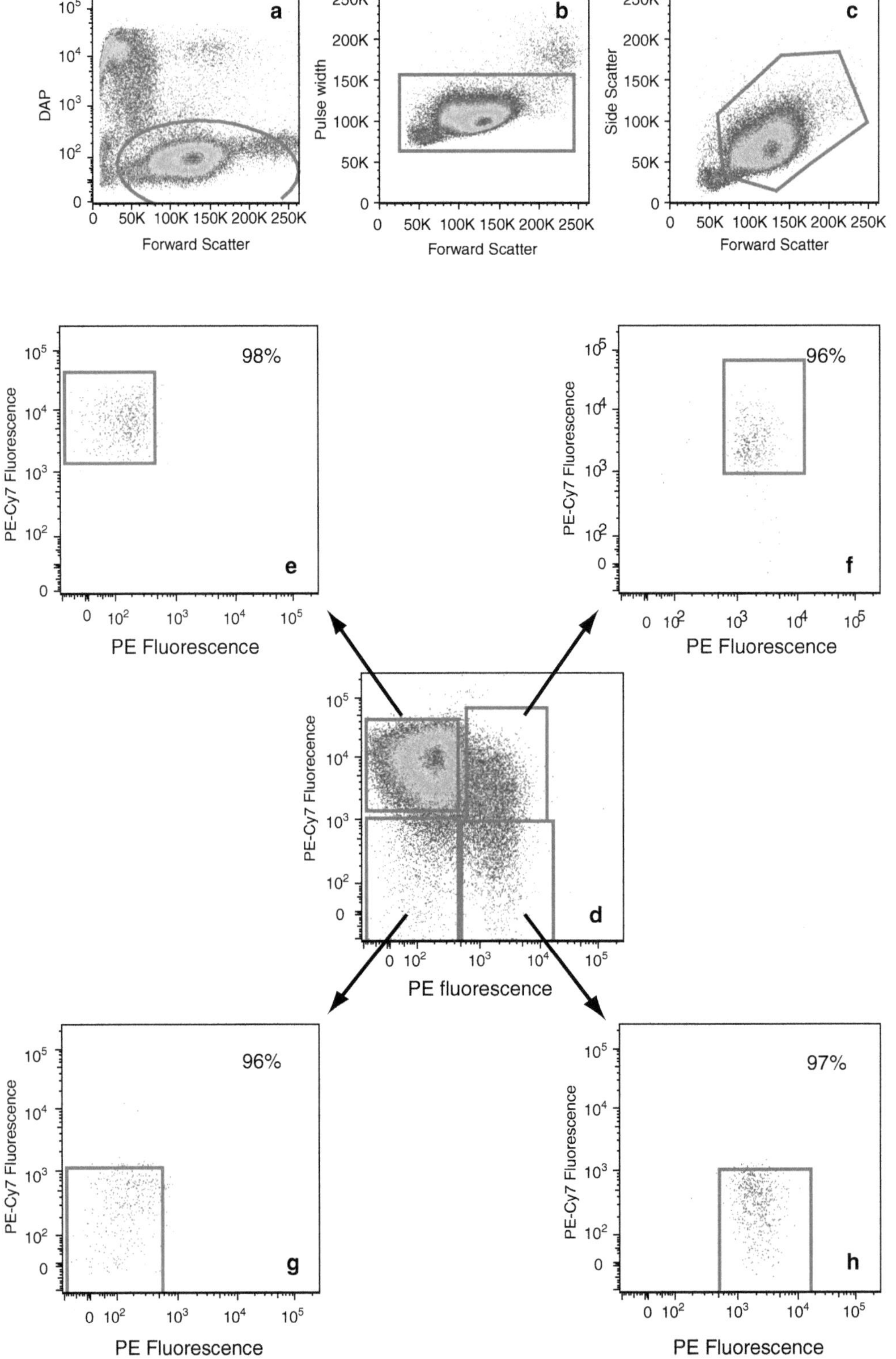

Fig. 3. A two-colour four-way sort. Cells have been stained with two fluorochrome-labelled antibodies—PE and PE-Cy7. The *top three panels* show the gating to exclude dead cells (**a**), doublets (**b**), and debris (**c**). After this, four regions have been set to sort populations on the basis of their fluorescence expression (**d**). The resultant sorted populations show single PE-Cy7 positive cells (**e**), double positive cells (**f**), double negative cells (**g**), and single PE positive cells (**h**). In each case the sorted populations have purities above 95%.

4. Notes

1. Phenol Red in media may interfere with subsequent procedures so it is generally advisable to use media without this when harvesting cells and subsequent antibody staining.
2. For some experiments, the use of trypsin may be contraindicated, for example, if the antigen under consideration is known to be cleaved by enzymatic action. Alternative enzymes may be used such as collagenase (0.1% w/v), dispase (0.1% w/v), or commercial preparations such as Accutase (Innovative Cell Technologies, San Diego, USA). If the antigen under investigation is sensitive to both enzymatic and chelating methods then scraping cells from the surface with a sterile cell scraper may be sufficient although this can lead to an increased number of cell clumps. This should be aseptic and only one cell scraper should be used for each vessel.
3. With all samples it is important to avoid cell loss so when spinning it is advisable to avoid braking as this can lead to cell loss. Additionally, aspirating supernatants rather than decanting may avoid cell loss. This also applies during subsequent antigen staining.
4. All pipettes, tips, and containers should either be purchased sterile or should be autoclaved.
5. Always wear gloves when handing potentially mutagenic chemicals such as propidium iodide.
6. If the cells still look clumped when examined microscopically, it is advisable to pass the cell suspension through a 21-gauge needle which will help to disperse these but should have minimal effect on cell viability. Samples may also be filtered through a stainless steel or nylon mesh. Various pore sizes are available, the most suitable being 30 μm when small cells are being used and 70 μm when larger or adherent cells are used.
7. At the end of processing, the cells should be assessed to get an accurate cell count. This may be achieved by the use of a haemocytometer or by an automated method such as a Vi-Cell (Beckman Coulter, Miami, USA) or equivalent. It is also useful to check the preparation microscopically to ensure that the preparation is free from excessive clumps and/or debris and, if unfixed, that the cell viability is good. Staining for antigenic determinants, functional state, or DNA content should be performed once processing is complete. If antibodies are used to identify the relevant cell population it is necessary to consider if these cause any downstream effects which may affect post-sorting processing—ideally antibodies should not alter or modulate cell function or behaviour.

8. The amount of antibody added will depend on the number of cells to be stained and the concentration of antibody in the staining solution. This is best determined empirically in positive control cells by a pilot experiment of test dilutions to determine the optimal concentration. Always do these experiments on equivalent numbers of cells and remember to scale up the amount of antibody used when doing bulk staining. The dilutions required for optimal staining using commercial antibodies will vary widely. Also, in general, the dilution for flow cytometry is lower than for slide-based immunofluorescence i.e. a higher antibody concentration is needed. Also it is important after washing steps to remove as much fluid as possible to avoid subsequent dilution of antibodies.
9. The length of time taken for antigen staining can vary—most antibody binding is very rapid (seconds), but some low-density, low-affinity antigens may take longer. The optimal temperature for staining is 37°C, unless using lymphocytes or other cells where antigen capping may occur in which case 4°C is preferable—in these cases the incubation time should be increased (doubled).
10. Immediately before flow sorting, cells should be suspended in low-serum or low-protein medium. Protein has a tendency to coat the sides of the sample lines in the cytometer and this can lead to blockages which are best avoided. Collection medium, however, should contain serum and antibiotics.
11. Phosphate buffered saline is generally used as a sheath fluid although any ionised fluid will be suitable. Obviously this needs to be sterile for sterile sorting; sterility is achieved by washing all fluid containers and fluid lines in the cytometer with alcohol, and sterility may also be improved by having a 0.22-μm filter in the sheath line.
12. Due to the possibility of coincidence of a wanted and an unwanted cell in either the same droplet or consecutive droplets, a decision as to whether these droplets should be sorted can be made by the operator depending on the desired purity and recovery for the cell fraction. Sorting a "one-drop envelope" will give high purity but will lead to reduced recovery; increasing the size of the deflection envelope will increase recovery at the expense of purity.
13. It is advisable to collect cells into medium, especially if they are to be re-cultured after sorting. This medium may also contain double-strength antibiotics—cells may be spun out of this medium after sorting and re-cultured in standard medium.
14. Cells, especially adherent cells or cells recovered from solid tumours, have a tendency to clump which will lead to machine blockages. This can be reduced by pre-filtering the cells to be sorted through a sterile gauze of a pore size appropriate to

the cells being used—35 μ for small cells (e.g. blood or bone marrow), 70 μ for epithelial or tumour cells. Gauze may be sterilised by autoclaving or by washing briefly in 70% v/v ethanol.

15. Clumping may also be due to high levels of cell death leading to nuclear break-up and free DNA. The addition of a small amount of DNase (5 KU/ml) to the cell suspension can be beneficial.
16. Viability of sorted cells is usually good (90%+)—there should be no significant reduction in viability of pre- and post-sorted cells. There may be loss of viability if the cells are not kept in optimal conditions and this may mean using temperature control for the duration of the sort.
17. It is important to maintain an optimal pH for the cells being sorted. This is particularly important in high-pressure sorters as increased pressure can lead to a reduction of pH if the cells are not adequately buffered. In these cases addition of 25 mM HEPES will help keep the cells in optimal condition.
18. Sterilisation will add to the preparation time of the cytometer and it is important to ask whether the sort will be sterile (if cells are to be re-cultured) or non-sterile (if cells are required for RNA extraction for example).
19. There are a number of practical considerations to be addressed before embarking on a sort. How numerous is the population of interest? How many cells are required at the end of the sort? How long will the sort take? The first two questions will enable the third to be determined. It may be that if a large number of cells of a minor population is required, flow sorting may actually be impractical. In these circumstances it is possible to pre-enrich for the population of interest by a prior step such as MACS-bead separation (36, 37).
20. Although the default sorting mode of most sorters is to maximise purity, this is user-changeable depending on whether high recovery and yield is needed at the expense of purity (for example when a small precious population needs to be recovered) or where high count accuracy is needed (for example in cloning experiments).
21. At the end of the sort a small aliquot of sorted cells should be re-analysed to determine the sort purity. Figs. 2 and 3 show examples of sorts based on triple- and single-colour staining.
22. Cell sorting and the operation of cell sorters is not a trivial task, and a trained and experienced operator is needed who is able to advise on the practicalities of any sort. Many institutions have set up specific core facilities for cytometry and other disciplines which have specific advantages in centralising both hardware and expertise.

References

1. Ormerod MG (ed) (1994) Flow cytometry: a practical approach. IRL, Oxford
2. Shapiro HM (2003) Practical flow cytometry, 4th edn. Wiley-Liss, New York, NY
3. Longobardi-Givan A (1992) Flow cytometry. First principles. Wiley-Liss, New York, NY
4. Ormerod MG (2008) Flow cytometry: a basic introduction. ISBN 978-0-9559812-0-3
5. Herzenberg LA, Sweet RG, Herzenberg LA (1976) Fluorescence activated cell sorting. Science 234:108–117
6. Duhnen J, Stegemann J, Wiecorek C, Mertens H (1983) A new fluid switching cell sorter. Histochemistry 77:117–121
7. Orfao A, Ruiz-Arguilees A (1996) General concepts about cell sorting. Clin Biochem 29:5–9
8. Carter NP (1990) Measurement of cellular subsets using antibodies. In: Ormerod MG (ed) Flow cytometry: a practical approach. IRL, Oxford, pp 45–67
9. Tough GF, Sprent J (1994) Turnover of naive- and memory-phenotype T cells. J Exp Med 179:1127–1135
10. Herzenberg LA, Tung J, Moore WA, Herzenberg LA, Parks DR (2006) Interpreting flow cytometry data: a guide for the perplexed. Nat Immunol 7:681–685
11. Krishan A (1975) Rapid flow cytofluorometric analysis of mammalian cell cycle by propidium iodide. J Cell Biol 66:188–193
12. Crissman HA, Steinkamp JA (1987) A new method for rapid and sensitive detection of bromodeoxyuridine in DNA-replicating cells. Exp Cell Res 173:256–261
13. Shapiro HM, Natale PJ, Kamentsky LA (1979) Estimation of membrane potentials of individual lymphocytes by flow cytometry. Proc Natl Acad Sci USA 76:5728–5730
14. Rijkers JT, Justement LB, Griffioen AW, Camber JC (1990) Improved method for measuring intracellular Ca++ with Fluo-3. Cytometry 11: 923–927
15. Wieder ED, Hang H, Fox MH (1993) Measurement of intracellular pH using flow cytometry with carboxy-SNARF 1. Cytometry 14:916–921
16. Darzynkiewicz Z, Bruno S, Del Bino G, Gorzyca W, Hotz MA, Lassota P, Traganos F (1992) Features of apoptotic cells measured by flow cytometry. Cytometry 13:795–808
17. Ormerod MG, Sun XM, Brown D, Snowden RT, Cohen GM (1993) Quantification of apoptosis and necrosis by flow cytometry. Acta Oncol 32:417–424
18. Nolan GP, Fiering S, Nicolas JF, Herzenberg LA (1985) Fluorescence activated cell analysis and sorting of viable mammalian cells based on ß-D-galactosidase activity after transduction of *Escherichia coli lacZ*. Proc Natl Acad Sci USA 85:2603–2607
19. Maftah A, Huet O, Gallet PF, Ratinaud MH (1993) Flow cytometry's contribution to the measurement of cell functions. Biol Cell 78: 85–93
20. Leonce S, Burbridge M (1993) Flow cytometry: a useful technique in the study of multidrug resistance. Biol Cell 78:63–68
21. Darzynkiewicz Z, Evenson D, Staiano-Coico L, Sharpless T, Melamed MR (1979) Relationship between RNA content and progression of lymphocytes through S phase of the cell cycle. Proc Natl Acad Sci USA 76:358–362
22. Bauman JGJ, Bayer JA, van Dekken H (1990) Fluorescent in situ hybridization to detect cellular RNA by flow cytometry and confocal microscopy. J Microsc 157:73–81
23. Wieckiewicz J, Krzeszowiak A, Ruggiero I, Pituch-Nowarolska A, Zembala M (1998) Detection of cytokine gene expression in human monocytes and lymphocytes by fluorescent in situ hybridization in cell suspension and flow cytometry. Int J Mol Med 1:995–999
24. Lauzon W, Sanchez Dardon J, Cameron DW, Badley AD (2000) Flow cytometric measurement of telomere length. Cytometry 42:159–164
25. Galbraith DW, Anderson MT, Herzenberg LA (1999) Flow cytometric analysis and FACS sorting of cells based on GFP accumulation. Methods Cell Biol 58:315–341
26. Shaner NC, Steinbach PA, Tsien RY (2005) A guide to choosing fluorescent proteins. Nat Methods 2:905–909
27. Goodell MA (2005) Stem cell identification and sorting using the Hoechst 33342 side population (SP). Curr Protoc Cytom Chapter 9: Unit 9.18
28. Johnson KW, Dooner M, Quesenberry PJ (2007) Fluorescence activated cell sorting: a window on the stem cell. Curr Pharm Biotechnol 8:133–139
29. Leary JF (1994) Strategies for rare cell detection and isolation. Methods Cell Biol 42: 331–357
30. Horan PK, Wheeless LL (1977) Quantitative single cell analysis and sorting. Science 198: 149–157
31. Williams C, Davies D, Williamson R (1993) Segregation of ΔF508 and normal CFTR alleles in human sperm. Human Mol Genet 2:445–448

32. Dunne JF, Thomas J, Lee S (1989) Detection of mRNA in flow-sorted cells. Cytometry 10:199–204
33. Green DK (1990) Analysis and sorting of human chromosomes. J Microsc 159:237–245
34. Davies DC, Monard SP, Young BD (2000) Chromosome analysis and sorting by flow cytometry. In: Ormerod MG (ed) Flow cytometry: a practical approach, 3rd edn. IRL, Oxford, pp 189–201
35. Berglund DL, Starkey JR (1989) Isolation of viable tumor cells following introduction of labelled antibody to an intracellular oncogene product using electroporation. J Immunol Methods 125: 79–87
36. Miltenyi S, Müller W, Weichel W, Radbruch A (1990) High gradient magnetic cell separation with MACS. Cytometry 11:231–238
37. Pickl WF, Majdic O, Kohl P, Stöckl J, Riedl E, Scheinecker C, Bello-Fernandez C, Knapp W (1996) Molecular and functional characteristics of dendritic cells generated from highly purified CD14+ peripheral blood monocytes. J Immunol 157:3850–3859

Chapter 13

Detection of Putative Cancer Stem Cells of the Side Population Phenotype in Human Tumor Cell Cultures

Matthias Christgen, Matthias Ballmaier, Ulrich Lehmann, and Hans Kreipe

Abstract

Human solid tumors and clonal tumor cell lines comprise phenotypically and functionally diverse subsets of cancer cells and also contain stem cell-like cancer cells. Side population (SP) cells, which pump out the fluorescent dye Hoechst 33342 (H33342) via the ABCG2 transporter, define a fraction of adult tissue stem cells in a wide variety of organs. Rare cancer cells with similar H33342 efflux capacity and delimited expression of *ABCG2* are present in various types of human tumors, such as breast cancer, lung cancer, and hepatocellular carcinoma. These cancer SP cells display several properties attributable to stem cell-like cancer cells and have been implicated in tumor growth, progression, and metastasis. Here we provide a detailed protocol for the detection of putative cancer stem cells of the SP phenotype in human adherent breast cancer cell cultures.

Key words: Tumor stem cells, SP cells, Breast cancer, Cell culture, Tumor growth, Metastasis, Flow cytometry, Hoechst 33342

1. Introduction

Human solid tumors and clonal tumor cell lines comprise phenotypically and functionally diverse subsets of cancer cells and also contain stem cell-like cancer cells (1, 2). Analogous to adult tissue stem cells, which maintain tissue homeostasis in various organs with high cell turnover, a subset of rare stem cell-like cancer cells plays a pivotal role for the maintenance of neoplastic growth in cancer (3–8). Aberrant differentiation of cancer stem cell-derived descendant cells rather than genetic instability accounts for the broad phenotypic heterogeneity among tumor cells of clonal

Miriam Dwek et al. (eds.), *Metastasis Research Protocols*, Methods in Molecular Biology, vol. 878,
DOI 10.1007/978-1-61779-854-2_13, © Springer Science+Business Media, LLC 2012

origin (1). Stem cell-like cancer cells have also been implicated in chemoresistance, tumor dormancy, and metastasis (9, 10). Hence, stem cell-like cancer cells represent a promising target for novel prognostic and therapeutic approaches in cancer medicine. Consequently, fractionation of clonal tumor cell populations into phenotypically and functionally diverse subsets, dissection of their interrelationships, and assessment of cancer stem cell properties, such as enhanced clonogenic growth in vitro, increased tumorigenicity in vivo, differentiation capacity, and asymmetric cell division, have become central issues in cancer research. Given the notorious difficulties associated with the fractionation of human primary tumor tissue into viable cell subsets (11), the identification of robust cancer stem cell pattern in human cancer cell lines has tapped an important alternative source for the systematic investigation of stem cell-like cancer cells (12–18). Several cell separation techniques have empirically been demonstrated to enrich stem cell-like cancer cells. These include the isolation of E-fraction cells by Percoll gradient centrifugation (19), the manual picking of holoclones from monolayer cancer cell cultures (13), and the flow cytometric isolation of tumor cells exhibiting the $CD44^{+}/CD24^{-/low}$ phenotype (4, 15). Based on the assumption that adult tissue stem cells and stem cell-like cancer cells might represent biologically related cell types with similar properties, several additional techniques, which were originally developed for the isolation of adult tissue stem cells, have successfully been adopted for the purification of stem cell-like cancer cells. This applies to the selective propagation of stem cell-like cancer cells as nonadherent tumor spheres (16, 17), the purification of tumor cells expressing the neuronal stem cell marker CD133 (5, 20–22), the isolation of cancer cells with high ALDH1 enzymatic activity (23, 24), and the purification of side population (SP) cells (12, 25–30), which is the subject of this chapter. The flow cytometric detection and isolation of SP cells takes advantage of their ability to actively pump out the fluorescent dye Hoechst 33342 (H33342) through the ATP-binding cassette (ABC) transporter ABCG2, which is a conserved feature of stem cells from a wide variety of sources (31, 32). However, as the expression of *ABCG2* has also been detected in some differentiated cell types, such as endothelial cells (33), the SP cell phenotype should be interpreted with caution. H33342 is a DNA-binding dye that is excited by UV wavelengths and exhibits an emission maximum at 450 nm (blue) (Fig. 1a). Importantly, H33342 also emits red light, although much dimmer than at the emission peak wavelength. Depending on the chromatin structure of the DNA that H33342 binds to, a slight red-shift of the emission spectrum occurs (34). H33342 labeling and detection are performed on viable cells (Fig. 1b). Simultaneous display of H33342-labeled cells at blue and red emission wavelengths (dual wavelength fluorescence analysis) reveals distinct cell populations, which are determined by their specific DNA content, chromatin structure, and H33342

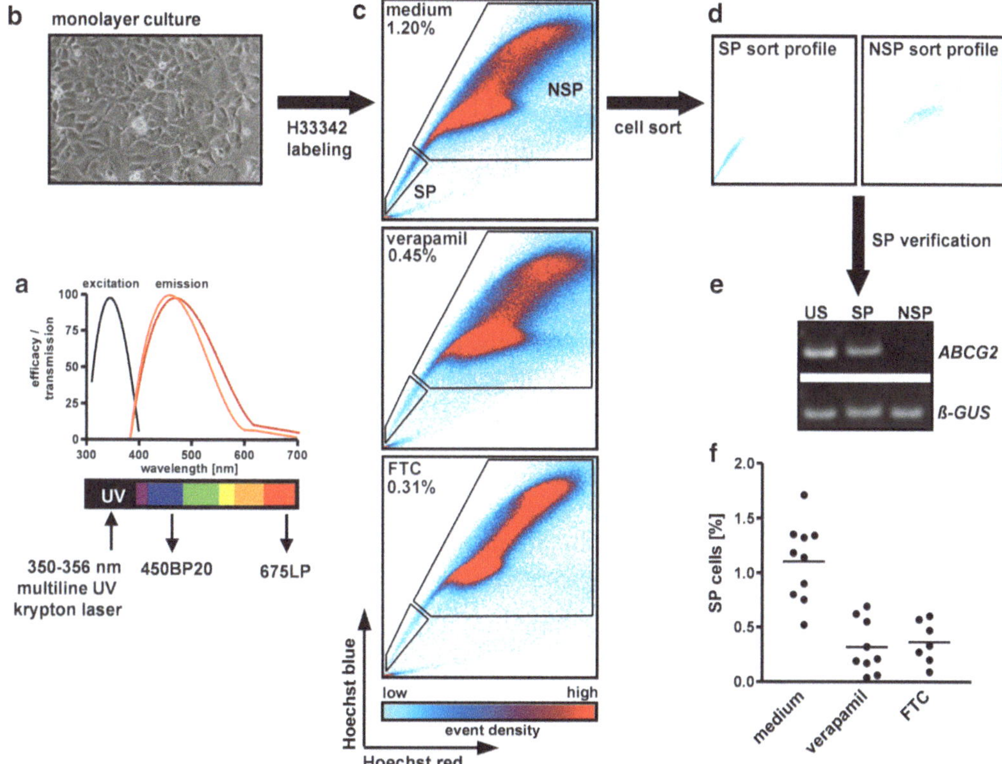

Fig. 1. Principle and work flow of the detection of putative cancer stem cells of the SP phenotype in human adherent tumor cell cultures. Data presented correspond to the CAL-51 human breast cancer cell line. (**a**) Schematic presentation of the general fluorescent dye properties of H33342. The H33342 dye is excited by UV wavelengths (350–356 nm multiline UV laser). Blue (450BP20) and red (675LP) fluorescence are collected. Depending on the chromatin structure of the DNA that H33342 binds to, a slight red-shift in the emission spectrum occurs (allegorized by a *light red* and a *dark red line*) (34). (**b**) Monolayer cancer cell culture. (**c**) H33342 dual wave length fluorescence analysis using verapamil and FTC for SP inhibition. Percentages of SP cells are given in the *upper left* corner of the plots. (**d**) Reanalysis of freshly sorted cancer SP and NSP cells. (**e**) SP verification by RT-PCR-based demonstration of a restriction of *ABCG2* mRNA expression to cancer SP cells. *US* unsorted cancer cells; *NSP* non-SP cancer cells. Amplification of *ß-GUS* serves as a control for the integrity of the cDNA preparations. (**f**) Repeated assessments of SP percentages in the same cancer cell culture by dual wavelength fluorescence analysis with verapamil and FTC. Each *dot* corresponds to an independent experiment.

efflux capacity (Fig. 1c) (34). Major populations regularly resolved by H33342 dual wavelength fluorescence analysis correspond to cells in the G_0/G_1-, S-, and G_2/M-phase of the cell cycle, which together compose the non-SP (NSP) cells, and the SP (Fig. 1c). Dead cells are excluded from the analysis. This is achieved by an additional propidium iodide (PI) staining of the H33342-labeled cell preparation, which allows the discrimination of dead cells by means of their extremely bright red PI fluorescence. Cells with an intrinsic capacity to actively efflux H33342 form a distinctly shaped tail of dimly stained cells, termed the side population (Fig. 1c). Inhibition of H33342 efflux by the broad range ABC transporter inhibitor verapamil or the ABCG2-specific inhibitor fumitremorgin c (FTC) results in a diminished appearance of SP cells (Fig. 1c, f)

(30, 32, 35). This serves as a provisional verification that the H33342 dimness of these cells is indeed due to active efflux and is not due to improper or inhomogeneous H33342 staining (see Note 1). In addition, RT-PCR-based verification of *ABCG2* mRNA restriction to the H33342 dim cell population is required as a definite proof of the genuine nature of the SP detected by dual wavelength fluorescence analysis (Fig. 1d, e) (30, 31). Originally, SP cells were discovered as a rare population of Sca-1^{+} hematopoietic stem cells in murine bone marrow (31, 35). Human bone marrow SP cells represent a fraction of $CD34^{-}$ hematopoietic stem cells (36). Interestingly, adult tissue stem cells from a wide variety of organs, such as the skeletal muscle, the liver, and the mammary gland, also display the SP cell phenotype (34, 37–40). These findings gained the notion that the SP cell phenotype is a universal stem cell feature (34). With respect to human solid tumors, Scharenberg and colleagues first demonstrated the presence of intrinsic SP cells among A549 non-small cell lung cancer cells (32). At the same time, Kim and colleagues reported that the MCF-7 breast cancer cell line harbors a very small fraction of intrinsic SP cells (41). However, Kondo and colleagues were the first to demonstrate that SP cells identified in the rat glioma cell line C6 represent a population of highly tumorigenic, undifferentiated cancer cells, which generate NSP cells as heterologous descendant cells with disparate differentiation attributes (12). Hence, cancer SP cells were defined as a population of cancer stem cells (12, 42). Although this understanding of cancer SP cells has partly been challenged (43–45), there are numerous confirmatory studies that support a cancer stem cell function of cancer SP cells (25–29, 46–55). Interestingly, human cancer cell lines void of SP cells, such as the Hep-G2 cell line, appear to be extraordinarily rare (25–29, 46–55). Consistently, SP cells have been detected in most cases of a yet limited number of primary tumor cell preparations analyzed for SP cells (25, 28, 48). In summary, cancer SP cells are a conserved phenomenon in human solid tumors and tumor cell lines and warrant further investigation regarding their stem cell-like features and their contribution to tumor growth, progression, and metastasis. Here we provide a detailed protocol for the detection and isolation of putative cancer stem cells of the SP phenotype in human adherent breast cancer cell cultures.

2. Materials

2.1. SP Detection by H33342 Dual Wavelength Fluorescence Analysis

1. Dulbecco's modified Eagle's medium (DMEM, Invitrogen/Gibco, Karlsruhe, Germany) supplemented with 5% fetal calf serum (FCS, Invitrogen/Gibco).
2. Phosphate-buffered saline (PBS, Invitrogen/Gibco).

3. PBS supplemented with 0.5% (w/v) bovine serum albumin (BSA, Sigma, Deisenhofen, Germany) and 2 mM EDTA (Invitrogen/Gibco).
4. 0.5 mg/ml trypsin/0.22 mg/ml EDTA (PAA, Pasching, Austria).
5. H33342 (Sigma) dissolved in autoclaved double-distilled (dd) water at a concentration of 1 mg/ml, stored in aliquots at 4°C in the dark. This is a 200× concentrated stock solution. The standard final H33342 concentration during H33342 labeling is 5 μg/ml.
6. Verapamil (Sigma) dissolved in autoclaved dd water at a concentration of 10 mM, stored in aliquots at 4°C. This is a 200× concentrated stock solution. The standard final verapamil concentration during H33342 labeling for SP inhibition studies is 50 μM.
7. FTC (Calbiochem, Darmstadt, Germany) dissolved in ethanol at a concentration of 1 mM. This is a 100× concentrated stock solution. The standard final FTC concentration during H33342 labeling for SP inhibition studies is 10 μM.
8. Propidium iodide (Sigma) dissolved in autoclaved dd water at a concentration of 5 mg/ml. This is a 100× stock solution. The final concentration during dual wavelength fluorescence analysis is 50 μg/ml.
9. 75-cm^2 adherent tissue culture flasks (Fisher Scientific, Schwerte, Germany).
10. 75 × 12 mm flow cytometry tubes with a 35-μm cell strainer in the cap (BD Bioscience, Erembodegem, Belgium).
11. Flow cytometer equipped with a UV laser for excitation of H33342. Near UV solid-state lasers (370 nm) have also been successfully used by some groups for the detection of SP cells. A 450/20-nm band pass filter is used in front of the detector for H33342 blue fluorescence and a 675-nm long pass filter for the detection of H33342 red fluorescence. A 610-nm short pass dichroic mirror separates the two emission wave lengths. In our studies, we use the multiline UV (350–365 nm) of a Coherent Innova 90C krypton laser mounted on a MoFlo cell sorter (Cytomation, Fort Collins, USA) (see Note 2).

2.2. SP Verification by RT-PCR-Based ABCG2 mRNA Expression Analysis

2.2.1. RNA Extraction

1. Trizol reagent (Invitrogen).
2. Chloroform.
3. Isopropanol.
4. Diethyl pyrocarbonate-treated water (DEPC water, Ambion, Austin, Texas, USA).
5. 3 M Na acetate pH 5.2.

6. 70% and 100% ethanol.
7. Glycogen, 20 μg/μl (Roche, Mannheim, Germany).

2.2.2. ABCG2 Reverse Transcription (RT)-PCR

1. 10× concentrated universal PCR reaction buffer without $MgCl_2$ (Invitrogen).
2. 25 nM $MgCl_2$ (Invitrogen).
3. 10 mM dATP, dGTP, dCTP, and dTTP (Fermentas, Leon-Rot, Germany).
4. 25 μM ROX reference dye (Invitrogen).
5. 12.5× concentrated Sybr Green I dye (Invitrogen).
6. Desalted oligonucleotide primers, 10 pmol/μl each, as indicated under Subheading 3.2.2.
7. Platinum Taq DNA polymerase (Invitrogen).
8. PCR thermal cycler.
9. PCR product gel loading dye (Fermentas).
10. Equipment and reagents for polyacrylamide (PAA) gel electrophoresis:
    - 30% PAA.
    - 1× TBE buffer (90 mM Tris-borate, pH 8.0/2 mM EDTA).
    - 10% (w/v) ammonium persulfate (APS).
    - Tetramethyethylenediamine (TEMED).

3. Methods

3.1. SP Detection by H33342 Dual Wavelength Fluorescence Analysis

1. Seed cells at low density into 75-cm² adherent tissue culture flasks (see Notes 1 and 3). For each tumor cell type, three tissue culture flasks are required (flask "A" for H33342 labeling, flask "B" for H33342 labeling in the presence of verapamil, and flask "C" for H33342 labeling in the presence of FTC) (see Note 4).
2. Prewarm 25 ml DMEM/5% FCS medium, 130 ml PBS, and 25 ml trypsin/EDTA in a 37°C water bath. Store another aliquot of 160 ml PBS and an aliquot of 2 ml PBS/0.5% BSA/2 mM EDTA on ice.
3. Prepare five individual 15 ml tubes according to the scheme presented in Table 1 using the H33342, verapamil, and FTC stock solutions and the prewarmed DMEM/5% FCS medium. Place these tubes in a 37°C water bath.
4. Remove the growth medium from tissue culture flask "B" and rinse cells once with 10 ml prewarmed PBS. Remove the PBS. Add 3.5 ml verapamil solution from tube "B1" into tissue

Table 1
Formulation for stock solutions of H33342, verapamil and FTC in DMEM/5% FCS

			Add stock solutions			Resulting concentration		
Tube number	Tube label	DMEM/5% FCS	H33342 stock solution (1 mg/ml)	Verapamil stock solution (10 mM)	FTC stock solution (1 mM)	H33342	Verapamil	FTC
1	A1	7.960 ml	40 μl	–	–	5 μg/ml	–	–
2	B1	3.960 ml	–	40 μl	–	–	100 μM	–
3	B2	3.960 ml	40 μl	–	–	10 μg/ml	–	–
4	C1	3.920 ml	–	–	80 μl	–	–	20 μM
5	C2	3.960 ml	40 μl	–	–	10 μg/ml	–	–

culture flask "B." Carefully tilt the flask several times and pre-incubate the cells in this 100 μM verapamil solution for 10 min at 37°C.

5. Meanwhile, remove the growth medium from tissue culture flask "C" and rinse cells once with 10 ml prewarmed PBS. Remove the PBS. Add 3.5 ml from tube "C1" into tissue culture flask "C." Carefully tilt the flask several times and pre-incubate the cells in this 20 μM FTC solution for 10 min at 37°C.
6. Meanwhile, remove the growth medium from tissue culture flask "A" and rinse cells once with 10 ml prewarmed PBS. Remove the PBS. Add 7 ml from tube "A1" into tissue culture flask "A." Carefully tilt the flask several times and incubate the cells in this 5 μg/ml H33342 solution for 90 min at 37°C.
7. When the preincubation of tissue culture flask "B" is completed, directly add 3.5 ml H33342 solution from tube "B2" into tissue culture flask "B," which already contains 3.5 ml verapamil solution from tube "B1." Carefully tilt the flask several times to mix the H33342 and the verapamil solutions and incubate the cells in the resulting 5 μg/ml H33342/50 μM verapamil solution for 90 min at 37°C.
8. When the preincubation of tissue culture flask "C" is completed, directly add 3.5 ml H33342 solution from tube "C2" into tissue culture flask "C," which already contains 3.5 ml FTC solution from tube "C1." Carefully tilt the flask several times to mix the H33342 and the FTC solution and incubate the cells in the resulting 5 μg/ml H33342/10 μM FTC solution for 90 min at 37°C.

Table 2
Formulation for stock solutions of H33342, verapamil and FTC in trypsin EDTA

			Add stock solutions			Resulting concentration		
Tube number	Tube label	Trypsin/ EDTA	H33342 stock solution (1 mg/ml)	Verapamil stock solution (10 mM)	FTC stock solution (1 mM)	H33342	Verapamil	FTC
6	AT	7.960 ml	40 μl	–	–	5 μg/ml	–	–
7	BT	7.920 ml	40 μl	40 μl	–	5 μg/ml	50 μM	–
8	CT	7.880 ml	40 μl	–	80 μl	5 μg/ml	–	10 μM

9. During the 90-min H33342 incubation period, prepare three individual 15 ml tubes according to the scheme presented in Table 2 using the H33342, verapamil, and FTC stock solutions and the prewarmed trypsin/EDTA. Place these tubes in a 37°C water bath.
10. When the 90-min H33342 incubation period of tissue culture flasks "A", "B," and "C" is completed, remove the H33342 labeling solutions and briefly rinse the cells three times with 10 ml prewarmed PBS.
11. Add 7 ml prewarmed trypsin/EDTA from tube "AT" into culture flask "A." Likewise, add 7 ml prewarmed trypsin/EDTA from tube "BT" and "CT" into tissue culture flasks "B" and "C," respectively. Incubate cells for 7 min at 37°C.
12. Meanwhile, label three individual 50 ml tubes as "A," "B," and "C" and add 30 ml ice-cold PBS per tube.
13. Directly transfer the trypsin/EDTA-released cells from tissue culture flask "A" into the 50-ml tube labeled as "A," which already holds 30 ml ice-cold PBS. Re-suspend the cells by pipetting up and down several times with a 5-ml pipette. Proceed with tissue culture flasks "B" and "C" in the same way.
14. Pellet the cells by centrifugation at 4°C. Re-suspend the cells in 10 ml ice-cold PBS. Transfer the cells in corresponding 15 ml tubes labeled as "A," "B," and "C".
15. Pellet the cells by centrifugation at 4°C and wash the cells once in 10 ml ice-cold PBS.
16. Pellet the cells by centrifugation at 4°C. Re-suspend the cells in 500 μl ice-cold PBS/0.5% BSA/2 mM EDTA. Filter the cells through 35-μm cell strainers and transfer the cells in

corresponding 75 × 12 mm flow cytometry tubes labeled as "A," "B," and "C".

17. Before flow cytometric analysis, add 5 μl propidium iodide stock solution to each flow cytometry tube and vortex well. Keep tubes chilled during analysis.

18. Analyze 1×10^5 cells from tube "A" on a UV or near-UV laser equipped flow cytometer using linear amplification for blue and red fluorescence signals. Adjust voltages of the photomultiplier tubes so that unstained cells are seen in the lower left corner of a bivariate dot plot with the red fluorescence signal on the abscissa and the blue fluorescence signal on the ordinate. PI-stained cells should appear on a vertical line at the right end of the plot. Center the main population of H33342-labeled cells. Depending on the cell type and the culture conditions you will be able to distinguish a G_0/G_1 population and a cloud of cells of higher fluorescence in the upper right corner representing S- and G_2/M cells with a higher DNA content. Use pulse width versus heights for discrimination of cell clusters. Gate out clustered cells (high pulse width) and dead cells (extremely bright red fluorescence). Define the distinctly shaped tail of H33342 dim cells, if present, as the SP and as region R_1. Assess SP cell frequency relative to all NSP cells.

19. Verify that the relative SP cell frequency drops significantly in cells from tube "B" (H333342 labeling in the presence of verapamil) and in cells from tube "C" (H33342 labeling in the presence of FTC) to preclude that the H33342 dim cells represent an artifact due to inhomogeneous dye distribution (see Note 5).

20. Sort 1×10^5 SP cells and 1×10^5 NSP cells into separate 15 ml tubes. Pellet the cells and carefully discard the supernatant. Next, re-suspend the cells in 100 μl ice-cold PBS and transfer them into separate 0.5 ml tubes (see Notes 6 and 7).

21. Pellet the cells and carefully discard the supernatant. Next, add 100 μl Trizol reagent and re-suspend by pipetting up and down several times with a 100-μl pipette tip. Store the Trizol samples at −80°C until further processing. Also prepare a 100-μl Trizol sample of the leftover unsorted (US) cells.

3.2. SP Verification by RT-PCR-Based ABCG2 mRNA Expression Analysis

3.2.1. RNA Extraction

1. Add 20 μl chloroform to each Trizol sample from step 21 under Subheading 3.1 and vortex well. Process each sample as outlined below.

2. Incubate at room temperature (RT) for 10 min.

3. Centrifuge for 15 min at 4°C and at 13,000 rpm (~17,900 × *g*) in a table-top microcentrifuge.

4. Transfer the upper, aqueous phase (~50 μl) to a fresh 0.5-ml tube.

5. Add 50 μl isopropanol and 1 μl glycogen (20 μg/μl).

6. Incubate at RT for 10 min.
7. Centrifuge for 10 min at 4°C and at 13,000 rpm (~17,900 × *g*) in a table-top microcentrifuge.
8. Carefully discard the supernatant without removing the precipitate pellet from the bottom of the tube (see Note 8).
9. Add 50 μl 70% ethanol and vortex well.
10. Centrifuge for 5 min at 4°C and at 13,000 rpm (~17,900 × *g*) in a table-top microcentrifuge.
11. Carefully discard the supernatant without removing the precipitate pellet from the tube.
12. Add 9 μl DEPC water, 1 μl 3 M Na acetate pH 5.2, and 25 μl ethanol and vortex well.
13. Incubate at –20°C overnight.
14. Centrifuge for 30 min at 4°C and at 13,000 rpm (~17,900 × *g*) in a table-top microcentrifuge.
15. Carefully discard the supernatant without removing the precipitate pellet from the tube.
16. Add 50 μl 70% ethanol and vortex well.
17. Centrifuge for 5 min at 4°C and at 13,000 rpm (~17,900 × *g*) in a table-top microcentrifuge.
18. Carefully discard the supernatant without removing the precipitate pellet from the tube.
19. Allow the precipitate RNA pellet to air-dry.
20. Resolve the precipitate RNA pellet in 7 μl DEPC water.
21. Optional: Measure OD_{260} and determine the concentration. Howsoever, immediately proceed with step 1 outlined under Subheading 3.2.2.

3.2.2. ABCG2 Reverse Transcription (RT)-PCR

1. Perform cDNA synthesis for each RNA sample (SP, NSP and US cells) with reagents of choice using the entire RNA yield in a final reaction volume of 20 μl.
2. For each cDNA template, two PCR reactions are carried out. These correspond to the *ABCG2* gene (forward, 5′-CTGAGAT CCTGAGCCTTTGG-3′; reverse, 5′-AAGCCATTGGTGTTT CCTTG-3′) and the reference gene *ß-GUS* (forward, 5′-CTC ATTTGGAATTTTGCCGATT-3′; reverse, 5′-CCGAGTGAA GATCCCCTTTTTA-3′) (see Note 9).
3. The constituents of each PCR reaction of 25 μl total volume are as follows: 15.5 μl dd water, 2.5 μl universal PCR reaction buffer, 2.5 μl $MgCl_2$, 0.5 μl dNTPs, 0.25 μl ROX reference dye, 2 μl Sybr Green I dye, 0.6 μl mixed (1:1) forward and reverse oligonucleotide primers, 0.15 μl Platinum Taq DNA polymerase, and 1 μl cDNA template from step 1 (see Note 10).

4. Perform the PCR with an initial polymerase activation (5 min at 95°C) followed by 40 two-step PCR cycles comprising denaturing (15 s at 95°C), annealing, and extension (1 min at 60°C).
5. Check PCR products on a 6% PAA-TBE gel. The *ß-GUS*, which serves as a control for the integrity of the cDNA preparation, is amplified as a 81-bp product and *ABCG2* as a 123-bp product. If SP detection and SP cell sort were carried out successfully, the *ABCG2* product is absent in the NSP cell sample and present in the SP cell and US cell sample.
6. Optional: Repeat steps 2–4 several times as independent PCR runs and use Sybr Green I fluorescence for real-time detection of PCR products. Calculate the relative *ABCG2/ß-GUS* mRNA ratio in US, SP, and NSP cells. In comparison to US cells, cancer SP cells show a 10- to 30-fold increased *ABCG2/ß-GUS* mRNA ratio.

4. Notes

1. Cell cohesiveness affects H33342 staining properties of epithelial cancer cells. Cell clustering during H33342 labeling in cell suspension may result in an inhomogeneous distribution of the fluorescent dye among individual cells and may blur the dual wavelength fluorescence analysis profile or may even imitate a disproportionally large SP. Contrary to SP analysis of bone marrow, it is therefore advantageous to perform the H33342 labeling of epithelial cancer cells in subconfluent monolayer cultures (41).
2. Cancer SP cell analyses have been conducted with different flow cytometers equipped with slightly different lasers, optical systems, and filter setups (12, 25–30, 32, 48, 50, 56). It has been suggested that this may explain the slight interlaboratory variation of SP cell percentages in some common cancer cell lines (26). However, interindividual variations in performing the H33342 labeling, the source of the used cell lines, and the cell culture conditions have an important influence on the results and may therefore also account for this phenomenon (see below).
3. It has clearly been demonstrated that cell culture density affects the SP cell prevalence in cancer cell lines (50). It has been suggested that cancer SP cells preferentially prevail at low culture density (50). Therefore, it is recommended to perform H33342 labeling on subconfluent monolayer cultures under standardized conditions. For excellent staining conditions, seed 0.4×10^6

cells into a 75-cm^2 culture flask and allow the cells to grow for 1-2 doubling times before H33342 labeling.

4. It is recommended to determine the sensitivity towards verapamil in the cells of interest prior to the SP analyses. Some tumor cell cultures are extraordinarily sensitive towards verapamil exposure and thus require the use of FTC for SP inhibition studies. Other cultures may tolerate exposure to high concentrations of verapamil without loss of cell viability (30). Thus, depending on the cell type, SP inhibition studies are conducted with verapamil concentrations ranging from 20 to 250 μM (40, 57).
5. Ultimately, this requires repeated SP analyses in independently processed cultures (Fig. 1f).
6. As the SP cell prevalence is extraordinarily low in most cancer cell cultures, this may require an upscaling of the experiment size. For instance, to obtain 1×10^5 SP cells from a culture with an SP prevalence of about 1%, at least ten individual 75-cm^2 culture flasks (each containing 1×10^6 cells at the day of H33342 labeling) have to be processed as a batch of flasks in parallel.
7. To assure correct cell separation, it is useful to perform a brief flow cytometric reanalysis of a small aliquot of the freshly sorted cells (Fig. 1d).
8. The precipitate pellet may be barely visible or invisible due to the small sample size (1×10^5 cells).
9. The *ABCG2* RT-PCR oligonucleotide primers indicated amplify a 123-bp product that spans over the *ABCG2* start codon and the intron 1 (~18.5 kb).
10. This RT-PCR may be carried out as either a qualitative or a quantitative RT-PCR. As cancer NSP cells entirely lack *ABCG2* mRNA, a qualitative RT-PCR is usually sufficient and demonstrates selective *ABCG2* expression in the corresponding SP cells fraction (26, 30).

References

1. Wicha MS, Liu S, Dontu G (2006) Cancer stem cells: an old idea—a paradigm shift. Cancer Res 66:1883–1890, discussion 1895–1896
2. Mackenzie IC (2005) Retention of stem cell patterns in malignant cell lines. Cell Prolif 38:347–355
3. Wulf GG, Wang RY, Kuehnle I, Weidner D, Marini F, Brenner MK, Andreeff M, Goodell MA (2001) A leukemic stem cell with intrinsic drug efflux capacity in acute myeloid leukemia. Blood 98:1166–1173
4. Al-Hajj M, Wicha MS, Benito-Hernandez A, Morrison SJ, Clarke MF (2003) Prospective identification of tumorigenic breast cancer cells. Proc Natl Acad Sci USA 100:3983–3988
5. Singh SK, Clarke ID, Terasaki M, Bonn VE, Hawkins C, Squire J, Dirks PB (2003) Identification of a cancer stem cell in human brain tumors. Cancer Res 63:5821–5828
6. Hemmati HD, Nakano I, Lazareff JA, Masterman-Smith M, Geschwind DH, Bronner-Fraser M, Kornblum HI (2003) Cancerous

stem cells can arise from pediatric brain tumors. Proc Natl Acad Sci USA 100:15178–15183

7. Collins AT, Berry PA, Hyde C, Stower MJ, Maitland NJ (2005) Prospective identification of tumorigenic prostate cancer stem cells. Cancer Res 65:10946–10951
8. O'Brien CA, Pollett A, Gallinger S, Dick JE (2007) A human colon cancer cell capable of initiating tumour growth in immunodeficient mice. Nature 445:106–110
9. Dean M, Fojo T, Bates S (2005) Tumour stem cells and drug resistance. Nat Rev Cancer 5:275–284
10. Balic M, Lin H, Young L, Hawes D, Giuliano A, McNamara G, Datar RH, Cote RJ (2006) Most early disseminated cancer cells detected in bone marrow of breast cancer patients have a putative breast cancer stem cell phenotype. Clin Cancer Res 12:5615–5621
11. Hill RP (2006) Identifying cancer stem cells in solid tumors: case not proven. Cancer Res 66:1891–1895, discussion 1890
12. Kondo T, Setoguchi T, Taga T (2004) Persistence of a small subpopulation of cancer stem-like cells in the C6 glioma cell line. Proc Natl Acad Sci USA 101:781–786
13. Locke M, Heywood M, Fawell S, Mackenzie IC (2005) Retention of intrinsic stem cell hierarchies in carcinoma-derived cell lines. Cancer Res 65:8944–8950
14. Matsui W, Huff CA, Wang Q, Malehorn MT, Barber J, Tanhehco Y, Smith BD, Civin CI, Jones RJ (2004) Characterization of clonogenic multiple myeloma cells. Blood 103:2332–2336
15. Phillips TM, McBride WH, Pajonk F (2006) The response of CD24(-/low)/CD44+ breast cancer-initiating cells to radiation. J Natl Cancer Inst 98:1777–1785
16. Ponti D, Costa A, Zaffaroni N, Pratesi G, Petrangolini G, Coradini D, Pilotti S, Pierotti MA, Daidone MG (2005) Isolation and in vitro propagation of tumorigenic breast cancer cells with stem/progenitor cell properties. Cancer Res 65:5506–5511
17. Fang D, Nguyen TK, Leishear K, Finko R, Kulp AN, Hotz S, Van Belle PA, Xu X, Elder DE, Herlyn M (2005) A tumorigenic subpopulation with stem cell properties in melanomas. Cancer Res 65:9328–9337
18. Zhou J, Wulfkuhle J, Zhang H, Gu P, Yang Y, Deng J, Margolick JB, Liotta LA, Petricoin E 3rd, Zhang Y (2007) Activation of the PTEN/mTOR/STAT3 pathway in breast cancer stem-like cells is required for viability and maintenance. Proc Natl Acad Sci USA 104: 16158–16163
19. Resnicoff M, Medrano EE, Podhajcer OL, Bravo AI, Bover L, Mordoh J (1987) Subpopulations of MCF7 cells separated by Percoll gradient centrifugation: a model to analyze the heterogeneity of human breast cancer. Proc Natl Acad Sci USA 84: 7295–7299
20. Ma S, Chan KW, Hu L, Lee TK, Wo JY, Ng IO, Zheng BJ, Guan XY (2007) Identification and characterization of tumorigenic liver cancer stem/progenitor cells. Gastroenterology 132: 2542–2556
21. Ma S, Lee TK, Zheng BJ, Chan KW, Guan XY (2008) CD133+ HCC cancer stem cells confer chemoresistance by preferential expression of the Akt/PKB survival pathway. Oncogene 27:1749–1758
22. Yin S, Li J, Hu C, Chen X, Yao M, Yan M, Jiang G, Ge C, Xie H, Wan D, Yang S, Zheng S, Gu J (2007) CD133 positive hepatocellular carcinoma cells possess high capacity for tumorigenicity. Int J Cancer 120: 1444–1450
23. Charafe-Jauffret E, Ginestier C, Iovino F, Wicinski J, Cervera N, Finetti P, Hur MH, Diebel ME, Monville F, Dutcher J, Brown M, Viens P, Xerri L, Bertucci F, Stassi G, Dontu G, Birnbaum D, Wicha MS (2009) Breast cancer cell lines contain functional cancer stem cells with metastatic capacity and a distinct molecular signature. Cancer Res 69:1302–1313
24. Ginestier C, Hur MH, Charafe-Jauffret E, Monville F, Dutcher J, Brown M, Jacquemier J, Viens P, Kleer CG, Liu S, Schott A, Hayes D, Birnbaum D, Wicha MS, Dontu G (2007) ALDH1 is a marker of normal and malignant human mammary stem cells and a predictor of poor clinical outcome. Cell Stem Cell 1:555–567
25. Hirschmann-Jax C, Foster AE, Wulf GG, Nuchtern JG, Jax TW, Gobel U, Goodell MA, Brenner MK (2004) A distinct "side population" of cells with high drug efflux capacity in human tumor cells. Proc Natl Acad Sci USA 101:14228–14233
26. Patrawala L, Calhoun T, Schneider-Broussard R, Zhou J, Claypool K, Tang DG (2005) Side population is enriched in tumorigenic, stem-like cancer cells, whereas ABCG2+ and ABCG2- cancer cells are similarly tumorigenic. Cancer Res 65:6207–6219
27. Haraguchi N, Utsunomiya T, Inoue H, Tanaka F, Mimori K, Barnard GF, Mori M (2006) Characterization of a side population of cancer cells from human gastrointestinal system. Stem Cells 24:506–513
28. Szotek PP, Pieretti-Vanmarcke R, Masiakos PT, Dinulescu DM, Connolly D, Foster R,

Dombkowski D, Preffer F, Maclaughlin DT, Donahoe PK (2006) Ovarian cancer side population defines cells with stem cell-like characteristics and Mullerian Inhibiting Substance responsiveness. Proc Natl Acad Sci USA 103: 11154–11159

29. Chiba T, Kita K, Zheng YW, Yokosuka O, Saisho H, Iwama A, Nakauchi H, Taniguchi H (2006) Side population purified from hepatocellular carcinoma cells harbors cancer stem cell-like properties. Hepatology 44:240–251
30. Christgen M, Ballmaier M, Bruchhardt H, von Wasielewski R, Kreipe H, Lehmann U (2007) Identification of a distinct side population of cancer cells in the Cal-51 human breast carcinoma cell line. Mol Cell Biochem 306:201–212
31. Zhou S, Schuetz JD, Bunting KD, Colapietro AM, Sampath J, Morris JJ, Lagutina I, Grosveld GC, Osawa M, Nakauchi H, Sorrentino BP (2001) The ABC transporter Bcrp1/ABCG2 is expressed in a wide variety of stem cells and is a molecular determinant of the side-population phenotype. Nat Med 7:1028–1034
32. Scharenberg CW, Harkey MA, Torok-Storb B (2002) The ABCG2 transporter is an efficient Hoechst 33342 efflux pump and is preferentially expressed by immature human hematopoietic progenitors. Blood 99:507–512
33. Diestra JE, Scheffer GL, Catala I, Maliepaard M, Schellens JH, Scheper RJ, Germa-Lluch JR, Izquierdo MA (2002) Frequent expression of the multi-drug resistance-associated protein BCRP/MXR/ABCP/ABCG2 in human tumours detected by the BXP-21 monoclonal antibody in paraffin-embedded material. J Pathol 198:213–219
34. Camargo FD, Ramos CA, Goodell MA (2004) Handbook of stem cells. Academic, Orlando, FL, pp 329–336
35. Goodell MA, Brose K, Paradis G, Conner AS, Mulligan RC (1996) Isolation and functional properties of murine hematopoietic stem cells that are replicating in vivo. J Exp Med 183:1797–1806
36. Goodell MA, Rosenzweig M, Kim H, Marks DF, DeMaria M, Paradis G, Grupp SA, Sieff CA, Mulligan RC, Johnson RP (1997) Dye efflux studies suggest that hematopoietic stem cells expressing low or undetectable levels of CD34 antigen exist in multiple species. Nat Med 3:1337–1345
37. Challen GA, Little MH (2006) A side order of stem cells: the SP phenotype. Stem Cells 24:3–12
38. Dontu G, Abdallah WM, Foley JM, Jackson KW, Clarke MF, Kawamura MJ, Wicha MS (2003) In vitro propagation and transcriptional profiling of human mammary stem/progenitor cells. Genes Dev 17:1253–1270
39. Behbod F, Xian W, Shaw CA, Hilsenbeck SG, Tsimelzon A, Rosen JM (2006) Transcriptional profiling of mammary gland side population cells. Stem Cells 24:1065–1074
40. Alvi AJ, Clayton H, Joshi C, Enver T, Ashworth A, Vivanco MM, Dale TC, Smalley MJ (2003) Functional and molecular characterisation of mammary side population cells. Breast Cancer Res 5:R1–R8
41. Kim M, Turnquist H, Jackson J, Sgagias M, Yan Y, Gong M, Dean M, Sharp JG, Cowan K (2002) The multidrug resistance transporter ABCG2 (breast cancer resistance protein 1) effluxes Hoechst 33342 and is overexpressed in hematopoietic stem cells. Clin Cancer Res 8:22–28
42. Setoguchi T, Taga T, Kondo T (2004) Cancer stem cells persist in many cancer cell lines. Cell Cycle 3:414–415
43. Zheng X, Shen G, Yang X, Liu W (2007) Most C6 cells are cancer stem cells: evidence from clonal and population analyses. Cancer Res 67:3691–3697
44. Burkert J, Otto WR, Wright NA (2008) Side populations of gastrointestinal cancers are not enriched in stem cells. J Pathol 214:564–573
45. Adamski D, Mayol JF, Platet N, Berger F, Herodin F, Wion D (2007) Effects of Hoechst 33342 on C2C12 and PC12 cell differentiation. FEBS Lett 581:3076–3080
46. Bar EE, Chaudhry A, Lin A, Fan X, Schreck K, Matsui W, Vescovi AL, Piccirillo S, Dimeco F, Olivi A, Eberhart CG (2007) Cyclopamine-mediated hedgehog pathway inhibition depletes stem-like cancer cells in glioblastoma. Stem Cells 25:2524–2533
47. Zen Y, Fujii T, Yoshikawa S, Takamura H, Tani T, Ohta T, Nakanuma Y (2007) Histological and culture studies with respect to ABCG2 expression support the existence of a cancer cell hierarchy in human hepatocellular carcinoma. Am J Pathol 170:1750–1762
48. Ho MM, Ng AV, Lam S, Hung JY (2007) Side population in human lung cancer cell lines and tumors is enriched with stem-like cancer cells. Cancer Res 67:4827–4833
49. Wang J, Guo LP, Chen LZ, Zeng YX, Lu SH (2007) Identification of cancer stem cell-like side population cells in human nasopharyngeal carcinoma cell line. Cancer Res 67:3716–3724
50. Mitsutake N, Iwao A, Nagai K, Namba H, Ohtsuru A, Saenko V, Yamashita S (2007) Characterization of side population in thyroid cancer cell lines: cancer stem-like cells are enriched partly but not exclusively. Endocrinology 148:1797–1803

51. Woodward WA, Chen MS, Behbod F, Alfaro MP, Buchholz TA, Rosen JM (2007) WNT/beta-catenin mediates radiation resistance of mouse mammary progenitor cells. Proc Natl Acad Sci USA 104:618–623
52. Harris MA, Yang H, Low BE, Mukherje J, Guha A, Bronson RT, Shultz LD, Israel MA, Yun K (2008) Cancer stem cells are enriched in the side population cells in a mouse model of glioma. Cancer Res 68:10051–10059
53. Engelmann K, Shen H, Finn OJ (2008) MCF7 side population cells with characteristics of cancer stem/progenitor cells express the tumor antigen MUC1. Cancer Res 68:2419–2426
54. Yin L, Castagnino P, Assoian RK (2008) ABCG2 expression and side population abundance regulated by a transforming growth factor beta-directed epithelial-mesenchymal transition. Cancer Res 68:800–807
55. Kabashima A, Higuchi H, Takaishi H, Matsuzaki Y, Suzuki S, Izumiya M, Iizuka H, Sakai G, Hozawa S, Azuma T, Hibi T (2009) Side population of pancreatic cancer cells predominates in TGF-beta-mediated epithelial to mesenchymal transition and invasion. Int J Cancer 124:2771–2779
56. Xu JX, Morii E, Liu Y, Nakamichi N, Ikeda J, Kimura H, Aozasa K (2007) High tolerance to apoptotic stimuli induced by serum depletion and ceramide in side-population cells: high expression of CD55 as a novel character for side-population. Exp Cell Res 313:1877–1885
57. Clarke RB, Spence K, Anderson E, Howell A, Okano H, Potten CS (2005) A putative human breast stem cell population is enriched for steroid receptor-positive cells. Dev Biol 277:443–456

Chapter 14

Genes Involved in the Metastatic Cascade of Medullary Thyroid Tumours

Caroline Schreiber, Kirsten Vormbrock, and Ulrike Ziebold

Abstract

The process of how a benign tumour turns invasive and capable to survive in distant organs remains poorly understood, despite the evidence that metastasis formation is the primary cause of cancer patient mortality. This ignorance is partly due to the lack of appropriate animal models from which to investigate this complex process. The retinoblastoma (Rb) tumour suppressor pathway (pRb/E2F) is mutated in almost all human tumours, and a number of laboratories have now established pRb- or E2F-deficient mouse models. Consistent with the role of mutation in retinoblastoma in cancer biology, *Rb* heterozygous mice are prone to develop tumours. Among the ensuing tumours, the medullary thyroid carcinomas (MTCs) have a lessened tendency to form secondary cancers and metastases. Intriguingly, if an *E2f3* mutation is introduced in this genetic background, more aggressive MTCs develop, which metastasize more frequently. Gene chip microarrays, however, provide an unbiased approach for examining the genome-wide expression levels and enable identification of a large set of metastasis-enriched gene sets. The identified genes may simply represent putative markers of the disease stage. Alternatively, genes may be identified that causally determine a link to the onset of metastasis. We describe the use of gene chip microarrays for identification of putative markers enriched in metastatic mouse MTCs. The chapter details how the most promising candidates are verified using additional methods, such as quantitative real-time PCR. In this case, co-transfection of the E2F-transcription factor using a heterologous reporter gene system is suggestive of E2Fs directly regulating putative metastasis markers.

Key words: Retinoblastoma tumour suppressor protein, E2F, Metastasis, Vimentin, Microarray, Luciferase assay, Quantitative real-time PCR

1. Introduction

The retinoblastoma tumour suppressor protein (pRb) was the first tumour suppressor to be identified (1–3). The E2F transcription factors are the most widely studied interaction partners of pRb.

Miriam Dwek et al. (eds.), *Metastasis Research Protocols*, Methods in Molecular Biology, vol. 878,
DOI 10.1007/978-1-61779-854-2_14,

E2F1–3 all bind pRb and are responsible for the control of the cell cycle at the G1/S transition (4). A combined *Rb* and *E2f3*-knockout mouse model demonstrated that an absence of E2F3 results in an alteration to the tumour spectrum developing in *Rb*+/– mice. Although E2F3 loss was capable of suppressing the development of pituitary tumours, its absence also promoted the development of medullary thyroid carcinomas (MTCs) yielding metastases at an increased frequency (5). It is, therefore, tempting to speculate that E2F3 may control an important switch to the onset of metastasis in MTCs.

MTCs that arise from calcitonin-secreting C-cells are a rare but a clinically relevant problem. MTCs are known to metastasize frequently and this phenomenon accounts for the low 5-year survival rate of 50% in human patients. Importantly, for the sporadic MTC subtype (which comprises more than 75% of all cases), there is little data available as to the underlying genetic defects of these tumours (6). We exploited a combined *Rb/E2f3*-knockout mouse to identify markers of the metastatic cascade in MTCs. Gene chip microarray analysis was performed and compared for distinct sets of tumours. Among the gene set that was found over-expressed, and enriched in the metastatic subgroup, was *vimentin*. This intermediate filament molecule is known to support migration and has been reported to be up-regulated in cells undergoing the epithelial to mesenchymal transition, both processes associated with metastasis. A mild up-regulation of *vimentin* in the metastatic mouse MTCs was verified using quantitative real-time PCR. As the pRB-deficient MTCs contain deregulated E2Fs, the direct regulation of vimentin by E2Fs seemed likely. Deploying a dual-specific *Luciferase* reporter gene assay (7) with various *vimentin*-reporter constructs, we established that the *vimentin* promoter is E2F responsive. This chapter describes a gene array-based approach used to determine genes involved in the metastatic process and methods by which to verify findings of such array-based work.

2. Materials

2.1. RNA Preparation

1. Trizol (Invitrogen, Catalogue number 15596018) is toxic, thus always handle under a fume hood.
2. Phenol: Phenol is toxic, so always handle under a fume hood.
3. Chloroform: Chloroform needs to be handled under a fume hood.
4. Isopropanol.
5. Ammonium acetate is dissolved in DEPC-treated H_2O at 3 M, and then adjusted to pH 5.5 using acetic acid.

6. 75% v/v ethanol: Absolute ethanol is diluted with DEPC-treated H_2O (Applichem, Catalogue number A0881).
7. DEPC-treated H_2O; double-distillate H_2O supplemented with 1:10,000 DEPC is stirred at room temperature and autoclaved the next day to destroy any remaining RNase activity. DEPC is toxic, so always handle with care under a fume hood.
8. RNeasy kit (Qiagen).
9. T7-(dT) 24primer: GGCCAGTGAATTGTAATACGACTCA CTATAGGGAGGCGG-(dT24).
10. 1× first-strand buffer; dissolve 5× first-strand buffer in DEPC-treated H_2O.
11. 1× second-strand reaction buffer (Invitrogen, Catalogue number 10812-014).
12. DTT is dissolved in DEPC-treated H_2O at 1 M and stored at −20°C.
13. dNTP mix is dissolved in DEPC-treated H_2O at 10 mM and stored at −20°C.
14. SuperScript II reverse transcriptase (Invitrogen, Catalogue number 18064-022).
15. *Escherichia coli* DNA ligase (New England Biolabs, Catalogue number M0205).
16. *Escherichia coli* DNA polymerase I (New England Biolabs, Catalogue number M0209).
17. *Escherichia coli* RNase H (New England Biolabs, Catalogue number M0297).
18. T4 DNA polymerase (New England Biolabs, Catalogue number M0203).
19. BioArray high yield RNA transcription labeling kit (Enzo Diagnostics, Farmingdale, NY).
20. 5× Fragmentation buffer: 200 mM Tris acetate, pH 8.1, 150 mM Mg acetate, 500 mM K acetate.

2.2. Quantitative Real-Time PCR

1. Random primer (5′-NNN NNN-3′).
2. 5× first strand buffer (Invitrogen, Catalogue number supplied with Superscript II).
3. RNase Out (Invitrogen, 10777-019).
4. Reverse transcriptase (Invitrogen, supplied with Superscript II).
5. 10 μM primer mix.
6. 2× qPCR master-mix (2× SYBR-green, ABgene, Thermo Scientific, Surrey, UK).
7. 96-well PCR plate.
8. Optical qPCR seal.

2.3. Cell Culture

1. Dulbecco's modified Eagle's medium (DMEM) supplemented with 10% v/v foetal bovine serum (FBS) and 1% v/v penicillin/streptomycin (Gibco, Catalogue number 15140-122).
2. Trypsin/EDTA (Gibco, Catalogue number 25200056).

2.4. $CaPO_4$ Transfection

1. TEN: 10 mM Tris–HCl, pH 7.8, 1 mM EDTA, 300 mM NaCl.
2. $CaCl_2$ is dissolved at 2 M in ddH_2O and sterilized through a 0.2-μM filtration device aliquotted and stored at −20°C.
3. 2× sHBS: 280 mM NaCl, 1.5 mM Na_2HPO_4, 50 mM HEPES, adjust pH to exactly 7.13 with HCl (the exact pH is critical). Filter as per step 2 and store at −20°C.

2.5. Preparation of Cell Lysate and Luciferase Assay

1. PBS: 140 mM NaCl, 2.6 mM KCl, 4.5 mM Na_2HPO_4, 1.4 mM KH_2PO_4. Adjust to pH 7.4 and autoclave.
2. Lysis buffer: 50 mM HEPES, pH 7.4, 250 mM KCl, 0.1% v/v NP-40, 10% v/v glycerol 1 mM DTT, 2 mM PMSF.
3. Sterile cell-scraper.
4. *Firefly* luciferase injection solution: 25 mM glycyl-glycin; 15 mM $KxPO_4$, prepare by mixing 15 mM KH_2PO_4 and K_2HPO_4 to pH 8.0; 4 mM EGTA (ROTH, Karlsruhe, Catalogue number 3054); 15 mM $MgSO_4$ 1 mM DTT; 1 mM ATP, dissolve in ddH_2O at 100 mM and store at −20°C. 100 μM CoEnzymA: Dissolve with ddH_2O at 10 mM and store in aliquots at −20°C. 75 mM luciferin (Applichem, Catalogue number A1006): Dissolve with ddH_2O, aliquot, and store at −20°C. Always keep luciferin-containing solutions in the dark (cover tubes with foil).
5. *Renilla* luciferase injection solution: 1.1 M NaCl; 2.2 mM EDTA, adjust to pH 8.0 with NaOH; 220 mM $KxPO_4$, mix KH_2PO_4 and K_2HPO_4 to pH 5.1; 0.44 mg/ml BSA: dissolve in ddH_2O at 50 mg/ml and store at −20°C. 0.008% w/v NaN_3, 1.43 μM Coelenterazin H (Applichem, Catalogue number A7508); always keep Coelenterazin H in the dark (foil-covered tubes).
6. Microtitre plates.

3. Methods

Searching for differentially expressed genes using gene-chip microarray analysis yields vast amounts of data and necessitates vigorous testing of the input data. The validity of the initial data set will depend heavily on the quality of the input RNA. Since RNases are ubiquitous, it is mandatory to use gloves and filter tips during all the handling steps. As far as possible, all aqueous solutions should be treated with DEPC first and then autoclaved. It is best to keep the samples cold at all times to reduce the activity of RNases and it is highly recommended that a person trained in bioinformatics and/or biostatistics is involved with the data analysis.

To reduce non-specific effects, we recommend that the cut-off for the gene expression is carefully set and that you verify the data obtained from such analysis; it is advisable to perform quantitative real-time PCR or similar control experiments (refer to Chapter 5). Amplification of RNA-specific products necessitates construction of primers that span an intron. If this is not possible, treat the RNA using DNase; this ensures that amplification from residual genomic DNA is avoided.

The results of gene-array analysis alone are insufficient to clarify if the gene of interest is regulated as predicted in silico and accordingly some differentially expressed genes should be verified. If feasible, perform additional controls to assess whether your gene of interest is a direct target of the process you are looking at. To evaluate the transcription factor action, a dual-specific luciferase assay may be performed. Here, the promoter of the gene of interest is cloned into a reporter system, such as the promoter-less *Firefly* luciferase version of pGL3. The activity of this reporter construct can then be examined by the co-transfection of a desired transcription factor construct. The efficiency of the transfection may be evaluated by the co-transfection of a *Renilla* luciferase reporter gene as this contains a constitutively active promoter, often the SV40 (see Note 1).

3.1. Preparation of Samples for Microarray Analysis

1. To isolate RNA from a resected tumour, add a minimum of 1 ml of Trizol reagent per 50–100 mg of tissue and homogenize in a 14-ml tube (see Note 2).
2. Incubate the homogenized sample at room temperature for 5 min and prepare a 1-ml aliquot in 2-ml micro-centrifuge tubes.
3. Add 0.2 ml chloroform per 1 ml of Trizol reagent used, vortex the samples vigorously for 15 s, and incubate for 2–3 min at room temperature.
4. Centrifuge samples at no more than 12,000 × *g* for 15 min at 2–8°C.
5. Transfer the aqueous upper phase to a new micro-centrifuge tube and add 0.5 ml isopropanol per 1 ml Trizol reagent used.
6. Mix by inversion and incubate for 10 min at room temperature.
7. Centrifuge at no more than 12,000 × *g* for 15 min at 2–8°C.
8. Remove the supernatant carefully and wash the RNA pellet once with 1 ml 75% v/v ethanol per 1 ml of Trizol reagent used.
9. Vortex and centrifuge at no more than 7,500 × *g* for 5 min at 2–8°C.

10. Remove the supernatant and air-dry the RNA pellet.
11. Re-suspend in 50–100 μl DEPC-treated H_2O.
12. Remove the contaminants by following the RNeasy kit according to the manufacturer's instructions.
13. For cDNA synthesis, mix 10 μg RNA with 100 pmol of T7-(dT)24 primer and incubate the samples at 70°C for 10 min (see Note 3).
14. Cool the samples on ice, add 1× first-strand buffer, 10 mM DTT, and 2 mM dNTP mix, and incubate at 42°C for 2 min.
15. Add 400 U of SuperScript II reverse transcriptase and incubate at 42°C for 1 h.
16. Briefly centrifuge the first-strand reaction mixture to drop all the liquid to the bottom of the Eppendorf tube and maintain it on ice.
17. Add 1× second-strand reaction buffer, 800 μM dNTP mix, 10 U *E. coli* DNA ligase, 40 U *E. coli* DNA polymerase I, and 2 U *E. coli* RNase H.
18. Top the reaction volume up to 150 μl with ddH_2O and incubate for 2 h at 16°C.
19. Add 10 U T4 DNA polymerase and incubate for a further 5 min.
20. Stop the reaction by adding 10 μl of 0.5 M EDTA.
21. To purify the double-stranded cDNA, use phenol/chloroform extraction and precipitate the cDNA with ammonium acetate and 100% ethanol.
22. To obtain biotin-labelled cRNA, use the BioArray high yield RNA transcription labelling kit (see Note 4).
23. To remove non-incorporated nucleotides from the cRNA, RNeasy columns are used. 15 μg of the cRNA is fragmented by adding fragmentation buffer and incubating at 94°C for 35 min. The RNA is now ready for labelling with the appropriate dye (normally Cy3/Cy5) and applying to gene expression array. This should be performed according to the supplier's recommendations. The results shown in Fig. 1a were obtained with the aid of the Genome Array Facility from the Max Delbrück Center who performed the hybridization step using the MOE430A Arrays (Affymetrix). The in silico analyses were performed in our laboratory using Genespring Software (see Note 5).

3.2. Sample Preparation for Quantitative Real-Time PCR

1. Isolate the RNA using the same protocol as for the microarray sample preparation until the cleaning step. If necessary, include a DNase treatment after RNA isolation (see above).
2. To transcribe the RNA into cDNA, use 2–5 μg RNA, add 0.5 μg of random primers, and dilute the sample with DEPC-H_2O to a final volume of 35 μl.

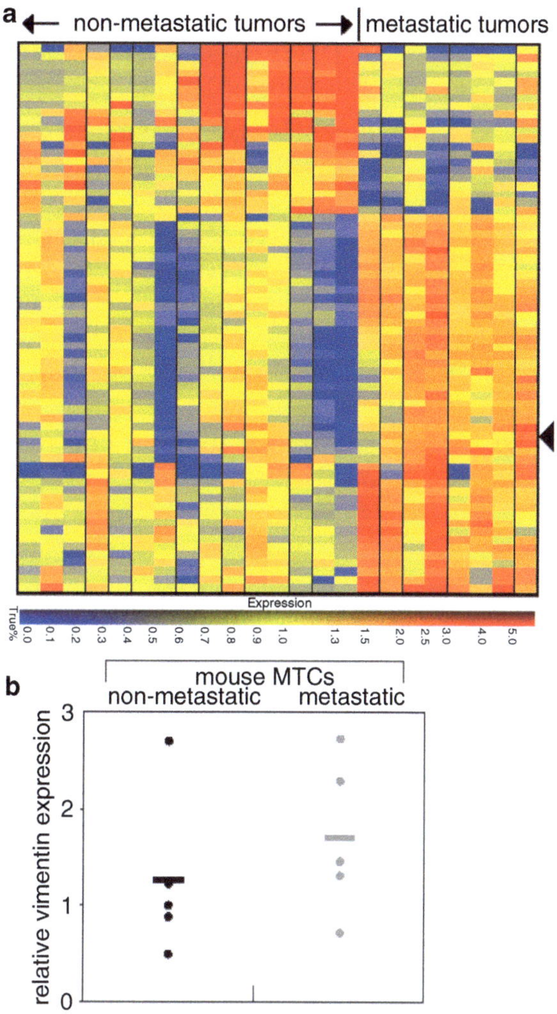

Fig. 1. Elevated *vimentin* expression levels were identified in a mouse MTC model. (**a**) Gene tree representation of the comparative mRNA expression levels discriminating non-metastatic and metastatic MTCs. *Vimentin* is highlighted by the *arrowhead* since it was classified as up-regulated in the metastatic tumour fraction. (**b**) The expression of *vimentin* in non-metastatic MTCs (*black dots*) is compared to metastatic MTCs (*grey dots*) using quantitative real-time PCR. Here, *vimentin* is detected with a reading on average 1.7 in the metastatic MTCs, whereas the lower average level of 1.2 was found in the non-metastatic tissues as shown by the *black* or *grey bars*, respectively. RNA from ten individual MTCs were analyzed.

3. Heat the samples at 65°C for 5 min and then quickly cool on ice.
4. Take out 10 μl; this will act as an "RT-minus" control (see Note 6).
5. To the remaining 25 μl RNA–random primer mix, add 10 μl 5× first-strand buffer, 2.5 μl 20 mM DTT, 1.25 μl RNase Out, 2.5 μl 10 mM dNTP mix, and 0.625 μl reverse transcriptase and DEPC-treated H_2O to a final volume of 50 μl.

6. Incubate the samples at 42°C for 1 h.
7. Dilute the samples with 50 µl DEPC-H_2O to result in a total volume of 100 µl.
8. Store the samples (and the control which includes the "RT-minus" control tube) at –20°C until ready for real-time PCR.

3.3. Quantitative Real-Time PCR

1. Prepare a Mastermix for the cDNA (cDNA-Mastermix) and also for the specific primers (primer-Mastermix). For each sample, calculate 0.5 µl cDNA diluted with 4.5 µl H_2O to make 5 µl cDNA-Mastermix. Likewise, calculate 0.5 µl of 10 µM primer mix diluted with 4.5 µl H_2O to yield 5 µl primer-Mastermix (see Note 7).
2. Pipette 5 µl of the cDNA-Mastermix and 5 µl of the primer-Mastermix into each well of the PCR plate.
3. Add 10 µl 2× qPCR Mastermix into each well.
4. Close the PCR plate with the optical qPCR seal. Be careful to close the seal tightly onto the plate; any small holes will lead to the evaporation of your sample and a failed reaction.
5. Collect the liquids in the PCR plate by centrifugation with a swing at 200 × *g*.
6. Start the PCR according to the manufacturer's instructions (see Note 8).
7. At the end of each run, the PCR machine should be set to calculate a melting curve of the products (see Note 9).
8. Analyze the real-time PCR result according to the delta-delta-CT method. For this, a reference gene (housekeeping gene) should be assayed. Ensure that the cycle threshold is not more than 41 cycles.
9. Figure 1b shows an example for a differential expression of *vimentin* in metastatic tumours as compared to non-metastatic tumours. The expression levels were normalized to beta-actin expression levels. In addition, the expression of one of the non-metastatic tumours was used as a reference and set to "1".

3.4. $CaPO_4$ Transfection for Luciferase Assays

1. Always passage only sub-confluent dishes of C33A cells using trypsin/EDTA. Keep a back-up for future experiments. It is best to seed cells intended for the $CaPO_4$ transfection into 6-well dishes approximately 24 h before transfection. Allow the cells to achieve about 50% confluence on the day of transfection.
2. A day ahead of the transfection, incubate the DNA with 500 µl of TEN and 300 µl of ethanol. Store overnight at –20°C; this step precipitates the DNA in an aseptic manner (see Note 10).
3. The next day, DNA is pelletted at 2–8°C by centrifugation at 16,000 × *g* for 15 min. Discard the supernatant without

disturbing the pellet containing the DNA. Allow the DNA to air-dry.

4. Under aseptic conditions, re-suspend the DNA in 109 µl ddH_2O and 16 µl 2 M $CaCl_2$.
5. Prepare 125 µl of 2× HBS in a 12-ml tube. Start the timer, and vortex the 12-ml tube while slowly adding the 125 µl DNA solution in a dropwise manner (see Note 11).
6. Incubate the samples for precisely 30 min at room temperature. Slowly add the 250 µl crystals into the culture dishes by dripping them onto the cells. The media will need to be changed the following morning (see Note 12).

3.5. Sample Preparation for the Luciferase Assay

1. 24–48 h after the transfection, the cells can be harvested. Remove and discard the media and wash the cells with ice-cold PBS. Perform all the subsequent steps on ice.
2. Add 500 µl of ice-cold PBS to the cells. Use a cell scraper to remove the cells from the culture dish. Collect the cells in a 1.5-ml micro-centrifuge tube and pellet the cells by centrifugation at 4°C.
3. Discard the supernatant. The cell pellets are re-suspended in a 10× volume of lysis buffer.
4. Incubate the cells for 20 min on ice. Clear the lysate by undertaking a centrifugation step at 16,000 × *g* for 15 min at 4°C.
5. Transfer the cell lysate into new tubes. The pellet may be discarded.

3.6. Dual-Specific Luciferase Assay

1. Pipette 10 µl of each of the cleared cell lysates into the wells of a 96-well microtitre plate, preparing duplicates, and maintain on ice.
2. Add 50 µl of *Firefly* luciferase injection solution and 50 µl of *Renilla* luciferase injection solutions into the well, and measure the light reaction immediately using a Luminometer.
3. The relative activity of the reporter gene product can be assessed by first establishing the activity of the responsive reporter and then the promoter of your interest. The established values for the *Firefly* light reaction are then divided by the values for *Renilla* light reaction, and are compared to the control values (*see* Note 10).
4. An example of the results of a luciferase assay is shown in Fig. 2. The basal activity of a wild-type *vimentin* reporter was measured, and compared to the activity when specific E2F transcription factors were co-expressed (Fig. 2b). To validate that the altered activity of the *vimentin* reporter is as a result of

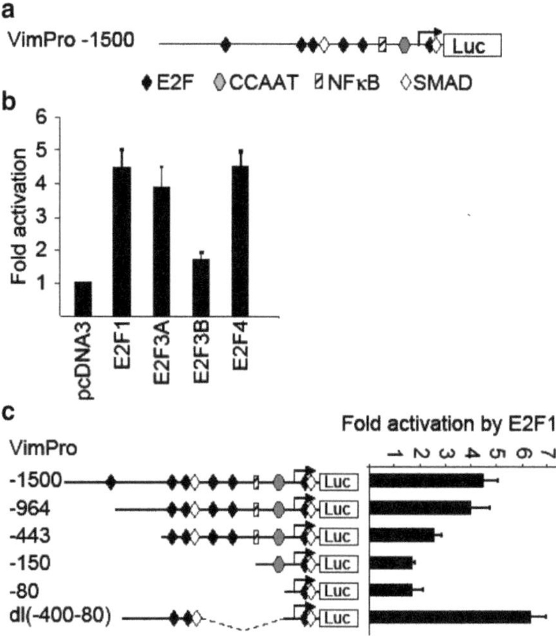

Fig. 2. The human *vimentin* promoter responds to E2F activity. (**a**) Schematic representation of a heterologous luciferase reporter gene. The construct consists of 1.5 kb of the human *vimentin* 5 region that contains five putative E2F-binding sites, two previously mapped SMAD, an NFκB-binding site, and the so-called CCAAT-box. (**b**) The *vimentin* promoter responds to E2F transcription factor activity. The VimPro-1500 *vimentin* reporter construct (*a kind gift from C. Gilles, Université de Liège, Belgium*) was co-transfected with specific E2F expression vectors or a control vector. In response to E2F1, E2F3A, and E2F4, the *vimentin* reporter efficiently activated luciferase, while the E2F3B was less capable of doing so. (**c**) Putative active or repressive E2F sites control *vimentin*. Three deletion constructs (VimPro-964, -443, and -150 construct) of the *vimentin* reporter were co-transfected with E2F1 leading to gradually lessened reporter gene activity. Loss of four E2F-binding sites (but also one SMAD and NFκB site) in VimPro-80 almost abolished E2F responsiveness, suggesting that these are critical "activating" E2F sites. It is likely that two more distal E2F sites (or an SMAD site) are critical for repression, since an internal deletion construct dl(-400-80) is highly active or "derepressed". It is twice as sensitive to E2F1 as the VimPro-443 construct, which contains two additional E2F sites, which are classified as putative "repressor" sites.

direct binding of the co-expressed E2F, a number of putative transcription factor-binding sites within the *vimentin* promoter were mutated by site-directed mutagenesis (Fig. 2c).

4. Notes

1. To assess the degree by which your reporter is activated by a specific transcription factor, it is incumbent upon you to include both a positive and a negative control in the assays.

2. Ensure that the tissue is completely homogenized using a tissue disruptor, such as the Ultra-Turrax (IKA, Staufen) or a similar device. Thorough homogenization will simplify the following steps. Clean the tissue disruptor thoroughly in between the treatment of different samples to avoid RNA carry-over from one sample to the next.
3. Depending on how much RNA is available, you may also use less RNA per reaction. But remember to use equal amounts of RNA in every reaction to prevent bias.
4. We have obtained good results using half of the suggested labelling reaction.
5. Many alternative programs that are less costly than Genespring are available online. We favour the BRB-ArrayTool from the National Cancer Institute, NCI (http://linus.nci.nih.gov/BRB-ArrayTools.html); it is easy to use and has many advantageous features, for example graphical display modes. Differentially expressed transcripts should always be identified using the "significance analysis of microarrays" (SAM) software.
6. You will need a control to determine if and how much genomic DNA is left. We call this the "RT-minus" reaction, and this can be performed alongside your cDNA of interest. If possible, use an intron–exon spanning primer pair at all times. This will allow you to differentiate between the PCR products from the cDNA and unwanted PCR products derived from genomic DNA.
7. To calculate the relative concentration of the gene of interest, you require an intrinsic control. The gene-encoding beta-actin is commonly used as a control. It is noteworthy that in specific cases other reference genes might be preferred. For cell cycle analysis, the S14 gene is often used. In contrast to actin, this gene remains relatively constant across the cell cycle. Include an "RT-minus" control with your intrinsic control primers and ensure that this appears at least ten cycles after the corresponding cDNA.
8. Our best primers work under the following parameters: melting temperature close to 60°C and a product size between 150 and 250 bp.
9. The melting curves should only display one amplicon. Analyze the primers by assessing the reaction product on an agarose gel. Verify that only one product at the appropriate size is visible on the gel.
10. Typically, ten times the amount of the responsive *Firefly* reporter is used as compared to the constitutive *Renilla* reporter. As a starting point, add the same amount of *Firefly* reporter as the amount for the expression vectors coding for the transcription

factor of interest; the precise ratio should, however, be established empirically. Be sure to include a number of controls, both negative and positive. Measuring the extracts of mock-transfected cells allows you to establish the background of reading on the Luminometer; this value may be subtracted from other measurements. Also, include combinations, where cells are transfected with a promoter-less reporter, various amounts of expression construct for the transcription factor of interest, and a known activator of your reporter.

11. Try to add the solution as slowly as possible, and aim to end up with really small crystals. The precipitation of the crystals may be assessed using the light microscope. Check the width of the 12-ml tubes beforehand to ensure that the pipettes fit into the tubes.
12. Strictly adhere to the time the crystals should be prepared. Aim to make and add them to the cells in precisely 30 min. Make duplicates or triplicates of every transfection.

References

1. Benedict WF, Murphree AL, Banerjee A, Spina CA, Sparkes MC, Sparkes RS (1983) Patient with 13 chromosome deletion: evidence that the retinoblastoma gene is a recessive cancer gene. Science 219:973–975
2. Godbout R, Dryja TP, Squire J, Gallie BL, Phillips RA (1983) Somatic inactivation of genes on chromosome 13 is a common event in retinoblastoma. Nature 304:451–453
3. Sparkes RS, Murphree AL, Lingua RW, Sparkes MC, Field LL, Funderburk SJ, Benedict WF (1983) Gene for hereditary retinoblastoma assigned to human chromosome 13 by linkage to esterase D. Science 219:971–973
4. Trimarchi JM, Lees JA (2002) Sibling rivalry in the E2F family. Nat Rev Mol Cell Biol 3:11–20
5. Ziebold U, Lee EY, Bronson RT, Lees JA (2003) E2F3 loss has opposing effects on different pRB-deficient tumors, resulting in suppression of pituitary tumors but metastasis of medullary thyroid carcinomas. Mol Cell Biol 23:6542–6552
6. Knostman KA, Jhiang SM, Capen CC (2007) Genetic alterations in thyroid cancer: the role of mouse models. Vet Pathol 44:1–14
7. Hampf M, Gossen M (2006) Protocol for combined *Photinus* and *Renilla* luciferase quantification compatible with protein assays. Anal Biochem 356:94–99

Chapter 15

High-Resolution Quantitative Methylation Analysis of MicroRNA Genes Using Pyrosequencing™

Ulrich Lehmann, Cord Albat, and Hans Kreipe

Abstract

MicroRNA (miRNA) genes have been shown to perform a crucial role in breast cancer metastasis. The epigenetic inactivation of such microRNA genes, as a result of aberrant DNA methylation, is frequently found in human tumours including those of the breast, and this is an area of considerable research activity.

Pyrosequencing™ is a new quantitative method for the assessment of DNA methylation, with single CpG site resolution. Pyrosequencing™ can easily be performed in a 96-well-plate format with a cost-effective medium-sized throughput.

This chapter provides a general outline of DNA methylation analysis, a detailed protocol of the Pyrosequencing™ procedure, and guidelines for the design of new assays. The strengths and limitations of this approach are discussed throughout the chapter.

Key words: Pyrosequencing™, Methylation, Analysis, CpG, Epigenetic, MicroRNA

1. Introduction

MicroRNAs (miRNAs) are a newly discovered class of very small, non-coding RNAs involved in the regulation of gene expression by inducing mRNA degradation or interfering with mRNA translation (1). miRNAs are typically 19–25 nucleotides in length and are synthesized via several intermediates and by cleavage of much longer precursor transcripts (2). Our current knowledge of the contribution of this class of regulatory transcripts in human development, physiology, and disease is still rapidly accumulating (3, 4).

Several studies have found mis-expression of various miRNAs in a diverse range of human tumours ((5) and references therein). Many genes coding for miRNAs are located near genomic breakpoints (6)

Miriam Dwek et al. (eds.), *Metastasis Research Protocols*, Methods in Molecular Biology, vol. 878,
DOI 10.1007/978-1-61779-854-2_15, © Springer Science+Business Media, LLC 2012

and are affected by copy number alterations (7). Protein factors involved in miRNA biosynthesis are also observed as dysregulated in human tumours (8, 9). Additionally, microRNA genes can be epigenetically deactivated by aberrant DNA methylation in a manner similar to that of classical tumour-suppressor genes (10, 11).

A pivotal role of several miRNAs in the process of metastasis has been described ((12) and references therein). In the case of breast cancer, microRNAs *hsa-mir-126* and *335* have been reported to suppress the invasive behaviour of breast cancer cells (13), whereas *hsa-mir-373* and *520c* are reported to promote metastasis (14). For this reason, the analysis of the DNA methylation patterning in breast cancer specimens is an active area of research (11, 15, 16).

Pyrosequencing™ is a new method for sequencing DNA that allows for the exact quantification of the amount of nucleotide incorporated into each position (see Fig. 1). This process enables the quantification of the allelic ratio should a polymorphic position be sequenced. Treatment of genomic DNA using bisulphite (17) creates a polymorphic position at every potential methylation site (TG in the case of completely unmethylated, CG for completely

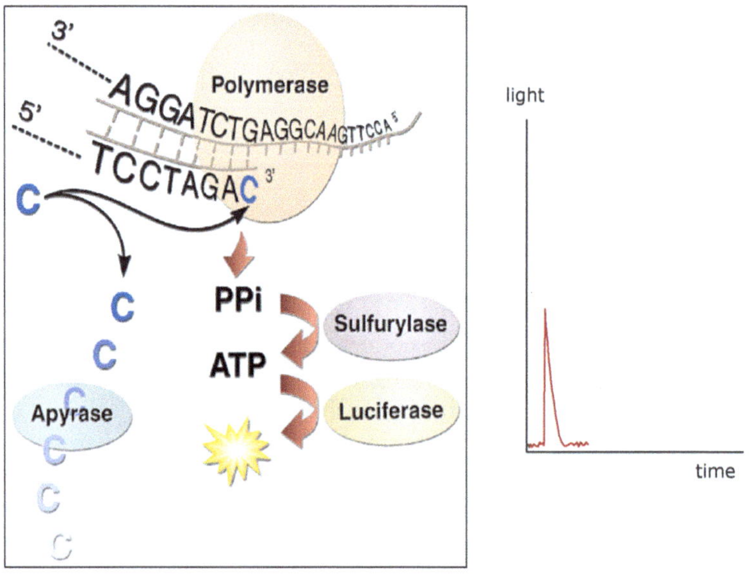

Fig. 1. The principle of Pyrosequencing™ (original kindly provided by Qiagen, slightly modified by the authors). To extend the product starting from the sequencing primer, a nucleotide is added separately and incorporated into the new DNA strand, thus releasing a pyrophosphate molecule. The latter induces an enzymatic cascade creating a light flash whose intensity is proportional to the amount of pyrophosphate released. The number of incorporated nucleotides can be measured directly using a camera inside the pyrosequencer. Before the next nucleotide is added, the enzyme apyrase degrades any remaining nucleotides.

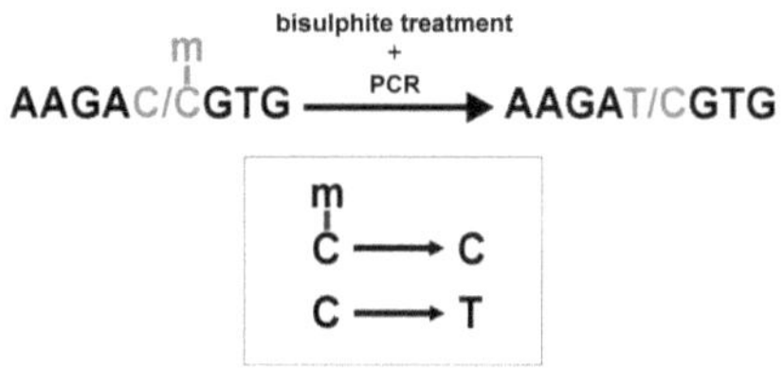

Fig. 2. Principle of bisulphite treatment. The treatment of genomic DNA with bisulphite converts the methylated cytosine (which is not detectable by conventional Sanger sequencing) into a different sequence (TG: unmethylated, CG: methylated). This occurs by deamination of the unmethylated cytosine into uracil; this is subsequently amplified as thymine. This sequence difference can be identified qualitatively or quantitatively by a variety of methods (see ref. 25 for an overview).

methylated, and YG, with Y=T or C, for methylation of only a portion of the DNA molecules under study; see Fig. 2). After bisulphite pretreatment, Pyrosequencing™ can determine the level of DNA methylation present at every CpG site in a given amplicon in a very precise manner. The first time the feasibility of this approach for a single CpG site was demonstrated was by Peter Nürnberg and colleagues in 2002 (18). The development of new software by Biotage (now Qiagen) allows for analysis of longer sections of DNA (up to 150 bp) that contain numerous methylation sites (19). This chapter provides an overview of the "work-flow", a detailed protocol for sample preparation and Pyrosequencing™ procedures, and guidelines and hints for the development of a new Pyrosequencing™ assay. The equipment required to perform Pyrosequencing™ analysis includes a special (and quite expensive) analyzer, set up exclusively to carry out Pyrosequencing™ reactions. This analyzer may not be available in every laboratory.

The exact quantification of the methylation level present in each CpG dinucleotide under study renders this methodology clearly superior to most methods currently in use in the field of DNA methylation. The limited reading length, inaccessibility of some sequences (due to extremely high GC content), and the difficulties the assay design sometimes poses (see Subheading 3.6 below) are the limitations of this approach.

2. Materials

2.1. DNA Isolation

1. DNeasy Blood and Tissue Kit (Qiagen, catalogue number 69506) (see Note 1).
2. Xylol.
3. Ethanol.
4. Tabletop centrifuge, room temperature.

5. Spectrophotometer.
6. Heating block.
7. Refrigerated tabletop centrifuge.
8. Aerosol filter tips (see Note 2).

2.2. Bisulphite Treatment

1. EZ DNA methylation kit from Zymo Research (HISS Diagnostics, catalogue number D5002) (see Note 3).
2. Refrigerated tabletop centrifuge.
3. Room-temperature tabletop centrifuge.
4. Heating block with dark hood.
5. Aerosol filter tips (see Note 2).

2.3. Pyro-PCR

1. Taq-Polymerase with buffer and $MgCl_2$ ("Platinum-Taq", Invitrogen, catalogue number 16966-034) (see Note 4).
2. Primers (see Note 5).
3. dNTPs.
4. Aerosol filter tips.
5. PCR tubes (Sarstedt, catalogue number 710900) or plates (Sarstedt, catalogue number 72.1979.202).
6. PCR machine (for example PersonalCycler from Biometra).

2.4. Pyrosequencing™

1. Pyrosequencing™ system (PyroMark MD, Qiagen, Hilden, Germany).
2. Pyrosequencing™ reagent kit (Qiagen, catalogue number 972812).
3. Reagent dispensing tips (Qiagen, catalogue number 979102).
4. Capillary dispensing tips (Qiagen, catalogue number 979104).
5. Q-CpG software (Qiagen).
6. Primers for Pyrosequencing™.
7. PCR plates (Sarstedt, catalogue number 72.1979.202).
8. Pyrosequencing™ plates (Qiagen, catalogue number 979001).
9. Streptavidin Sepharose HP beads (GE Healthcare, Uppsala, Sweden, catalogue number 1-5113-01).
10. Vacuum workstation (Qiagen, catalogue number 9001529) with the appropriate filter tips (Qiagen, catalogue number 979010) and troughs (Qiagen, catalogue number 979011).
11. Binding buffer: 10 mM Tris–HCl, pH 7.6, 2 M NaCl, 1 mM EDTA, 0.1% v/v Tween 20.
12. Denaturation solution: 0.2 N NaOH.
13. Washing buffer: 10 mM Tris acetate, pH 7.6.

14. Annealing buffer: 20 mM Tris acetate, pH 7.6, 2 mM Mg acetate.
15. Plate mixer (H+P Labortechnik AG, catalogue number 50094537).
16. Heating plate or thermoblock for the denaturation step.
17. Thermoplate for sample preparation (Qiagen, catalogue number 9019071).

3. Methods

3.1. DNA Isolation

The isolation of DNA from formalin-fixed, paraffin-embedded human tissue samples is routinely performed in our laboratory using the DNeasy Blood and Tissue Kit from Qiagen following the protocol supplied by the manufacturer (see Note 1).

3.2. Bisulphite Treatment

The treatment of genomic DNA with bisulphite is performed using the EZ DNA methylation kit from Zymo Research following the protocol supplied by the manufacturer (see Note 3).

Usually, 0.5–1 μg genomic DNA is treated with bisulphite. Depending on the quality of the DNA (and the efficiency of subsequent PCR amplifications), much less DNA can be used.

To maximize the yield of the elution, the total volume is split into two or more steps with fractions of the desired elution volume (for example for 50 μl: 2×25 μl).

3.3. Pyro-PCR

10–20 ng of bisulphite-treated DNA are amplified in a total volume of 25 μl using 0.5 unit Platinum-Taq (see Note 4), 10 pmol of each primer, and 200 μM dNTPs. The optimal $MgCl_2$ concentration and annealing temperature have to be determined empirically (see Note 6). After completion of the PCR step, 5 μl of the reaction mixture is resolved on a polyacrylamide gel and stained with ethidium bromide in order to verify the identity and purity of the PCR product.

3.4. Pyrosequencing™

1. Fill the troughs at the vacuum preparation station. Trough 1: 70% v/v ethanol; trough 2: 0.2 M NaOH; trough 3: 1× washing buffer; and trough 4: distilled water. Switch on heating block (set to 80°C) and place thermoplate onto the heating block.
2. Dispense into a 96-well plate the following mixture for every sample:
   (a) 47 μl binding buffer.
   (b) 3 μl streptavidin-Sepharose beads (mix well before use).
   (c) Up to 20 μl PCR product.
   (d) Add HPLC water to a total volume of 80 μl.

3. Shake the plate for 5 min at room temperature.
4. Meanwhile, prepare the Pyrosequencing™ plate by adding 11.5 μl annealing buffer for each reaction well and 0.5 μl (5 pmol) Pyrosequencing™ primer.
5. The order of samples within the wells of the PCR plate must be identical to that of the order in the binding reaction and the Pyrosequencing™ plate.
6. Turn on the pump of the vacuum preparation station, close the valve to create a vacuum, and wash the tips for 10 s in trough 4. Next, check the filter tips of the aspiration device by aspirating water from a separate PCR plate containing 100 μl HPLC water in each well. The water should be aspirated from every well within a few seconds. If this is not the case, all corresponding tips must be replaced.
7. Aspirate the Sepharose beads from the sample plate for at least 10 s and carefully remove the aspiration device. Take care not to lose any Sepharose beads!
8. Immerse the aspiration device for 5 s in trough 1, and then place the device in troughs 2 and 3 for 10 s each (see above; step 1).
9. Hold the aspiration device above the Pyrosequencing™ plate, turn off the vacuum, and check that the air pressure has reached a normal level taking care to ensure that the display has reached zero! Immerse the tips in the annealing mix afterwards. Shake gently for 15–30 s to release the beads into the wells. Make sure that all beads are shaken off by holding the plate against a light source. If the aspiration device is lowered too fast, the annealing mixture might be drawn off.
10. Place the Pyrosequencing™ plate on top of the heating block (80°C), cover with the thermoplate, and incubate for 2 min. Allow to cool for 5–10 min.
11. In the meantime, start the Q-CpG software system selecting the appropriate assay set-up file. Register all samples in the 96-well plate in the preset worksheet. When selecting "Volume Information", the program calculates the required amount of enzyme, substrate, and nucleotides (depending on the number of samples and the reading length for every individual sample).
12. Dispense an appropriate amount of enzyme, substrate, and nucleotide into the cartridges. Avoid the ingress of air bubbles, as this may cause an obstruction in the fine dispensation needles or dispensation errors in the pyrogram. Biotage (now Qiagen) recommends centrifugation of the nucleotides before use for 5 min, at full speed at room temperature, to avoid the transfer of any particles into any of the six cartridges.

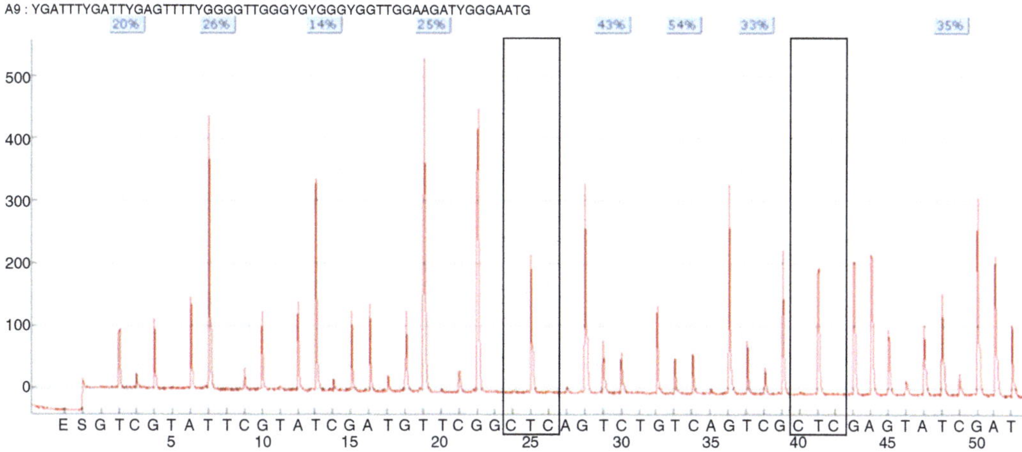

Fig. 3. An example pyrogram, the primary output of the Q-CpG software. The numbers above the potential methylation sites indicate the percent methylation. The "bisulphite controls" (see Subheading 3.5, step 4) are *boxed*. No signal is generated by dispensing a cytosine, indicating complete conversion of the original DNA at this position. Longer sequences (up to 150 bp) can be analyzed, but then a nearly linear decrease in signal intensity during the Pyrosequencing™ reaction is observed due to exhaustion of the enzyme mixture. Depending on the quality of the Pyrosequencing™ primer, a concomitant increase in background signal can sometimes be observed. These two effects determine the readable length of a program (signal-to-noise ratio). The example shows methylation analysis of the microRNA gene *has-mir-148a*.

13. Place the holder with the filled cartridges and the Pyrosequencing™ plate into the analyzer, close the reaction chamber via the software, close the lid of the analyzer manually, and start the run via the software (see Note 7).
14. After finishing the run, prompt and careful cleaning of the cartridges is of utmost importance.

3.5. Guidelines for Pyrogram Evaluation

1. If the substrate is dispensed at the very beginning, it should not generate a signal (or only a peak much smaller than the nucleotide signal). A large substrate peak indicates pyrophosphate contamination, and should be avoided.
2. Peaks should be sharp and slender, and should reach the baseline after less than 30 s (halfway before the next dispensation).
3. Single-nucleotide peaks should be very similar for C, G, and T and slightly larger for A.
4. The "bisulphite controls" (introduced automatically by the software or manually in addition) should be "empty" (i.e. no signal greater than background noise after dispensing a cytosine outside the context of a CpG dinucleotide, Fig. 3).
5. No additional peaks should appear between the regularly spaced nucleotide peaks.
6. The average peak height should be clearly above background, at least 25 units above. 50 units above the background is clearly

much better and achievable with the Pyro Mark MD system we use. Values differ for the other two systems from Qiagen.

7. The measured peaks should fit quite well into the calculated histogram.

3.6. Assay Design

The primers utilized for the amplification of the sequence to be analyzed should not contain potential methylation sites. In addition, the sequencing primer for the Pyrosequencing™ reaction should not contain any potential methylation sites, and should not overlap completely with any of the flanking PCR primers; this is to ensure that specific binding takes place only to the PCR product (and not to any side product, which also contains the PCR primer sequences at the ends). In our experience, a complete overlap of the sequence primer with one of the PCR primers (see Fig. 4) very often increases the background signal. The three primers should not contain long homopolymer stretches or, conversely, short repeats, as this will reduce specificity. Longer complementary stretches induce primer dimer formation and should also be avoided. In order to minimize secondary structure formation during the Pyrosequencing™ reaction, the amplicon size should not exceed 200–250 bp. Longer amplicons might be analyzable, but the success rate (yield of PCR product and quality of the pyrogram) will be clearly reduced.

Due to the reduced complexity of bisulphite-treated DNA (three instead of four bases, with the exception of methylated CpG dinucleotides), the fulfilment of these requirements often represents a major challenge. In addition, many sequences of interest (classical "CpG islands") are extremely CpG rich and have a very high GC content. This sometimes makes the design of primer sequences without a potential methylation site, i.e. no CpG dinucleotide, and a GC content not exceeding 60% impossible because the amplicon size should not exceed a certain length (see below) and the annealing temperature should not exceed 65°C.

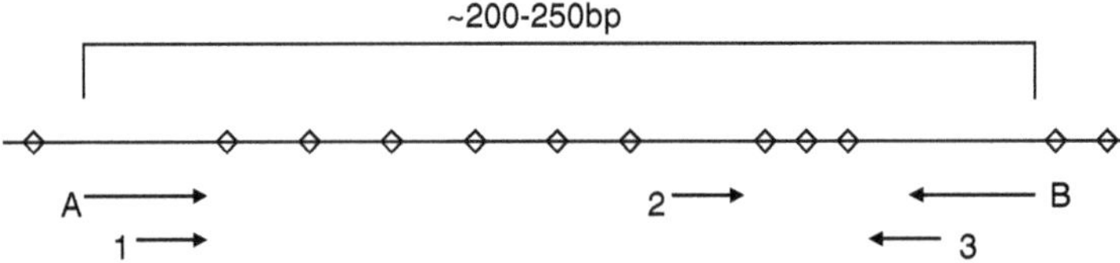

Fig. 4. Schematic view of primer location. A and B are the flanking primers, neither of which contains potential methylation sites in its recognition sequence. Methylation sites are symbolized by a rhombus. The numbers 1, 2, and 3 show the Pyrosequencing™ primers. Primer 1 completely overlaps with primer A. This should be avoided (see text for details).

In order to ensure the specificity of the amplification towards fully converted DNA, the primer-binding sites should contain several cytosine residues which have been converted to thymine during the bisulphite treatment. If it is not possible to avoid a potential methylation site in the primer-binding sites (as a C/T in the forward primer or a G/A in the reverse primer), this should be placed towards the 5′ end of the primer.

After bisulphite conversion, the two DNA strands will no longer be complementary. This offers the possibility that one might analyze one of the two strands, but it can also lead to confusion when trying to localize published primers. In rare cases (for example, for the analysis of hemimethylation), both strands have to be analyzed. See ref. 20 for more details.

Several freely available software tools help in the selection of primers for the amplification of bisulphite-treated DNA: MethPrimer (21), Bisearch (22), MethylPrimer Express (Applied Biosystems, Darmstadt, Germany), CpGWARE (Chemicon International, Billerica, Massachusetts, USA), and PerlPrimer(23). However, in our experience, the manual selection of primers supported by a conventional primer design program (to calculate T_{Ann} and GC content and identify potential primer dimers) is equally efficient and sometimes even better. We routinely use PrimerExpress (Applied Biosystems) or Primer 3 (http://frodo.wi.mit.edu/) for this purpose.

The following negative controls should be included in order to check a new assay for unwanted self-priming reaction components that may create strong background signals:

1. Inclusion of only the sequencing primer in the Pyrosequencing™ reaction
2. Inclusion of only the biotinylated primer in the Pyrosequencing™ reaction
3. Inclusion of only the sequencing primer and the biotinylated primer in the Pyrosequencing™ reaction
4. Inclusion of only the PCR product

4. Notes

1. Numerous DNA isolation kits are commercially available. We did not perform a comprehensive comparison of all the available products but gained positive experiences with the Qiagen kit. However, the majority of the other kits would most likely yield comparable results. Alternatively, DNA can be isolated by conventional Proteinase K digestion, organic extraction, and subsequent ethanol extraction. If performed well, this

procedure yields high amounts of pure DNA, but requires more practical skills and is quite time consuming.

2. Repeated amplification of the same target sequence from limited amounts of starting material poses a serious risk of the cross-contamination of samples. For this reason, strict guidelines concerning the handling of samples before and after amplification and the cleaning of all instruments must be enforced and carried out by all personnel involved. We perform all pre-amplification steps in a separate laboratory ("pre-PCR area"). Everything used in this laboratory (including lab coats, pens, and notepads) are dedicated exclusively to this room and are strictly separated from the post-PCR area. Plastic lab-ware and benches are cleaned regularly using a 3% v/v hypochlorite solution. PCR products are subsequently analyzed in a separate laboratory ("post-PCR area"). Under no circumstances should amplified samples or equipment from this working area be brought back to the pre-PCR area. Aerosol filter tips are also an absolute necessity in the pre-PCR area. Also refer to Chapter 5 for further guidance on PCR-based assays.
3. As with DNA isolation, many different kits are available for the bisulphite treatment of genomic DNA. Here, we gather the most extensive experience using a product from Zymo Research (EZ DNA methylation kit); we also carried out a few comparisons of other products. In our opinion, this one works best, but other kits may work comparably well.
4. In our experience, a hot-start enzyme is absolutely essential for specific and efficient amplification of bisulphite-treated DNA. We gained extensive positive experiences with "Platinum-Taq" from Invitrogen (Carlsbad, California, USA), but other enzymes might work as well.
5. Biotinylated primers are more labile than normal primers. Repeated thawing and freezing should be avoided by the preparation of aliquots containing the stock solution. High purity (HPLC grade) is also required since unbound biotin competes for binding to the streptavidin-agarose beads and, thereby, reduces the efficiency of the purification step. For most assays, we use a universal biotinylated primer and a complementary tag at one of the two PCR primers (24). This saves quite a lot of money as it avoids the ordering of dozens of biotinylated primers, which are sometimes not used up within the shelf life of the reagent. Under most circumstances, the tag and the universal primer do not interfere with the amplification of the sequence of interest. In only a few cases for frequently used assays and if the tag interferes with the specificity and/or efficiency of the PCR we order assay-specific biotinylated primers.

6. Primer optimization, that is to say, testing different annealing temperatures and $MgCl_2$ concentrations, is strongly recommended. We routinely test three to six different temperatures around the calculated optimal T_{Ann} at two different $MgCl_2$ concentrations and select the reaction parameters giving the strongest band and no or only spurious side products.
7. Before each run, perform a dispensation test: Place a Pyrosequencing™ plate sealed with a plastic foil inside the reaction chamber and start a dispensation test using the software. Six discrete droplets should be clearly visible and each should be positioned at the centre of a well. This ensures that the dispensation is working at the beginning of the analysis. The test does not, however, prevent occasional obstruction of one or several dispensation needles during the Pyrosequencing™ run, but it does reduce the risk of a complete failure of the sequencing run.

Acknowledgements

The authors would like to thank Britta Hasemeier for expert assistance in preparing the figures and Dr. Florian Puls for critically reading the manuscript.

References

1. Mattick JS, Makunin IV (2006) Non-coding RNA. Hum Mol Genet 15:R17–R29
2. Sevignani C, Calin GA, Siracusa LD, Croce CM (2006) Mammalian microRNAs: a small world for fine-tuning gene expression. Mamm Genome 17:189–202
3. O'Rourke JR, Swanson MS, Harfe BD (2006) MicroRNAs in mammalian development and tumorigenesis. Birth Defects Res C Embryo Today 78:172–179
4. Shivdasani RA (2006) MicroRNAs: regulators of gene expression and cell differentiation. Blood 108:3646–3653
5. Esquela-Kerscher A, Slack FJ (2006) Oncomirs – microRNAs with a role in cancer. Nat Rev Cancer 6:259–269
6. Calin GA, Sevignani C, Dumitru CD, Hyslop T, Noch E, Yendamuri S, Shimizu M, Rattan S, Bullrich F, Negrini M, Croce CM (2004) Human microRNA genes are frequently located at fragile sites and genomic regions involved in cancers. Proc Natl Acad Sci USA 101:2999–3004
7. Zhang L, Huang J, Yang N, Greshock J, Megraw MS, Giannakakis A, Liang S, Naylor TL, Barchetti A, Ward MR, Yao G, Medina A, O'Brien-Jenkins A, Katsaros D, Hatzigeorgiou A, Gimotty PA, Weber BL, Coukos G (2006) MicroRNAs exhibit high frequency genomic alterations in human cancer. Proc Natl Acad Sci USA 103:9136–9141
8. Chiosea S, Jelezcova E, Chandran U, Acquafondata M, McHale T, Sobol RW, Dhir R (2006) Up-regulation of dicer, a component of the MicroRNA machinery, in prostate adenocarcinoma. Am J Pathol 169: 1812–1820
9. Sugito N, Ishiguro H, Kuwabara Y, Kimura M, Mitsui A, Kurehara H, Ando T, Mori R, Takashima N, Ogawa R, Fujii Y (2006) RNASEN regulates cell proliferation and affects survival in esophageal cancer patients. Clin Cancer Res 12:7322–7328
10. Lujambio A, Ropero S, Ballestar E, Fraga MF, Cerrato C, Setién F, Casado S, Suarez-Gauthier A, Sanchez-Cespedes M, Git A, Spiteri I, Das PP, Caldas C, Miska E, Esteller M (2007) Genetic unmasking of an epigenetically silenced microRNA in human cancer cells. Cancer Res 67:1424–1429

11. Lehmann U, Hasemeier B, Christgen M, Müller M, Römermann D, Länger F, Kreipe H (2008) Epigenetic inactivation of microRNA gene hsa-mir-9-1 in human breast cancer. J Pathol 214:17–24
12. Nicoloso MS, Spizzo R, Shimizu M, Shimizu M, Rossi S, Calin GA (2009) MicroRNAs – the micro steering wheel of tumour metastases. Nat Rev Cancer 9:293–302
13. Tavazoie SF, Alarcon C, Oskarsson T, Padua D, Wang Q, Bos PD, Gerald WL, Massagué J (2008) Endogenous human microRNAs that suppress breast cancer metastasis. Nature 451:147–152
14. Huang Q, Gumireddy K, Schrier M, le Sage C, Nagel R, Nair S, Egan DA, Li A, Huang G, Klein-Szanto AJ, Gimotty PA, Katsaros D, Coukos G, Zhang L, Puré E, Agami R (2008) The microRNAs miR-373 and miR-520c promote tumour invasion and metastasis. Nat Cell Biol 10:202–210
15. Lujambio A, Calin GA, Villanueva A, Ropero S, Sánchez-Céspedes M, Blanco D, Montuenga LM, Rossi S, Nicoloso MS, Faller WJ, Gallagher WM, Eccles SA, Croce CM, Esteller M (2008) A microRNA DNA methylation signature for human cancer metastasis. Proc Natl Acad Sci USA 105:13556–13561
16. Lujambio A, Esteller M (2009) How epigenetics can explain human metastasis: a new role for microRNAs. Cell Cycle 8:377–382
17. Frommer M, McDonald LE, Millar DS, Collis CM, Watt F, Grigg GW, Molloy PL, Paul CL (1992) A genomic sequencing protocol that yields a positive display of 5-methylcytosine residues in individual DNA strands. Proc Natl Acad Sci USA 89:1827–1831
18. Uhlmann K, Brinckmann A, Toliat MR, Ritter H, Nürnberg P (2002) Evaluation of a potential epigenetic biomarker by quantitative methyl-single nucleotide polymorphism analysis. Electrophoresis 23:4072–4079
19. Brakensiek K, Wingen LU, Langer F, Kreipe H, Lehmann U (2007) Quantitative high-resolution CpG island mapping with Pyrosequencing™ reveals disease-specific methylation patterns of the CDKN2B gene in myelodysplastic syndrome and myeloid leukemia. Clin Chem 53:17–23
20. Laird CD, Pleasant ND, Clark AD, Sneeden JL, Hassan KM, Manley NC, Vary JC Jr, Morgan T, Hansen RS, Stöger R (2004) Hairpin-bisulfite PCR: assessing epigenetic methylation patterns on complementary strands of individual DNA molecules. Proc Natl Acad Sci USA 101:204–209
21. Li LC, Dahiya R (2002) MethPrimer: designing primers for methylation PCRs. Bioinformatics 18:1427–1431
22. Tusnady GE, Simon I, Varadi A, Aranyi T (2005) BiSearch: primer-design and search tool for PCR on bisulfite-treated genomes. Nucleic Acids Res 33:e9
23. Marshall OJ (2004) PerlPrimer: cross-platform, graphical primer design for standard, bisulphite and real-time PCR Quantitative real-time PCR for equine cytokine mRNA in nondecalcified bone tissue embedded in methyl methacrylate. Bioinformatics 20:2471–2472
24. Issa JP, Gharibyan V, Cortes J, Jelinek J, Morris G, Verstovsek S, Talpaz M, Garcia-Manero G, Kantarjian HM (2005) Phase II study of low-dose decitabine in patients with chronic myelogenous leukemia resistant to imatinib mesylate. J Clin Oncol 23:3948–3956
25. Brena RM, Huang TH, Plass C (2006) Quantitative assessment of DNA methylation: potential applications for disease diagnosis, classification, and prognosis in clinical settings. J Mol Med 84:365–377

Chapter 16

RNAi Technology to Block the Expression of Molecules Relevant to Metastasis: The Cell Adhesion Molecule CEACAM1 as an Instructive Example

Daniel Wicklein

Abstract

Specific gene silencing using small hairpin RNA (shRNA) constructs offer researchers the possibility to study the influence of a single protein in the metastatic process. The role of the cellular adhesion molecule CEACAM1 on tumour formation and metastasis is of some interest. The human melanoma cell line FemX-1 was transfected with an shRNA construct directed against the CEACAM family. Stable clones were obtained and characterized via puromycin selection, single-cell dilution, and subsequent FACS analysis. The cell line showed a knock-down of CEACAM1 of more than 85%. This knock-down remained stable when examined in an SCID mouse xenograft experiment over 40 days.

Key words: RNAi, Specific gene silencing, siRNA, shRNA, Stable knock-down, CEACAM1

1. Introduction

A recent approach to gene silencing has been developed based on RNA interference (RNAi) (1). The system involves the introduction of double-stranded RNA (dsRNA) into the cells of interest. The RNAi sequence is homologous to a part of the target mRNA and it is digested in vitro to 21–23-bp small interfering RNA (siRNA) fragments which are bound by cellular proteins and form RNA-induced silencing complexes (RISCs). The RISCs specifically bind to the target transcript and this results in its degradation (2, 3). For stable knock-down of a target gene, it is essential that the specific dsRNA remains present within the cells. To achieve this, cells are transfected with vectors enabling constitutive expression

Miriam Dwek et al. (eds.), *Metastasis Research Protocols*, Methods in Molecular Biology, vol. 878, DOI 10.1007/978-1-61779-854-2_16, © Springer Science+Business Media, LLC 2012

of small hairpin RNA (shRNA) containing the dsRNA specific for the gene of interest. A method for obtaining cell lines with stable knock-down of a target gene is described in this chapter and the stable knock-down of mRNA for the cell adhesion molecule CEACAM1 in a human melanoma cell line is presented as an example.

2. Materials

2.1. Cells and Cell Culture

1. FemX-1 is a melanoma line established from a metastatic lymph node which has been shown to exhibit elevated expression levels of *CEACAM1* in cell culture (4, 5).
2. RPMI 1640 medium supplemented with 10% v/v fetal calf serum (FCS) and 2 mM l-glutamine, 100 U/ml penicillin, and 100 μg/ml streptomycin.
3. Enzyme-free cell dissociation buffer (Invitrogen).
4. Puromycin (Clontech, Saint-Germain-en-Laye, France), 1 mg/ml in PBS, aliquots frozen at –20°C.

2.2. Molecular Biology

1. LB medium and LB agar prepared according to the manufacturer's instructions.
2. Ampicillin, 100 mg/ml, aliquots frozen at –20°C.
3. Restriction enzymes MluI and XhoI, corresponding buffers supplemented with bovine serum albumin (BSA).
4. T4-Ligase, ligation buffer, supplemented with BSA.
5. QIAprep Miniprep Kit (Qiagen, Hilden, Germany) for plasmid purification.
6. 65 or 66 bp DNA oligonucleotides.
7. FuGENE transfection reagent (Roche Diagnostics, Mannheim, Germany).
8. RNAi-Ready pSIREN-RetroQ Vector (Clontech), prelinearized (BamHI, EcoRI digestion).
9. Sequencing facility.

2.3. Cytometry

1. FACS buffer: 1% w/v BSA in PBS with 0.05% v/v NaN_3.
2. Propidium iodide solution: 50 μg/ml propidium iodide in PBS.
3. Mouse anti-human CEACAM1 monoclonal antibody, phycoerythrin-labelled (R&D Systems, Wiesbaden, Germany) and corresponding mouse IgG2b-PE isotype control (Miltenyi Biotech, Bergisch-Gladbach, Germany).

3. Methods

Vectors for shRNA constructs are commercially available. Some even have the short hairpin sequences inserted, thereby allowing the specific knock-down of a target protein. For our experiment, we chose the RNAi-Ready pSIREN-RetroQ Vector system (Clontech), as this allows easy insertion of short hairpin sequences by using short commercially synthesized oligonucleotides. The shRNA is expressed under the control of the RNA polymerase III-dependent human U6 promotor. This is ideal for short sequences since it allows non-viral, as well as retroviral, transfection after production of infectious particles using a "packaging" cell line (e.g. HEK293 cells). However, we found that all the tumour cell lines we have used so far could be transfected without the addition of viral particles. The pSIREN-RetroQ Vector contains a puromycin resistance sequence, thereby allowing selection of stable tranfectants. It contains a sequence for ampicillin resistance for selection of transformed *Escherichia coli* cells (for cloning and later production of plasmid DNA). Before starting the experiment, the knock-down system has to be validated. The knock-down of the mRNA for the cell adhesion molecule CEACAM1 in the melanoma cell Line FemX-1 (4, 5) and its validation by flow cytometry are presented here.

3.1. Planning the Knock-Down Experiment

1. The mRNA sequence can be obtained from the NCBI/ Genbank (e.g. *Homo sapiens* carcinoembryonic antigen-related cell adhesion molecule 1 or biliary glycoprotein, CEACAM1, transcript variant 1, mRNA, accession number NM_001712).
2. The FASTA sequences (or the accession number) can be used for selection of potential siRNA sequences with the Clontech RNAi Target Sequence Selector using the default settings in the first instance (http://bioinfo.clontech.com/rnaidesigner/sirnaSequenceDesignInit.do); see Note 1.
3. The siRNA sequences can be verified and checked for potential off-target effects using the Blastn programme at the NCBI site (Human genomic + transcript, if planning the knock-down of a human gene). Only sequences showing no potential knock-down (possible off-target effects) apart from for the chosen protein—in this case the CEACAM—family targeted with the oligonucleotide CTCACAGCCTCACTTCTAA should be used in the later stages of the experiment.
4. 65 or 66 bp sequences of shRNA oligonucleotides with suitable single-strand overhangs and—for later insert verification—an additional restriction site not present in the vector are constructed using the Clontech shRNA Sequence

Designer tool for pSIREN insertion (http://bioinfo.clontech.com/rnaidesigner/oligonucleotideDesigner.do). In the case of the pSIREN-RetroQ vector, the restriction sites are BamHI and EcoRI with an additional MluI site (see Note 2).

3.2. Building the shRNA Vector for Transfection

Cloning the shRNA oligonucleotides into the chosen vector can usually be undertaken according to the manufacturer's instructions and is described, briefly, for the pSIREN-RetroQ vector here.

1. Order the oligonucleotides for shRNA expression and resuspend in nuclease-free water to a final concentration of 100 μM.
2. Mix the top and bottom strand at a 1:1 ratio and anneal by heating to 95°C for 30 s followed by incubation at 72, 37, and 25°C for 2 min each. This yields double-stranded oligonucleotides (50 μM).
3. Dilute the double-stranded oligonucleotides 1:100 with water to obtain a concentration of 0.5 μM. Use 1 μl of each of the diluted oligonucleotides for ligation reactions of 15 μl total volume. For this, 1 μl T4 DNA ligase, 0.5 μl BSA solution, and 2 μl of the BamHI/EcoRI pre-digested pSIREN-RetroQ Vector (25 ng/ml) are mixed with the oligonucleotide. Incubate the ligation reactions for 3 h at room temperature. Alternatively, 1 μl of an anti-luciferase control oligonucleotide or other negative control oligonucleotide may be used for ligation to generate a vector for negative control tranfectants.
4. For the transformation, thaw the chemo-competent *E. coli* cells (one vial per ligation reaction) on ice, add the whole ligation reaction mixture, and mix carefully by pipetting. Transform the bacteria by incubating the tubes for 30 min on ice, followed by a 42°C heat shock for 90 s and incubation on ice for an additional 10 min.
5. Add 200 μl LB-medium (without antibiotics) to each tube and incubate for 30 min at 37°C.
6. Plate the *E. coli* cells on 1 ml LB agar containing 100 μg/ml ampicillin (taken from a 1-μl stock solution), and incubate at 37°C overnight.
7. Pick single colonies, seed in 5 ml LB medium (100 μg/ml ampicillin), and incubate overnight at 37°C with vigorous shaking.
8. Prepare plasmids from 1 ml of the overnight cultures using, for example, the QIAprep Miniprep Kit (Qiagen) according to the manufacturer's instructions. In this system, the elution volume of the plasmid is 50 μl (the elution buffer is provided in the kit).
9. Perform a restriction enzyme analysis utilizing the additional cutting site introduced into the system via the ordered

oligonucleotides (there is no cutting site in the original vector) and a second enzyme with one cutting site in the original vector. In the case of the pSIREN-RetroQ vector, the insert can be verified using MluI and XhoI: Only clones with an insert will contain the MluI restriction site (a fragment of 1.2 kb is cut out), whereas the vector without the insert is simply linearized by XhoI. This can be evaluated using standard agarose gel electrophoresis.

10. It is important that the DNA concentration of the clones containing the insert is determined. It is also desirable to sequence the oligonucleotides by DNA sequencing. In the case of pSIREN-RetroQ, the U6 forward sequencing primer 5′-CAGGAAGAGGGCCTAT-3′ can be used.

3.3. Transfection

1. Before the cells can be transfected, the optimal concentration of the selection antibiotic (puromycin) has to be determined. The optimal seeding densities and concentrations have to be determined individually (see Note 3), but for most cell lines the following will be adequate: seed cells into two 6-well cell culture plates at 2.5×10^4 cells per well in 5 ml culture medium and, after 1 day, add 2, 1.8, 1.5, 1, 0.8, 0.6, 0.5, 0.4, 0.3, 0.2, or 0.1 μg/ml, respectively. Also include a well without puromycin. Incubate the cells under their standard culture conditions (usually 37°C, in a 5% CO_2 atmosphere) and change the medium every 3–4 days for 2 weeks. Evaluate the viability of the cells every 1–2 days by viewing under a standard inverted light microscope. Use the lowest concentration of antibiotic that ensures that all the cells are dead after 2 weeks.
2. Seed 1×10^5 to 1.5×10^5 cells into 6-well plates 18–24 h before transfection using 5 ml of cell culture medium.
3. Add 6 μl Fugene transfection reagent to 94 μl of medium (without serum and antibiotics) in 1.5-ml reaction tubes. Vortex for 2 s. Add 2 μg of vector containing knock-down shRNA oligonucleotide or control oligonucleotide (for example, anti-CEACAM or anti-Luciferase oligonucleotides); this will be in the region of 10 μl of the plasmid preparation. Vortex the tubes again for 2 s and incubate at room temperature for at least 15 min (up to 45 min) (see Note 4).
4. After the incubation step, remove the medium from the cells and replace with 1 ml medium without serum and antibiotics. Add the transfection reaction in a dropwise manner and gently swirl the plates. Incubate for at least 6 h, preferably overnight, at 37°C in a 5% CO_2 atmosphere.
5. Remove the transfection medium, replace with 5 ml of cell culture medium, and incubate for 48 h under standard conditions (see Note 5).

3.4. Selection of Stable Transfectants

1. Replace the medium with 5 ml selection medium: fresh culture medium containing the optimal concentration of selection antibiotic, e.g. 1 μg/ml puromycin. Change the medium every 3–4 days. Non- or weakly transfected cells should detach from the surface and die (see Note 6) and these can conveniently be removed in the medium. With time and depending on the proliferation rate of the cells, foci of antibiotic-resistant cells should appear.
2. Grow the resistant cells to near confluence and then split the cultures. Great care must be taken to obtain a single-cell suspension at this stage, preferably by the use of an enzyme-free dissociation buffer. Use part of the culture to evaluate the success of the knock-down (see below) and another part for a limiting dilution experiment (see step 5, below). Also reseed the original cultures.
3. If the protein target of the knock-down can be detected either on the surface or intracellularly by FACS, perform the FACS analysis here (for further details on FACS analysis, refer to the Chapter 12). For the FACS analysis, pellet the cells in a bench-top centrifuge, usually for 5 min at about 2,500 × *g*. Typically, up to 10^6 cells can be stained in 1.5-ml reaction tubes in 100 μl FACS buffer containing 1 μg/ml of the fluorescent labelled antibody (for example: anti-CEACAM1-PE or mouse IgG2b-PE isotype control). Incubate the cells on ice for 30 min, wash with 1 ml of FACS buffer, pellet the cells, and resuspend in 300 μl of FACS buffer. Transfer the cells to tubes suitable for use in the flow cytometer and perform the FACS analysis. Add 3 μl of the propidium iodide solution directly before each measurement. The instrument settings depend on the cell line and the flow cytometer used and have to be determined individually (see Note 7).
4. Compare the control transfection with the knock-down transfection (using the software of the cytometer or, for example, the freeware software WinMDI 2.9). Two distinct populations of cells should be detected: one population showing fluorescence similar to the control and another population showing reduced fluorescence (according to the level of knock-down; Fig. 1). If an isotype control is included, this can be used to estimate whether the knock-down efficiency is adequate. If successful, a number of cells should display a fluorescence signal not much higher than the isotype control.
5. Perform a limiting dilution experiment to obtain monoclonal cell lines from the knock-down cells and the control tranfectants. Dilute the cells in selection medium (depending on the cell line—typically 10 cells/ml, i.e. 1 cell/100 μl) and seed a number of

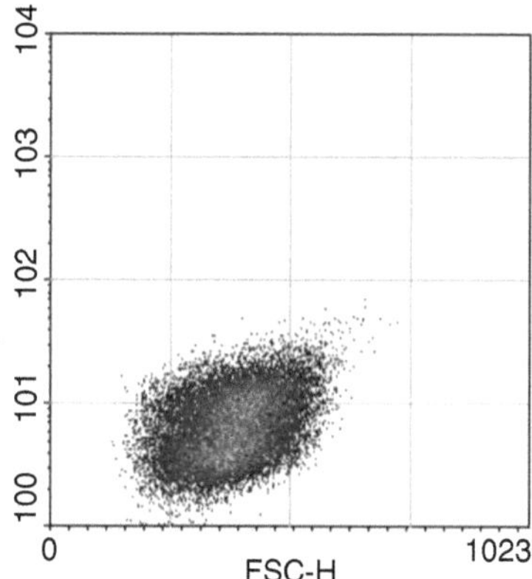

Fig. 1. FACS analysis of melanoma cell line FemX-1 transfected with pSIREN-RetroQ with anti-CEACAM sh-insert, showing the density plot. Cells were stained with an Anti-CEACAM monoclonal antibody. FSC-H: Height of forward scatter signal. The lower of the two populations contains transfected cells with different levels of CEACAM knock-down.

96-well cell culture plates (usually five to ten 96-well plates will be needed); place 100 μl of cell suspension in each well (see Note 8). Evaluate the control wells optically by light microscopy to ensure that only one cell is present per well and mark all wells that either do not have cells or have more than one cell per well.

6. After 1 week, check all the unmarked wells (that contained only one single cell in step 5) and mark those wells that now contain proliferating cells with another colour marker pen.
7. Grow the cells to near confluence, then transfer all clones to 24-well plates, and grow to near confluence again. This step should yield enough cells for the evaluation of the knock-down clones. Also prepare cells from the original, non-transfected cell line as a reference.
8. Prepare a staining solution containing the fluorescence labelled antibody against the protein of choice (for example, anti-CEACAM1-PE); typically you will require 50 μl per clone at 1 μg/ml in FACS buffer.
9. Remove the cells from the plate surface using an enzyme-free buffer, reseed part of the cell suspension into new plates, and transfer the rest of each clone into 1.5-ml reaction tubes.
10. Pellet the cells in a benchtop centrifuge (usually 5 min at about 2,500 × *g*). Resuspend the cells in 50 μl staining solution and incubate on ice for 30 min. After the incubation step, wash with 500 μl FACS buffer. Centrifuge as above, resuspend the cells in 300 μl FACS buffer, and finally transfer to tubes suitable for the cytometer. As a control, incubate cells of the

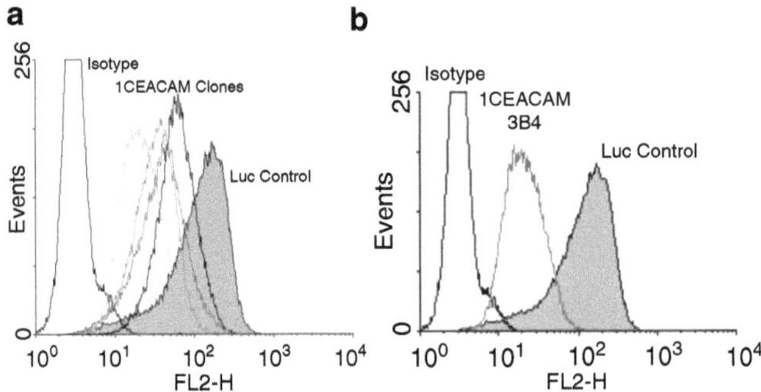

Fig. 2. Panel (**a**) FACS analysis of four CEACAM knock-down clones of FemX-1 (1CEACAM Clones) and one clone transfected with pSIREN-RetroQ with a control Luciferase sh-insert (Luc Control), histogram. Cells were stained with mouse anti-human CEACAM1-PE. Isotype: Isotype control mouse IgG2b-PE of Luc Control; FL2-H: Height of signal for fluorescence channel 2 (PE-Signal). The four clones show various levels of knock-down. Panel (**b**) as Panel (**a**), but only clone 1CEACAM 3B4 with the least CEACAM1 expression (highest knock-down) is shown. Median fluorescence of 1CEACAM 3B4 was 13% of Luc Control (note that FL-H is in logarithmic scale).

original cell line with the staining solution and a second aliquot with a suitable isotype control (for example mouse IgG2b-PE, 1 μg/ml) at the same concentration.

11. Perform FACS analysis of the clones. Include an isotype control of non-transfected cells as a reference. Shortly before the measurement of each clone, add 3 μl of the propidium iodide solution. Clones of the control transfection (for example those expressing anti-Luc shRNA) should show no reduction in fluorescence intensity when compared with the non-transfected cells.
12. Select clones (usually about five different clones) with the lowest fluorescence activity as well as some clones from the control transfection step. Propagate, and perform another FACS analysis with an isotype control for every clone.
13. Compare the median fluorescence of the antibody used above with the isotype control for each clone (for example, the fluorescence intensity of the anti-CEACAM1-PE is compared with the fluorescence intensity of the IgG2b-PE isotype control). The clone with the lowest ratio of antibody against isotype control is the prime candidate for further experiments (Figs. 2 and 3).

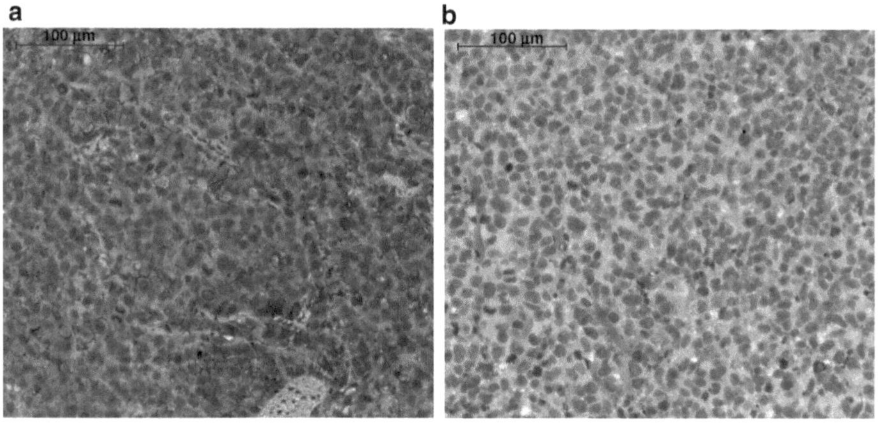

Fig. 3. Panel (**a**) Immunohistochemistry of primary tumours from FemX-1 melanoma cells subcutaneously injected into SCID mice (40 days after injection), stained for human CEACAM1. 10^6 FemX Luc control cells injected. Panel (**b**) 10^6 FemX 1CEACAM 3B4 cells injected.

4. Notes

1. Other companies offer similar Web sites.
2. If vectors other than pSIREN are used, other antibiotics might be required for the selection of transfected cells. Usually, information related to the required concentration of antibiotic can be obtained from the manufacturers.
3. Optimal seeding densities might differ, especially if particularly fast- or slow-growing cell lines are used.
4. If other transfection reagents are used, the protocol might differ. Refer to the manufacturer's information sheet. If the cell lines do not tolerate serum-free medium even for short periods, reagents such as Fugene (Roche) allow transfection in the presence of serum (for example FCS) containing media. However, transfection efficiency is usually higher when serum-free medium is used.
5. The incubation time of the cells without addition of the selection antibiotic after transfection depends on the growth rate of the cell line under study. Usually, 48 h is sufficient, but for particularly fast- or slow-growing cell lines this time may have to be shortened or lengthened. Depending on the cell line used, only a fraction of the seeded single cells will give rise to monoclonal colonies. The smaller this fraction, the more 96-well plates have to be seeded. As some tumour cell lines depend on secreted growth factors, medium conditioned by the particular cell line might improve the number of colonies or even be essential for the growth of the clones.

6. If non-adherent cells (suspension cultures) are transfected, cells should be centrifuged and reseeded each time the medium is changed.
7. If there is no labelled antibody available against the target protein, then the cells should be washed after incubation with the primary antibody and an appropriate secondary antibody with fluorescent label should be used in the second step. Propidium iodide stains only dead or dying cells and, thus, allows exclusion of these cells from the analysis. If the protein target of the knock-down is intracellular, the cells should be fixed and permeabilized. This can be done by incubating the cells with FACS buffer mixed 1:1 with PBS, 4% v/v formaldehyde (final concentration of 2% v/v formaldehyde) for 15 min at RT, followed by washing and incubation with FACS buffer diluted 1:1 with PBS and 1% v/v saponin (final concentration of 0.5% v/v saponin) for an additional 15 min at RT. Also use this 0.5% v/v saponin buffer for staining with the antibody (and second antibody if necessary).
8. "Play safe" and also reseed (and/or store away in liquid nitrogen) the original transfection culture as this includes all the possible clones. If no clones with sufficient knock-down are obtained or if the parameters of the limiting dilution are incorrect and there are insufficient clones (or none at all) to use to grow new colonies, then the experiment can be repeated using the original transfectants (Fig. 1).

References

1. Siomi H, Siomi MC (2009) On the road to reading the RNA-interference code. Nature 457:396–404
2. Hammond SM, Bernstein E, Beach D, Hannon GJ (2000) An RNA-directed nuclease mediates post-transcriptional gene silencing in Drosophila cells. Nature 404:293–296
3. Zamore PD, Tuschl T, Sharp PA, Bartel DP (2000) RNAi: double-stranded RNA directs the ATP-dependent cleavage of mRNA at 21 to 23 nucleotide intervals. Cell 101:25–33
4. Fodstad O, Kjønniksen I, Aamdal S, Nesland JM, Boyd MR, Pihl A (1988) Extrapulmonary, tissue-specific metastasis formation in nude mice injected with FEMX-I human melanoma cells. Cancer Res 48:4382–4388
5. Thies A, Mauer S, Fodstad O, Schumacher U (2007) Clinically proven markers of metastasis predict metastatic spread of human melanoma cells engrafted in SCID mice. Br J Cancer 96:609–616

Chapter 17

Galectin-3 Binding and Metastasis

Pratima Nangia-Makker, Vitaly Balan, and Avraham Raz

Abstract

Galectin-3 is a member of a family of carbohydrate-binding proteins. It is present in the nucleus, the cytoplasm, and also the extracellular matrix (ECM) of many normal and neoplastic cell types. Reports show an upregulation of this protein in transformed and metastatic cell lines (Raz and Lotan Cancer Metastasis Rev 6: 433–452, 1987; Raz et al. Int J Cancer 46: 871–877, 1990). Moreover, in many human carcinomas, an increased expression of galectin-3 correlates with progressive tumor stages (Lotan et al. Int J Cancer 56: 474–480, 1994; Bresalier et al. Gastroenterology 115: 287–296, 1998; Nangia-Makker et al. Int J Oncol 7: 1079–1087, 1995; Xu et al. Am J Pathol 147: 815–822, 1995).

Several lines of analysis have demonstrated that the galectins participate in cell–cell and cell–matrix interactions by recognizing and binding complementary glycoconjugates and thereby play a crucial role in normal and pathological processes. Elevated expression of the protein is associated with an increased capacity for anchorage-independent growth, homotypic aggregation, and tumor cell lung colonization (Lotan et al. Cancer Res 45: 4349–4353, 1985; Lotan and Raz J Cell Biochem 37: 107–117, 1988; Meromsky et al. Cancer Res 46: 5270–5275, 1986). In this chapter we describe the methods of purification of galectin-3 from transformed *Escherichia coli* and some of the commonly used functional assays for analyzing galectin-3 binding.

Key words: Galectin-3, Purification, Homotypic aggregation, Heterotypic aggregation, Anchorage-independent growth, Wound healing

1. Introduction

Human galectin-3 is approximately 30 kDa in size, it is a unique chimeric gene product belonging to the family of non-integrin β-galactoside binding lectins, and has conserved amino acid sequences in the carbohydrate-binding motif (10). Clinical investigations have shown a correlation between expression of galectin-3 and malignant properties of several types of cancers (11), consequently, galectin-3 has been described as a cancer-associated protein.

Miriam Dwek et al. (eds.), *Metastasis Research Protocols*, Methods in Molecular Biology, vol. 878,
DOI 10.1007/978-1-61779-854-2_17,

Galectin-3 has three structural domains. Each domain is associated with at least one specific function: (a) an N-terminal domain containing a serine phosphorylation site that is important in regulating cellular signaling (12); (b) a collagen-α-like sequence cleavable by matrix metalloproteinases (13); and (c) a COOH terminal containing a single carbohydrate recognition domain (CRD) and the NWGR anti-death motif (14).

Galectin-3 is a cytosolic protein, but it can also traverse the intracellular and plasma membranes to translocate into the nucleus, mitochondria and to be externalized, and these properties suggest that galectin-3 is a shuttling protein with multiple functions (15–17). Galectin-3 lacks the classical signal sequence for secretion and does not pass through the endoplasmic reticulum/Golgi apparatus pathways (18); however, it can be transported into the extracellular milieu via a nonclassical pathway (19), where it interacts with a myriad of partners regulating a number of biological functions (11, 20, 21). Many of the functions of galectin-3 are dependent on its carbohydrate-binding properties and therefore can be inhibited by a specific disaccharide inhibitor, lactose.

The analysis of the properties of galectin-3 often requires the use of recombinant galectin-3. Currently, a few methods are available for the purification of galectin-3 based on the affinity of galectin-3 to its substrate or by using affinity tags fused to galectin-3. Glutathione S-transferase (GST) is an affinity tag that is commonly used for the production of recombinant proteins. The method is based on the affinity of GST to the glutathione ligand coupled to a matrix. The binding of GST-tagged proteins to the matrix is reversible, and the fused protein can be eluted under mild, non-denaturating conditions by adding reduced glutathione to the elution buffer. If desired, cleavage of the protein from GST can be achieved using a site-specific protease whose recognition sequence is located immediately upstream of N-terminus of galectin-3.

GST-tagged galectin-3 is constructed by inserting the cDNA sequence encoding the full length or a fragment of galectin-3 to multiple cloning sites of one of the pGEX vectors or any other vector expressing GST protein under the regulation of a bacterial or a mammalian promoter. In the case of the pGEX vector, the expression is under control of the *tac* promoter, which is induced by the addition of isopropyl β-D thiogalactoside (IPTG). All pGEX vectors also contain the *lac*I*q* gene, which produces a repressor protein preventing expression until induction by IPTG. Although all pGEX vectors have a range of protease cleavage recognition sites we use the pGEX-6P vectors that contain a unique cleavage site recognized by the PreScission Protease (PP).

To analyze the binding of galectin-3 to its receptors the most commonly used protocol involves labeling the protein either with biotin or by ^{125}I (22, 23). We discuss, here, the biotinylation protocol because of its relative simplicity and avoidance of radioactive reagents. After binding of biotin-labeled protein to the cell surface receptors, the binding efficiency can be measured in terms of color development by the use of a substrate–chromogen mixture.

Laminin, fibronectin, and collagen type IV are ECM proteins with affinity for galectin-3 (24). We describe an assay for binding of the cell surface proteins to ECM ligands. Since a number of surface proteins can bind to a given ligand, this is generally performed as an indirect assay with a number of cell lines varying in their galectin-3 expression that are derived from the same origin.

It has been presumed that tumor cell surface lectins might play a role in cellular interactions in vivo that are important for the formation of emboli and for the arrest of circulating tumor cells (25). Homotypic aggregation is an assay that reflects the formation of tumor cell emboli in circulation. This assay is performed using asialofetuin, which is a glycoprotein possessing several branched oligosaccharide side chains with terminal nonreducing galactosyl residues. It binds to the lectins present on the cell surface of tumor cells and induces homotypic aggregation by serving as a cross-linking bridge between adjacent cells (26). Heterotypic interactions between the tumor cells and endothelial cells can be measured in a wound healing assay and as three-dimensional cocultures on Matrigel™ (27). These assays assist in the analysis of the interactive properties of different variants of galectin-3, thus elucidating the role of the various mutations or protein fragments in tumor biology.

One in vitro property of tumorigenic cells is their ability to grow progressively in semi-solid medium, this is a function of the cancer-associated autonomy from growth regulatory mechanisms (28). Anchorage-independent growth is an assay in which the cells are seeded on soft agar and allowed to grow. The cells, which divide and form colonies over a period of 10–15 days usually exhibit a higher metastatic potential in in vivo studies (5).

2. Materials

2.1. Purification of Galectin-3 Using Affinity Chromatography

1. Luria-Bertani (LB) broth: 1% w/v tryptone, 0.5% w/v yeast, and 1% w/v NaCl.
2. Galectin-3 expressing bacterial clone: *Escherichia coli* transformed with a suitable vector containing cDNA encoding the human galectin-3 transcript.
3. Ampicillin: 1% w/v ampicillin. The ampicillin stock should be stored at –20°C.
4. Lysis buffer: 150 mM NaCl; 1% w/v NP-40; 0.5% w/v DOC; 0.1% w/v SDS; 50 mM Tris–HCl, pH 8.0; 0.1% w/v leupeptin; and 1 mM PMSF. Leupeptin and PMSF stocks should be stored at –20°C and added prior to use.

5. IPTG: 0.1 M IPTG.
6. Phosphate buffer: 8 mM Na_2HPO_4; 2 mM NaH_2PO_4; 1 mM $MgSO_4$; 1 mM PMSF; 0.2% w/v NaN_3; and 5 mM DTT. PMSF and DTT should be added just prior to use.
7. Elution buffer: phosphate buffer containing 0.3 M lactose.
8. MOPS buffer: 0.5 M; pH 7.5.
9. Asialofetuin column: prepare as shown below and store at 4°C.

2.1.1. Preparation of Asialofetuin

1. Dissolve 100 mg fetuin in 40 ml 0.025 N H_2SO_4.
2. Incubate the solution at 80°C for 1 h.
3. Dialyze against 40 L of distilled H_2O to remove the SO_4 ions.
4. Lyophilize overnight and resuspend in 13 ml water.
5. Ensure that the fetuin has been converted into asialofetuin by running it on a reducing SDS–polyacrylamide gel (for more information on SDS–PAGE and Western blotting refer to the chapter by Blancher and McCormick in this volume). Fetuin has a molecular mass of approximately 66 kDa, whereas asialofetuin has a lower mass commensurate with the removal of the sialic acid residues. Run both the untreated and treated protein on the gel to check the reduction in mass associated with the removal of the sialic acid residues (see Note 1).

2.1.2. Preparation of an Asialofetuin Chromatography Column

1. Dissolve asialofetuin in 0.5 M MOPS buffer to a concentration of 10 mg/ml.
2. Approximately 15 min before the coupling reaction take 5 ml of Affigel-15 slurry (Bio-Rad, Catalogue no. 153-6051) and wash with 15 volumes of cold deionized water.
3. Combine 4 ml of 10 mg/ml cold asialofetuin/buffer solution with 4 ml of prepared Affigel-15 slurry. Mix at 4°C for 2 h.
4. Centrifuge the slurry, pour off the supernatant, and resuspend in deionized water. To block the unreacted sites on the gel matrix add 1 M ethanolamine, pH 8.0. Use 0.1 ml of ethanolamine per milliliter of gel slurry. Agitate gently at 4°C for 1 h.
5. Pack a column with the slurry, leave overnight in a cold room, and wash with 0.1 M MOPS buffer until eluant is free of protein.
6. Equilibrate the column using the 10 mM phosphate buffer containing 0.2% w/v sodium azide (see Note 2).

2.1.3. Purification of GST-Tagged Galectin-3

1. 2× YT broth: 1.6% w/v tryptone, 1% w/v yeast, and 0.5% w/v NaCl.
2. Galectin-3 expressing bacterial clone: *E. coli* transformed with a suitable vector containing cDNA encoding the human galectin-3 transcript.
3. Ampicillin: 1% w/v ampicillin. The ampicillin stock should be stored at –20°C.

4. Lysis buffer: 150 mM NaCl; 1% v/v NP-40; 0.5% w/v DOC; 0.1% w/v SDS; 50 mM Tris–HCl, pH 8.0; 0.1% leupeptin; and 1 mM PMSF. Leupeptin and PMSF stocks should be stored at –20°C and added prior to use.
5. IPTG: 0.1 M IPTG.
6. Phosphate buffer: 8 mM Na_2HPO_4; 2 mM NaH_2PO_4; 1 mM $MgSO_4$; 1 mM PMSF; 0.2% w/v NaN_3; and 5 mM DTT. The PMSF and DTT should be added just prior to use.
7. Cleavage buffer (CB) 100 ml: 50 mM Tris–HCl, pH 7; 150 mM NaCl; 1 mM EDTA; and 1 mM DTT.
8. PreScission Protease: For a 1 ml bed volume column, mix 80 μl (160 U) of PP with 920 μl of CB.
9. Glutathione Sepharose (GS) 4B beads (GE Healthcare Bioscience).

2.2. Preparation of Recombinant Galectin-3 for Binding to Cell Surface Receptors

1. EZ-link, Sufo-NHS-Biotinylation Kit (Pierce, IL, USA).
2. Substrate chromogen mixture, this should be prepared immediately before use: dissolve 0.5 mg/ml ABTS (2,2′-Azinothiazoline sulfonic acid) in 0.1 M citrate buffer, pH 4.2; containing 0.03% v/v hydrogen peroxide. Alternatively, the ABTS substrate kit for horseradish peroxidase (HRP) (Zymed Laboratories, Inc., San Francisco, CA, USA) can be used.
3. Lactose: 1 M in ddH_2O.
4. 96-well microtiter plates.
5. ELISA plate reader, with filters enabling readings to be taken at 405 nm.

2.3. Preparation of Recombinant Galectin-3 for Binding to Soluble Extracelluar Matrix Proteins

1. 96-well microtiter plates.
2. EHS laminin: 100 μg/ml, store at –20°C.
3. Collagen type IV: 100 μg/ml, store at –20°C.
4. Fibronectin: 100 μg/ml, store at –20°C.
5. Phosphate buffered saline (PBS).
6. Bovine serum albumin (BSA): 30% w/v in distilled water, store at 4°C.
7. Alamar blue (Biosource International).
8. ELISA plate reader, with filters enabling readings at 405 and 570 nm.

2.4. Homotypic Aggregation

1. EDTA: 0.02% w/v in CMF-PBS.
2. Asialofetuin: as described in Subheading 2.1.
3. Formaldehyde.
4. Trypan blue: 0.4% w/v in normal saline.

5. Glass tubes coated with Sigmacote (Sigma Chemical Co. USA): pour 1 ml Sigmacote in the glass tube; rotate the tube, so that the entire area inside the tube is covered with sigmacote. Let it dry and wash the excess with distilled water.
6. Orbital shaker.
7. Hemocytometer.

2.5. Heterotypic Aggregation

1. Matrigel™.
2. 8-Chamber slide.
3. Endothelial cells (HUVEC or any other).
4. Tumor cells (galectin-3 expressing).
5. Trypsin (0.25% w/v).
6. Trypan blue: 0.4% w/v in normal saline.
7. Hemocytometer.
8. Viable stains DiI and DiO (Invitrogen).
9. Tissue culture incubator.

2.6. Wound Healing Assay

1. Cell Culture inserts (ibidi GmbH, Germany).
2. Endothelial cells.
3. Tumor cells (galectin-3 expressing).
4. Trypsin (0.025% w/v).
5. Hemocytometer.
6. Viable stains DiI and DiO (Invitrogen).
7. Tissue culture incubator.

2.7. Anchorage-Independent Growth

1. Complete medium: appropriate medium with 10% w/v fetal bovine serum (FBS).
2. 1% w/v Agar solution: add 1 g of sea-plague agarose to 10 ml of distilled H_2O and autoclave. Allow to cool to 45°C and add 90 ml of complete medium and mix well. The agar solution prepared in medium can be stored for 2–3 days.
3. 2.5% w/v Glutaraldehyde.
4. 6-Well plates

3. Methods

3.1. Isolation of Recombinant Galectin-3

1. Inoculate 10 ml of LB Broth (containing 100 μg/ml of ampicillin) with the galectin-3 expressing bacterial clone.
2. Incubate overnight at 37°C with constant shaking at 225 rpm.

3. Use the overnight bacterial culture to inoculate a fresh 10 × 200 ml of LB, again, containing 100 μg/ml of ampicillin.
4. Incubate for 3 h at 37°C with shaking at 225 rpm (see Note 3).
5. Add ampicillin again and 2.5 ml of IPTG stock to induce the protein synthesis. It is important to add more ampicillin as this will select for the growth of resistant clones.
6. Incubate for 4 h at 37°C with shaking at 225 rpm.
7. Transfer the contents to centrifuge bottles and pellet the cells by centrifugation at 1,000 × *g* for 15 min at 4°C.
8. Discard the supernatant and resuspend the pellet (this contains the bacteria) in 20 ml PBS. Combine into one bottle and centrifuge again for 15 min at 1,000 × *g* at 4°C.
9. Discard the supernatant and store the pellet at −70°C overnight. The pellet can be stored for up to 1 week without a reduction in the protein yield.
10. Resuspend the pellet in 80 ml of ice-cold lysis buffer.
11. Disrupt the cells with a probe-type sonicator using multiple short bursts at maximum intensity for 4 × 30 s, ensure that the cells are maintained on ice during this process to prevent protein denaturation. Do not sonicate for more than 30 s at a time as otherwise the temperature of the lysate will increase. Between each sonication cycle cool the lysate on ice.
12. Centrifuge the lysate for 20 min at 18,000 × *g* and retain the supernatant. The remaining steps should be performed in a cold room.
13. Equilibrate the asialofetuin column with three bed volumes of phosphate buffer.
14. Load the supernatant from step 12, above. Allow the supernatant to flow through the column at the rate of 10–12 drops/min.
15. Wash the column with three bed volumes of phosphate buffer.
16. Elute the protein with 15 ml of elution buffer; add the buffer in 1-ml aliquots in order to avoid compacting the chromatography column.
17. Wash the column with three bed volumes of phosphate buffer. The column can be reused if stored at 4°C.
18. Measure the protein content of the samples collected in step 16 by using standard protein estimation methods. Pool the samples containing the protein and dialyze against PBS until all the lactose has been removed from the samples. The purified protein can be stored at −70°C (see Notes 4–7).

3.2. Purification of Galectin-3 Using the GST Affinity Tag

1. Propagate BL21-CodonPlus (DE3)-RIPL cells (see Note 8) transformed with pGEX-6p-Galectin-3 in 25 ml 2× YT/AMP in a 100 ml flask, at 37°C overnight (O/N).
2. Transfer the bacteria to 1 L 2× YT/AMP media (5 L flask) and grow at 37°C until the OD_{600nm} reaches 0.5–0.8 (usually after 5–6 h). Transfer the flask to 20°C and wait about 30 min.
3. Add IPTG to a final concentration of 0.5 mM and continue incubation 4–5 h or O/N at 20°C.
4. Harvest the bacteria by centrifuging them at 6,000×*g* for 10 min. Maintain the bacteria at 4°C during this process. You can now freeze the bacterial pellet, or proceed to the cell-lysis step.
5. Cell lysis: resuspend the bacteria using 10 ml/1 g of pellet cold PBS supplemented with protease inhibitors.
6. Transfer the bacteria to 50-ml centrifuge tubes, place the tubes on ice, and sonicate the bacteria to lyse them. The bacteria should be in a volume of not less than 30 ml during the lysis step. Use a sonic probe for 2 min, with a 10 s pulse with 50 s off at 65% amplitude. Check the number of intact bacteria under the phase contrast microscope to ensure adequate cell lysis. Retain a sample of the supernatant. Immediately add Triton X-100 to a final concentration of 1% v/v and incubate for 30 min with gentle agitation at 4°C.
7. Centrifuge the sample at 15,000×*g* for 10 min at 4°C and then transfer to new tubes. Save a sample of the supernatant.
8. Incubate the supernatant with 0.5 ml of GS 4B beads (see Note 9). The beads should be pre-washed with 1× PBS. Use 0.5-ml GS beads per 10 ml bacterial lysate and incubate for 30–60 min at 4°C with gentle agitation. Centrifuge the GS bead—lysate mixture at 500×*g* for 1 min, aspirate the supernatant and wash the beads thrice with 5 ml PBS.
9. Centrifuge again at 500×*g* for 10 min, remove supernatant and wash the beads with 5 ml of CB.
10. During the wash (step 9, above), prepare the PP mixture in CB (see Note 10). For this, use 80 μl PP (160 U) and 920 μl of CB.
11. Add the PP mixture to the GS beads and incubate for 4 h to O/N at 4°C.
12. Centrifuge the beads at 500×*g* for 5 min and collect the supernatant.
13. Wash the beads three times with 500 μl PBS. Save aliquots for analysis and estimation of protein concentration. The eluate will contain galectin-3, while the GST moiety and PP (also GST-tagged) will remain bound to the beads.

3.3. Binding of Galectin-3 to Cell Surface Receptors via Biotinylation of Galectin-3

1. Isolate recombinant galectin-3 as in Subheading 3.1.
2. Dissolve 2 mg protein in 1 ml of PBS solution and calculate the number of millimoles dissolved using the following formula:

$$\frac{\text{mg protein}}{\text{MW protein}} = \text{mmol of protein.}$$

3. Dissolve 2 mg sulfo-NHS-Biotin in 100 μl distilled water and add 30 μl of this solution to the protein solution to give a 20-fold molar excess of biotin over galectin-3.
4. Incubate on ice for 2 h for the biotinylation of galectin-3 to be completed.
5. To remove excess salt from the protein, equilibrate the 10 ml desalting column with 30 ml PBS and apply the protein sample. Allow the sample to permeate the gel. Add buffer to the column in 1-ml aliquots and collect 1-ml fractions of the purified eluent protein in separate test tubes.
6. Monitor the protein content in the tubes by measuring the absorbance at 280 nm. Pool the fractions containing protein (see Note 11).
7. Plate cells in a 96-well plate at a density of 1×10^4 cells per well and incubate overnight.
8. Adjust the concentration of biotinylated galectin-3 such that it is in the range 0–20 μg/50 μl.
9. Remove the medium from the cells.
10. Add fresh medium containing 0.5% v/v FCS and include 50 μl biotin-labeled galectin-3 at a range of different concentrations (see Note 12).
11. Incubate the plates for 2 h at 37°C.
12. Wash the plate three times with PBS to remove any unbound material.
13. Add a 1:1,000 dilution of HRP conjugated streptavidin to the wells. The HRP complex binds to biotin-labeled galectin-3. Incubate for 30 min at room temperature.
14. Wash with PBS three times to remove any unbound material.
15. Add 100 μl of the substrate chromogen mixture and incubate for 1 min to allow the color to develop.
16. Measure the optical density at 405 nm (see Note 13).

3.4. Binding of Galectin-3 to Extracellular Matrix Proteins

1. Coat the 96-well microtiter plates with serially diluted (0–10 μg) EHS laminin, collagen type IV, or fibronectin.
2. Incubate the plates for 1 h at 37 or 4°C overnight to dry the ECM protein onto the surface of the plate (see Note 14).

3. Block the nonspecific sites in the wells of the 96-well plate using sterile 1% w/v BSA in PBS, for 1 h at 37°C.
4. Wash the wells with sterile PBS three times to remove any unbound protein.
5. Detach the cells from their plate using 0.02% w/v EDTA. Count the viable cells using the trypan blue exclusion method and using a hemocytometer. Seed the cells at 4×10^4 cells per well.
6. Allow the cells to adhere to the plates. Depending on the cell line used for these studies, the time taken for this step may range between 15 min and 24 h.
7. Wash off any nonadherent cells using fresh medium. Repeat twice (see Note 15).
8. To count the number of cells attached to the ECM proteins, add 200 μl of fresh medium containing a 1:10 dilution of Alamar blue dye (the live cells create a reducing environment, which changes the color of dye from blue to pink).
9. Incubate the cells with the medium/Alamar dye mixture for 3–4 h at room temperature.
10. Read the absorbance at 570 nm.

3.5. Homotypic Aggregation

1. Detach the cells from the monolayer with 0.02% w/v EDTA in CMF-PBS. The use of trypsin should be avoided because it may proteolytically digest the cell surface proteins that are of interest in this experiment.
2. Suspend the cells at 1×10^6 cells/ml in CMF-PBS with or without asialofetuin (20 μg/ml).
3. Transfer 0.5-ml aliquots, of the cell suspensions prepared above, into siliconized glass tubes. Agitate at 80 rpm for 1 h at 37°C.
4. Terminate the aggregation experiment by fixing the cells with 1% w/v formaldehyde in CMF-PBS.
5. Count the number of single cells that remain using a hemocytometer (see Notes 16–18).
6. Calculate the percent aggregation using the formula $(1\text{-Nt}/\text{Nc}) \times 100$, where Nc is the number of single cells in the control and Nt is the number of single cells in the presence of test compound.

3.6. Heterotypic Aggregation (for Further Information on the Matrigel™ Assay System, See the Chapter by Brooks SA in Volume 2 of This Series)

1. Thaw the Matrigel™ material on ice (see Note 19).
2. Cool an 8-chamber slide on ice.
3. Transfer 200 μl of the Matrigel™ material into each chamber of the 8-chamber slide. Carefully remove any air bubbles and place the slide in a 37°C tissue culture incubator for approximately 15 min until the Matrigel™ has solidified, forming a uniform layer.

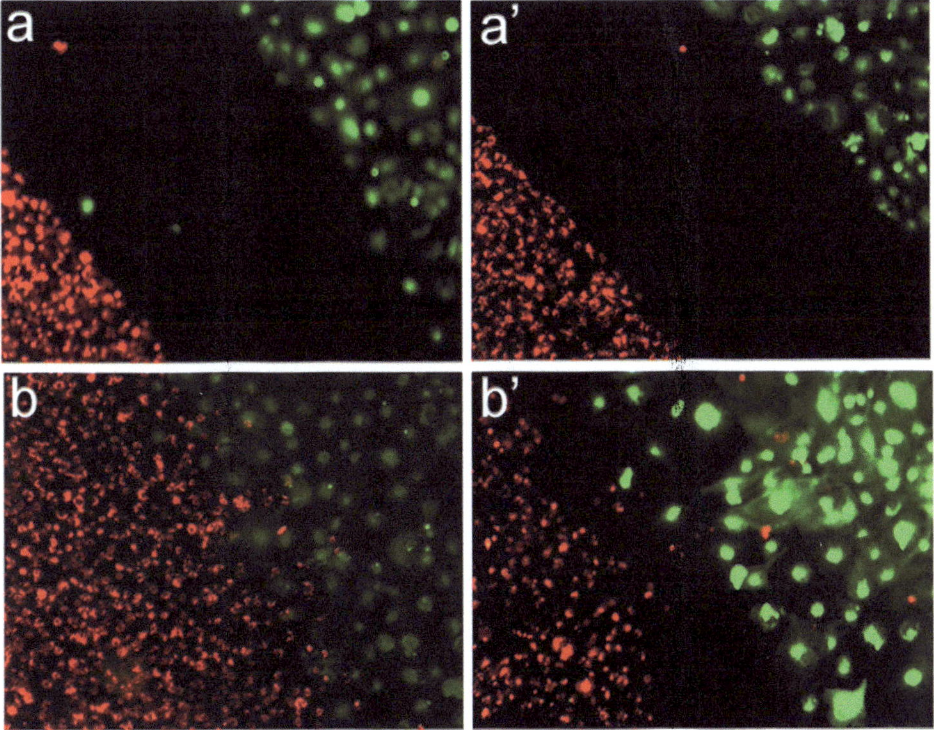

Fig. 1. Cell migration using the wound healing assay: endothelial cells BAMEC and a galectin-3 variant H64- or P64-transfected BT-549 cells were prelabeled with DiO (*green*) and DiI (*red*), respectively, and seeded in each chamber of the cell culture insert. After 24 h, the *inserts* were removed and cell migration was studied. (**a–a′**) Migration of BAMEC and BT-549-H64; (**b–b′**) migration of BAMEC and BT-549-P64; (**a**, **b**): 0 h; (**a′**, **b′**): 24 h.

4. Trypsinize the cells of interest and calculate the number of cells required for the experiment. You will need 50,000 cells per chamber for each cell line.
5. Suspend the cells at a concentration of 1×10^6/ml in complete medium in a tube. Incubate with 5 μl of DiI or DiO as desired at 37°C for 15 min in the dark.
6. Centrifuge the cells at $1,000\times g$ for 5 min and wash with complete medium.
7. Resuspend the cells at a density of 1×10^6/ml.
8. Mix 50×10^4 of the epithelial and 50×10^4 of the endothelial cells (each stained with a different dye) in an Eppendorf tube. Pellet the cells by centrifugation. Remove the medium and resuspend in 250 μl medium specific for the endothelial cells (see Note 20).
9. Seed the cells onto the Matrigel™ matrix (see Note 21).
10. Observe and photograph the cells after 24 h (Fig. 1).

3.7. Wound Healing Assay

1. Trypsinize cells and calculate the number of cells required for the experiment.

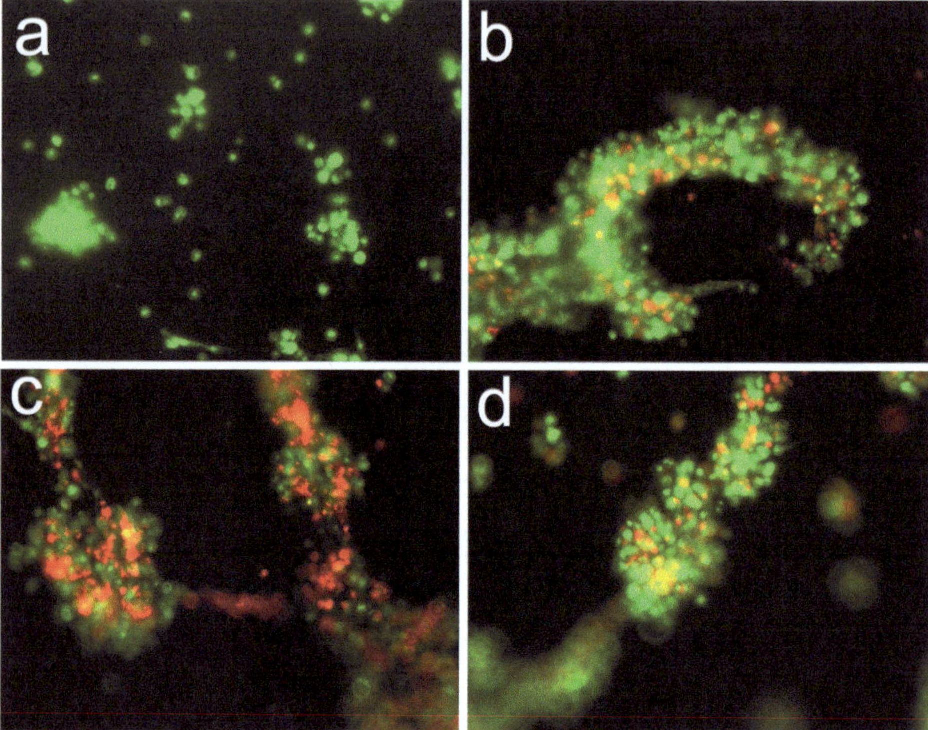

Fig. 2. Endothelial cell morphogenesis and interactions with H64- or P64-transfected BT-549 cells. Endothelial cells BAMEC and the galectin-3 variants H64- or P64-transfected BT-549 cells were prelabeled with DiO (*green*) and DiI (*red*) respectively. 50×10^3 cells were seeded in each chamber on top of gelled Matrigel™. The three-dimensional cultures were observed after 24 h. (**a**) BAMEC alone; (**b**) BAMEC and BT-549 cells; (**c**) BAMEC and BT-549 H64; and (**d**) BAMEC and BT-549 P64.

2. Pellet and resuspend the cells at a concentration of 1×10^6/ml complete medium.
3. Incubate the cells with 5 μl of DiI or DiO, as desired, at 37°C for 15 min.
4. Centrifuge cells at $1,000 \times g$ for 5 min.
5. Wash the cells once with complete medium and resuspend the cells at a density of 1×10^6/ml.
6. Seed 2.4×10^4 cells in each chamber of the cell culture insert. If the migration of cells towards each other is to be analyzed, two different kinds of cells may be seeded. We perform this assay to study wound healing and cell migration of endothelial cells towards epithelial cells harboring various galectin-3 variants.
7. After 24 h incubation in the tissue culture incubator, remove the insert carefully, using sterile forceps and wash the chamber 2–3 times with endothelial cell medium to remove any floating cells. Add fresh endothelial cell growth medium.
8. Observe the cell migration under the light or fluorescent microscope at different time intervals to analyze the rate of migration, Fig. 2 (see Note 21).

3.8. Anchorage-Independent Growth

1. Pour 2 ml of 1% w/v agar solution into each well of a 6-well plate.
2. Allow the agar to solidify at room temperature for 15–20 min. Plates can be wrapped with Parafilm and stored at 4°C for about 2 weeks.
3. Suspend 500 or 1,000 cells in 1 ml of complete medium and immediately add 1 ml of the 1% w/v agar solution. Mix and pour gently into the wells of the plates (see Note 16).
4. Allow the cell-containing top layer to solidify at room temperature for 15 min and 2 h at 4°C.
5. Transfer to a 37°C, 5% v/v CO_2 incubator O/N.
6. The following morning add 1 ml of complete medium to each of the wells and allow the colonies to grow for a further 2 weeks. The medium should be changed two times per week during this growth period (see Note 22).
7. Fix the colonies with 2.5% w/v glutaraldehyde solution and compare both the number and size of the colonies in the different wells.

4. Notes

1. In the dialyzed form asialofetuin can be stored at 4°C for months.
2. The binding efficiency of asialofetuin to the column can be determined by calculating the protein content in the flow through.
3. Determine the OD at 600 nm. It should be 0.5.
4. In our hands the yield of galectin-3 varies between 1 and 5 mg/L of bacterial culture. If the yield is lower than this we suggest troubleshooting by going through Notes 5–7.
5. After step 13, resuspend the pellet in fresh lysis buffer (20 ml/tube), repeat the sonication and centrifugation steps, again retaining the supernatant.
6. After step 15, save the flow through and reload onto the column.
7. At no stage let the column dry out as this will reduce the binding efficiency.
8. Although a variety of *E. coli* host strains can be used for cloning and expression work with pGEX vectors we recommend a specially engineered strain that is more suitable for maximizing the expression of mammalian proteins: BL21-CodonPlus (DE3)-RIPL.
9. A variety of GS matrices are available for affinity purification of GST-galectin-3. However, based on our experience we would

recommend GS 4B (GE Healthcare) as the first choice. The total volume of Sepharose beads required depends on the amount of galectin-3 that needs to be purified, the capacity of the matrix is supplied by the manufacturers in the data sheet for the material.

10. The choice of protease is based on a few factors including the presence, in galectin-3, of recognition sequences of other proteases as well as the protease specificity.
11. Store the biotinylated protein at 4°C until use. The protein can be stored under these conditions for 1 week.
12. In some wells 50–100 mM lactose can be added along with the protein for specific inhibitory studies.
13. The biotin–streptavidin complex cleaves H_2O_2 which is coupled to the oxidation of substrate ABTS giving a green end product.
14. Ensure that the plates are stored horizontally so that there is a uniform coating across the well.
15. The wells should be washed thoroughly to remove any unbound protein. However, the washes should be gentle so the cells remain unaffected.
16. Ensure that the cells are in a single cell suspension after detaching them from the monolayer.
17. It is important to siliconize the glass tubes in which aggregation is performed so that the cells adhere to the glass.
18. The aggregates are usually very fragile and can be disrupted by harsh pipetting.
19. Matrigel™ should be thawed at 4°C, because it has a tendency to gel at higher temperatures.
20. In our experience, the endothelial cells are more sensitive to changes in medium composition, while tumor cells tend to grow in endothelial cell specific medium. Therefore, we use endothelial cell medium for the cocultures.
21. The cells should be added very carefully in the center of the chamber so as not to disturb the Matrigel™ layer.
22. The agarose layers should be allowed to gel completely to prevent the top layer sliding and to prevent the passage of the cells to the bottom of the plates.

Acknowledgements

This work was supported by NIH R37CA46120-19 (A. Raz).

References

1. Raz A, Lotan R (1987) Endogenous galactoside-binding lectins: a new class of functional tumor cell surface molecules related to metastasis. Cancer Metastasis Rev 6:433–452
2. Raz A, Zhu DG, Hogan V, Shah N, Raz T, Karkash R, Pazerini G, Carmi P (1990) Evidence for the role of 34-kDa galactoside-binding lectin in transformation and metastasis. Int J Cancer 46:871–877
3. Lotan R, Ito H, Yasui W, Yokozaki H, Lotan D, Tahara E (1994) Expression of a 31-kDa lactoside-binding lectin in normal human gastric mucosa and in primary and metastatic gastric carcinomas. Int J Cancer 56:474–480
4. Bresalier RS, Mazurek N, Sternberg LR, Byrd JC, Yunker CK, Nangia-Makker P, Raz A (1998) Metastasis of human colon cancer is altered by modifying expression of the beta-galactoside-binding protein galectin 3. Gastroenterology 115:287–296
5. Nangia-Makker P, Thompson E, Hogan V, Ochieng J, Raz A (1995) Induction of tumorigenicity by galectin-3 in a non-tumorigenic human breast carcinoma cell line. Int J Oncol 7:1079–1087
6. Xu XC, el-Naggar AK, Lotan R (1995) Differential expression of galectin-1 and galectin-3 in thyroid tumors. Potential diagnostic implications. Am J Pathol 147:815–822
7. Lotan R, Lotan D, Raz A (1985) Inhibition of tumor cell colony formation in culture by a monoclonal antibody to endogenous lectins. Cancer Res 45:4349–4353
8. Lotan R, Raz A (1988) Endogenous lectins as mediators of tumor cell adhesion. J Cell Biochem 37:107–117
9. Meromsky L, Lotan R, Raz A (1986) Implications of endogenous tumor cell surface lectins as mediators of cellular interactions and lung colonization. Cancer Res 46:5270–5275
10. Barondes SH, Castronovo V, Cooper DN, Cummings RD, Drickamer K, Feizi T, Gitt MA, Hirabayashi J, Hughes C, Kasai K et al (1994) Galectins: a family of animal beta-galactoside-binding lectins. Cell 76:597–598
11. Dumic J, Dabelic S, Flögel M (2006) Galectin-3: an open-ended story. Biochim Biophys Acta 1760:616–635
12. Gong HC, Honjo Y, Nangia-Makker P, Hogan V, Mazurak N, Bresalier RS, Raz A (1999) The NH2 terminus of galectin-3 governs cellular compartmentalization and functions in cancer cells. Cancer Res 59:6239–6245
13. Ochieng J, Fridman R, Nangia-Makker P, Kleiner DE, Liotta LA, Stetler-Stevenson WG, Raz A (1994) Galectin-3 is a novel substrate for human matrix metalloproteinases-2 and -9. Biochemistry 33:14109–14114
14. Akahani S, Nangia-Makker P, Inohara H, Kim HR, Raz A (1997) Galectin-3: a novel antiapoptotic molecule with a functional BH1 (NWGR) domain of Bcl-2 family. Cancer Res 57:5272–5276
15. Davidson PJ, Li SY, Lohse AG, Vandergaast R, Verde E, Pearson A, Patterson RJ, Wang JL, Arnoys EJ (2006) Transport of galectin-3 between the nucleus and cytoplasm. I. Conditions and signals for nuclear import. Glycobiology 16:602–611
16. Li SY, Davidson PJ, Lin NY, Patterson RJ, Wang JL, Arnoys EJ (2006) Transport of galectin-3 between the nucleus and cytoplasm. II. Identification of the signal for nuclear export. Glycobiology 16:612–622
17. van den Brule FA, Waltregny D, Liu FT, Castronovo V (2000) Alteration of the cytoplasmic/nuclear expression pattern of galectin-3 correlates with prostate carcinoma progression. Int J Cancer 89:361–367
18. Hughes RC (1999) Secretion of the galectin family of mammalian carbohydrate-binding proteins. Biochim Biophys Acta 1473:172–185
19. Nickel W (2005) Unconventional secretory routes: direct protein export across the plasma membrane of mammalian cells. Traffic 6:607–614
20. Nangia-Makker P, Balan V, Raz A (2008) Regulation of tumor progression by extracellular galectin-3. Cancer Microenviron 1: 43–51
21. Ochieng J, Furtak V, Lukyanov P (2004) Extracellular functions of galectin-3. Glycoconj J 19:527–535
22. Dong S, Hughes RC (1997) Macrophage surface glycoproteins binding to galectin-3 (Mac-2-antigen). Glycoconj J 14:267–274
23. Raz A, Meromsky L, Zvibel I, Lotan R (1987) Transformation-related changes in the expression of endogenous cell lectins. Int J Cancer 39:353–360
24. Warfield PR, Makker PN, Raz A, Ochieng J (1997) Adhesion of human breast carcinoma to extracellular matrix proteins is modulated by galectin-3. Invasion Metastasis 17:101–112
25. Lotan R, Raz A (1983) Low colony formation in vivo and in culture as exhibited by metastatic melanoma cells selected for reduced homotypic aggregation. Cancer Res 43:2088–2093

26. Inohara H, Raz A (1994) Effects of natural complex carbohydrate (citrus pectin) on murine melanoma cell properties related to galectin-3 functions. Glycoconj J 11:527–532
27. Nangia-Makker P, Wang Y, Raz T, Tait L, Balan V, Hogan V, Raz A (2010) Cleavage of galectin-3 by matrix metalloproteases induces angiogenesis in breast cancer. Int J Cancer 127:2530–2541
28. Cifone MA, Fidler IJ (1980) Correlation of patterns of anchorage-independent growth with in vivo behavior of cells from a murine fibrosarcoma. Proc Natl Acad Sci USA 77:1039–1043

Chapter 18

Lectin Array-Based Strategies for Identifying Metastasis-Associated Changes in Glycosylation

Simon Fry, Babak Afrough, Anthony Leathem, and Miriam Dwek

Abstract

Since 2005, lectin microarray technology has emerged as a relatively simple yet powerful technique for the comprehensive analysis of glycoprotein glycosylation. Lectin microarrays represent a new analytical method that can be used to explore the human glycome, a unique source of markers of diseases including cancer. The lectin microarray technology is a sensitive tool with the potential to allow high-throughput analysis of cancer-associated changes in glycosylation. This chapter describes the generation of a lectin-binding signature associated with metastatic primary breast tumours that have been resected, fixed, and embedded in paraffin. Procedures concerning sample and lectin microarray preparation are explained, alongside experimental considerations and approaches to data analysis.

Key words: Metastasis, Lectin, Microarray, Glycome, Glycosylation

1. Introduction

The human glycome is of particular interest in cancer research given that aberrations in glycosylation are a common feature of a broad variety of tumour types (1–6). Progress in exploiting the human glycome as a source of disease markers has been hampered due to the complexity of glycans and a lack of analytical methods to decipher their structures. Capillary electrophoresis, liquid chromatography, and mass spectrometry are all commonly used as glycome profiling tools (7). However, these techniques may not be suited to the initial detection of glycosylation differences between samples due to time-consuming/low-throughput methods of analysis, sensitivity, sample contaminants, requirements for complex equipment and the generation of complex data.

The twenty-first century has seen the development of high-throughput technologies in the area of both genomics and

Miriam Dwek et al. (eds.), *Metastasis Research Protocols*, Methods in Molecular Biology, vol. 878, DOI 10.1007/978-1-61779-854-2_18,

proteomics. The field of glycomics followed these developments, and in 2005 the first publications emerged detailing the use of lectin microarrays facilitating high-throughput, sensitive, and relatively simple simultaneous analysis of *N*- and *O*-linked glycosylation. The advantage of these techniques included the minimal sample preparation and the ability to analyse glycosylation of proteins without the need for protein deglycosylation steps.

Lectin microarrays may be fabricated and analysed using a selection of the approximately 100 commercially available lectins, and widely available laboratory equipment (8, 9). Lectin microarrays are generated by the immobilisation of several lectins to a solid phase (typically a microscope slide) using chemical methodologies such as the attachment of lysine side chain amine groups to the solid phase through epoxy-functionalised or *N*-hydroxysuccinimidyl-derived esters (10). Such microarrays are then interrogated with fluorescently labelled glycoproteins, and the resultant pattern of fluorescence is interpreted using known binding specificities of lectins in order to annotate the glycan epitopes present in a sample of interest. Lectin arrays have been employed in this manner to analyse bacterial (11), viral (12), and mammalian (13) glycomes.

Since their first appearance lectin microarray technologies have advanced rapidly. When a complex sample of glycoproteins (e.g., tissue extract, serum) and a biological reference (used as an internal standard) are labelled with different fluorescent tags, mixed and used to probe a lectin microarray, the resultant competitive binding between analytes gives ratiometric binding data that can reveal subtle changes in glycosylation (14).

Covalent attachment of lectins to a solid phase occurs in a randomly orientated fashion and may reduce structural flexibility of lectins, altering the binding properties. Using recombinant lectins it has been demonstrated that site-directed orientation of lectins improves the overall sensitivity and increases the limit of detection of lectin microarrays (15).

Lectin microarrays may also being used as a quick and convenient means of gaining information on the glycosylation status of a specific glycoprotein in a mixture. Once an unlabelled mixture has bound to a lectin microarray a fluorescently labelled antibody can then be used to reveal which lectins on the array have bound a glycoprotein of interest (16).

It should be noted that there are inherent drawbacks to the use of lectin microarrays. Firstly, glycan recognition is limited to the combined substrate specificities of the arrayed lectins. In addition, glycan structures must be inferred from known lectin-binding specificities. As these specificities will usually be based on structural epitopes rather than the entire glycan, structural elucidation is often impossible. However, it is these epitopes that are thought to encode biological information and they can be targeted for additional analysis in other applications, such as lectin affinity chromatography.

In this chapter we describe a method for identifying metastasis-associated changes in glycosylation associated with metastatic breast cancer. Formalin-fixed paraffin-embedded (FFPE) primary tumours are described, but the technique can be applied to fresh tissue protein extracts, as well as serum or urine with minimal sample preparation (17).

2. Materials

2.1. Sample Preparation

1. Radio immune precipitation assay (RIPA) buffer: 50 mM Tris–HCl, 150 mM NaCl, SDS 0.1%, sodium deoxycholate 0.5%, and NP40 1%.
2. Phosphate buffered saline (PBS): PBS tablets (Sigma) yield 0.01 M phosphate buffer, 2.7 mM KCl, and 137 mM NaCl when dissolved in the appropriate volume of deionised water.

2.2. Sample Labelling

Three nanomolar Cy3 monoreactive dye: one vial of Cy3 monoreactive dye (GE Healthcare) contains sufficient dye to label 30 nM of amino groups. Dissolve one vial in 50 μl dimethylformamide and then divide in to 10×5 μl aliquots. Vacuum centrifuge each aliquot to dryness (this takes about 4 h). Store the vials in the dark with desiccant at 4°C.

2.3. Microarray Scanning

1. Probing buffer: Tris-buffered saline (TBS) containing 1% v/v Triton X-100 and 500 mM glycine. Note: some lectins require divalent cations such as calcium, magnesium, or manganese. Ensure these ions are present at a concentration of 1 mM if required.
2. Lectin microarray: for example LecChip™, GP BioSciences, lectin microarrays these contain 45 lectins in each of seven wells on a glass slide. Each lectin is spotted in triplicate.

3. Methods

3.1. Removal of Paraffin Wax from Formalin-Fixed Paraffin-Wax-Embedded Tumours (See Note 1)

1. Cut 5 μm × 10 μm sections from each FFPE primary breast tumour and add to 1 ml xylene. Incubate for 15 min at 40°C.
2. Centrifuge at $7{,}550 \times g$ for 10 min. Remove excess xylene, taking care not to disturb the pellet.
3. Repeat steps 1 and 2 twice.

4. Add 1 ml 95% v/v ethanol to each tissue pellet. Incubate for 10 min at room temperature and then centrifuge at 7,550 × *g* for 10 min.
5. Repeat step 4 twice. Reduce sample volume by vacuum centrifugation (do not allow sample to completely dry).

3.2. Protein Extraction

1. Add 580 μl RIPA buffer to dewaxed breast tumour tissue and incubate at 100°C for 20 min, and then 60°C for 120 min.
2. Centrifuge samples at 10,080 × *g* for 15 min at 4°C. Collect supernatant for analysis.

3.3. Protein Labelling

1. Quantify protein content of supernatant. In our laboratory we use the Micro BCA Protein Assay Kit (Thermo Scientific). Adjust protein concentration to 50 μg/ml with PBS.
2. Take 20 μl (1 μg) of each sample and add to 3 nM Cy3 monoreactive dye. Incubate for 60 min at room temperature in the dark.
3. Dilute Cy3 labelled proteins to 2 μg/ml by making up to 500 μl with probing buffer. This step should be performed immediately before the sample is applied to the lectin microarray.

3.4. Data Acquisition

We use a GlycoStation™ Reader 1200 (GP Biosciences) to detect fluorescent signals with the Array Pro analyser version 4.5 (Media Cybernetics) to analyse data, but a range of systems are available for this.

1. Prior to use, wash the lectin microarray by adding 100 μl of probing solution to each well. Remove the liquid by inverting the microarray and briefly shaking. Repeat twice taking care to ensure the surface of the microarray does not dry out.
2. Immediately add 100 μl of Cy3-labelled sample to each well of the lectin microarray slide. Incubate microarray in a chamber (>80% humidity) for 150 min at room temperature in the dark. It is important to ensure that an adequate number of biological and analytical replicates are prepared (see Note 2).
3. Remove sample from each well and wash five times with probing solution, taking care not to contaminate the adjacent wells of the microarray. After fifth wash, do not remove probing solution.
4. Acquire fluorescent images of lectin microarrays.

3.5. Data Analysis

1. Calculate the net fluorescent intensity for each spot on the lectin microarray by subtracting the background value from the raw signal intensity value (see Note 3).
2. Average the replicate net lectin-binding values.

3. Normalise the data by making the total fluorescence detected in each well (i.e. for all 45 lectins) equivalent for all samples being compared.

When identifying metastasis-associated changes in glycosylation, metastasis to the lymph node has been used as a surrogate for distant recurrences of the tumour. In this case the lectin-binding responses from the protein extracts from FFPE primary breast cancer tumours are grouped according to the lymph node status of the patients. Data normality is assessed using the D'Agostino-Pearson omnibus test. The non-parametric Mann–Whitney *U*-test is used to identify lectins showing statistically significant differences in median binding to glycoproteins prepared from FFPE primary breast tumours with lymph node metastasis compared to those where there has been no lymph node metastasis. Lectins demonstrating significant differences may then be selected as candidate markers for lymph node metastasis and used in further analyses (see Note 4).

4. Notes

1. Whilst the analysis of extracts from FFPE primary breast tumours has been described here, the technique is equally applicable to any source of glycoproteins. We have analysed metastasis-associated glycosylation using serum, urine, and cell line preparations.
2. In order to analyse the statistical significance of data a sufficient number of replicates should be measured. Typically each sample is applied to two wells containing each lectin in triplicate, to give six replicate readings per sample.
3. In order to make meaningful comparisons between samples it is essential to ensure that binding signals are produced in the linear binding range of each lectin. The linear binding range can be determined empirically by producing binding curves for each lectin. Binding curves are made by measuring lectin-binding signal intensities at a range of sample total protein concentrations. The highest concentration of sample that falls on the linear part of every arrayed lectin (rather than at the plateau) should be selected to ensure that all lectin-binding data is generated under non-saturating conditions.
4. When developing the lectin microarray it is advisable to ensure that all observed binding is specifically mediated through lectin-glycan interactions. Pure preparations of a glycoprotein in a native and deglycosylated form can be used to demonstrate the requirement of protein glycosylation in a sample for lectin

microarray binding. Prior to the analysis described here the signals produced by Cy3-labelled human serum albumin (HSA, a non-glycosylated protein) and GalNAc conjugated HSA upon binding to the lectin microarray were measured. As expected, native HSA produced negligible binding allowing the minimum net fluorescent intensity of a positive binding signal to be defined. GalNAc-HSA binding to specific lectins on the array demonstrated specific binding. Monosaccharide inhibition or exoglycosidase digestions can also be used to demonstrate the specificity of arrayed lectin binding.

References

1. Dennis JW, Granovsky M, Warren CE (1999) Glycoprotein glycosylation and cancer progression. Biochim Biophys Acta 1473:21–34
2. Dennis JW, Laferte S, Waghorne C, Breitman ML, Kerbel RS (1987) Beta 1-6 branching of Asn-linked oligosaccharides is directly associated with metastasis. Science 236:582–585
3. Dube DH, Bertozzi CR (2005) Glycans in cancer and inflammation-potential for therapeutics and diagnostics. Nat Rev Drug Discov 4:477–488
4. Feizi T (1985) Demonstration by monoclonal antibodies that carbohydrate structures of glycoproteins and glycolipids are onco-developmental antigens. Nature 314:53–57
5. Fuster MM, Esko JD (2005) The sweet and sour of cancer: glycans as novel therapeutic targets. Nat Rev Cancer 5:526–542
6. Hakomori S (1996) Tumor malignancy defined by aberrant glycosylation and sphingo(glyco) lipid metabolism. Cancer Res 56:5309–5318
7. Vanderschaeghe D, Festjens N, Delanghe J, Callewaert N (2010) Glycome profiling using modern glycomics technology: technical aspects and applications. Biol Chem 391:149–161
8. Angeloni S, Ridet JL, Kusy N, Gao H, Crevoisier F, Guinchard S, Kochhar S, Sigrist H, Sprenger N (2005) Glycoprofiling with micro-arrays of glycoconjugates and lectins. Glycobiology 15:31–41
9. Kuno A, Uchiyama N, Koseki-Kuno S, Ebe Y, Takashima S, Yamada M, Hirabayashi J (2005) Evanescent-field fluorescence-assisted lectin microarray: a new strategy for glycan profiling. Nat Methods 2:851–856
10. Pilobello KT, Krishnamoorthy L, Slawek D, Mahal LK (2005) Development of a lectin microarray for the rapid analysis of protein glycopatterns. Chembiochem 6:985–989
11. Zheng T, Peelen D, Smith LM (2005) Lectin arrays for profiling cell surface carbohydrate expression. J Am Chem Soc 127:9982–9983
12. Pilobello KT, Mahal LK (2007) Lectin microarrays for glycoprotein analysis. Methods Mol Biol 385:193–203
13. Hsu KL, Mahal LK (2006) A lectin microarray approach for the rapid analysis of bacterial glycans. Nat Protoc 1:543–549
14. Gupta G, Surolia A, Sampathkumar SG (2010) Lectin microarrays for glycomic analysis. OMICS 14:419–436
15. Hsu KL, Pilobello KT, Mahal LK (2006) Analyzing the dynamic bacterial glycome with a lectin microarray approach. Nat Chem Biol 2:153–157
16. Krishnamoorthy L, Bess JW Jr, Preston AB, Nagashima K, Mahal LK (2009) HIV-1 and microvesicles from T cells share a common glycome, arguing for a common origin. Nat Chem Biol 5:244–250
17. Pilobello KT, Slawek DE, Mahal LK (2007) A ratiometric lectin microarray approach to analysis of the dynamic mammalian glycome. Proc Natl Acad Sci USA 104:11534–11539

Index

Miriam Dwek et al. (eds.), *Metastasis Research Protocols*, Methods in Molecular Biology, vol. 878,
DOI 10.1007/978-1-61779-854-2, © Springer Science+Business Media New York 2012

MIX
Papier aus verantwortungsvollen Quellen
Paper from responsible sources
FSC® C105338

If you have any concerns about our products,
you can contact us on
ProductSafety@springernature.com

In case Publisher is established outside the EU,
the EU authorized representative is:
Springer Nature Customer Service Center GmbH
Europaplatz 3, 69115 Heidelberg, Germany

Printed by Libri Plureos GmbH
in Hamburg, Germany